Oil and Gas Field
Development Techniques

BASICS OF RESERVOIR ENGINEERING

FROM THE SAME PUBLISHER

INSTITUT FRANÇAIS DU PÉTROLE PUBLICATIONS

René COSSÉ
Reservoir Training Manager
at ENSPM-Formation Industrie
Professor at IFP School

Oil and Gas Field Development Techniques

BASICS OF RESERVOIR ENGINEERING

1993

Translation of
Le gisement, R. Cossé

ISBN 2-7108-0630-4
ISSN 1271-9048

FOREWORD

Oil and gas field development and production involve a range of techniques that may be grouped together under the following four headings:

(a) *Reservoir engineering.*
(b) *Drilling.*
(c) *Downhole production.*
(d) *Surface production.*

These four disciplines are interrelated, and the studies and operations involved call for a number of different specialists from operating and service companies and from equipment suppliers who may often require general information relating to the techniques as a whole.

These disciplines are covered in four volumes, under the general title:

OIL AND GAS FIELD DEVELOPMENT TECHNIQUES

The aim is to provide an understanding of the techniques and constraints related to each of these four disciplines and thereby facilitate communication between the different professions involved.

This information should also be of use to those with more specialized skills, such as computer and legal experts, economists, research workers, equipment manufacturers, etc. It should help them to relate their work to development techniques as a whole and to integrate them into the overall professional context.

These four volumes partly reproduce the content of introductory courses designed by *ENSPM-Formation Industrie* to meet specific needs and course material included in the curriculum of the École Nationale Supérieure du Pétrole et des Moteurs.

A. Leblond
Director
Drilling - Production - Reservoir Engineering
ENSPM-Formation Industrie

CONTENTS

FOREWORD. V

Main Symbols and Industry Units Used. XXI

Equivalent of Principal British/US Units in Metric Units. XXVII

INTRODUCTION

I.1 What is a Reservoir?. 3

I.2 What is Reservoir Engineering?. 3

I.2.1 Reservoir Image 6
I.2.2 Well Characteristics 7
I.2.3 Recovery Mechanisms 7
I.2.4 Uncertainties 9
I.2.5 Simulation of Reservoir Development 10
I.2.6 Development Project, Optimization 10

I.3 The Place of Reservoir Studies in Development. 11

Chapter 1
RESERVOIR GEOLOGY AND GEOPHYSICS

1.1 Goals: Image of the Reservoir (Geological Model) 15

1.2 Hydrocarbon Generation and Migration. 18

1.2.1 Generation 18
1.2.2 Migration 19

1.2.2.1 Primary Migration 19
1.2.2.2 Secondary Migration 20

1.3 Reservoirs. 21
1.3.1 Definitions 21
1.3.2 Reservoir Rocks 21
1.3.2.1 Sandstone Reservoirs 21
1.3.2.2 Carbonate Reservoirs 22

1.4 Traps. 23
1.4.1 Classification of Traps 23
1.4.2 Characteristics 24

1.5 Seismic Development Survey. 25
1.5.1 Principles 25
1.5.2 3D Seismic Prospecting 26
1.5.3 Vertical Seismic Profile (VSP) and Offset Vertical Seismic Profile (OSP) 27
1.5.4 Detection of Fluids 27

1.6 Examples of Reservoirs. 30

Chapter 2
CHARACTERIZATION OF RESERVOIR ROCKS

2.1 Porosity. 33
2.1.1 Definition 33
2.1.2 Determination of Porosity 35
2.1.2.1 Core Analysis 35
2.1.3 Effect of Pressure 38

2.2 Permeability. 38
2.2.1 Definition, Darcy's Law 38
2.2.2 Laws of Horizontal Flow in Steady-State Conditions 40
2.2.2.1 Liquids 41
2.2.2.2 Gas 42

2.2.3 Associations of Formations of Different Permeabilities 42
2.2.4 Specific Permeability, Effective and Relative Permeabilities ... 43
2.2.5 Measurement of Specific Permeability 44
2.2.5.1 Measurement Using the Variable Pressure Feed Permeameter (Fig. 2.9), Initial Apparatus, Measurement Principle 45
2.2.5.2 Constant Pressure Feed Air Permeameter 45
2.2.6 Porosity/Permeability Relationship 46
2.2.7 Porosity/Permeability Exercice 47

2.3 Saturations 49

2.3.1 Definition 49
2.3.2 Distribution of the Different Fluids in a Virgin Field, Capillary Mechanisms 50
2.3.2.1 Capillary Properties of Rocks 50
2.3.2.2 Equilibrium of a Virgin Field 54
2.3.2.3 Capillary Migration 56
2.3.3 Capillary Pressure Curves, Curves Representing the Average Capillary Properties of a Reservoir 57
2.3.3.1 Capillary Pressure Curve of a Sample 57
2.3.3.2 Passage from a $(P_C - S)_{lab}$ Curve to a $(P_C - S)_{reservoir}$ Curve 58
2.3.3.3 Curves Showing the Average Capillary Properties of a Reservoir 59
2.3.4 Determination of *in Situ* Saturation 60
2.3.4.1 Direct Method (Core Analysis) 60
2.3.4.2 Indirect Method by Analysis of Capillary Mechanisms 60
2.3.4.3 Well Log Method 61
2.3.5 Typical Section of Saturations in a Reservoir 61
2.3.6 Surveys with a Scanner: Porosity, Pore Morphology, Sweeping by Another Fluid 61

2.4 Well Logging 62

2.4.1 General 62
2.4.2 Electric Logs 65
2.4.2.1 Spontaneous Potential SP 65

2.4.2.2 Resistivity Logs 65
2.4.2.3 Microresistivity Logs 66
2.4.3 Radioactivity Logs 66
2.4.3.1 Gamma Ray Logs (GR) 66
2.4.3.2 Neutron Logs N 68
2.4.3.3 Density Logs (gamma/gamma) D 68
2.4.3.4 Neutron Relaxation TDT 68
2.4.4 Sonic (or Acoustic) Logs 68
2.4.5 Auxiliary Logs 69
2.4.5.1 Caliper Log 69
2.4.5.2 Dipmeter Log 69
2.4.5.3 Cement Bond Log (CBL) 69
2.4.6 Determination of Lithology, Porosity and Saturations 70
2.4.6.1 Lithology 70
2.4.6.2 Porosity 70
2.4.6.3 Saturations 71
2.4.6.4 Composite Log 71
2.4.6.5 Overall Interpretation 71
2.4.6.6 Cost 73
2.4.7 Production Logs 74

Chapter 3
FLUIDS AND PVT STUDIES

3.1 General Behavior 77
3.1.1 Pure Substances 78
3.1.1.1 Pressure/Specific Volume Diagram (Clapeyron Diagram) (Fig. 3.1) 78
3.1.1.2 Pressure/Temperature Diagram (Fig. 3.2) 79
3.1.2 Mixtures 80
3.1.2.1 Pressure/Specific Volume Diagram (Fig. 3.3) 80
3.1.2.2 Pressure/Temperature Diagram (Fig. 3.4) 81
3.1.2.3 Different Types of Reservoir 82

3.1.3 Behavior of Oils and Gases between the Reservoir and the Surface 84

3.1.4 Physical Components of Hydrocarbons and Other Components 86

3.2 Natural Gases. 86

3.2.1 Practical Equations of State 86

3.2.2 Volume Factor of a Gas B_g 88

3.2.3 Determination of Z 89

3.2.3.1 Experimental 89

3.2.3.2 Calculations and Charts 89

3.2.4 Condensables Content of a Gas (GPM) 89

3.2.5 Viscosity of a Gas 90

3.2.6 Effluent Compositions of Gas Wells 90

3.3 Oils. 91

3.3.1 Behavior in the One-Phase Liquid State and the Two-Phase State (Oil and Gas) 91

3.3.2 Formation Volume Factor and Gas/Oil Ratio 91

3.3.2.1 Variation between Reservoir and Stock Tank Oil at a Given Reservoir Pressure 91

3.3.2.2 Variation in FVF B_o and Solution GOR R_s with Reservoir Pressure, Production GOR R 94

3.3.3 Viscosity 99

3.3.4 Effluent Compositions of Oil Reservoir Wells 100

3.3.5 Exercice 100

3.4 Formation Water. 101

3.4.1 Composition 101

3.4.2 Compressibility 102

3.4.3 Viscosity 102

3.4.4 Water and Hydrocarbons 102

3.5 Charts. 102

3.6 Liquid/Vapor Equilibria, Equation of State. 104

Chapter 4
VOLUMETRIC EVALUATION OF OIL AND GAS IN PLACE

4.1 General. 115

4.2 The Different Categories of Oil and Gas in Place. 116

4.3 Volumetric Calculation of Oil and Gas in Place. 118

4.3.1 Principle of Volumetric Methods 118
4.3.2 Calculation of Impregnated Rock V_r 119
4.3.2.1 Calculation of the Volume of Rocks from Isobaths. Area/Depth Method 122
4.3.2.2 Rapid Calculation Method 123
4.3.3 Calculation of the Volume of Oil from Isovol Maps 123

4.4 Choice of Average Characteristics, Uncertainties. 125

4.4.1 Choice of Average Characteristics 125
4.4.2 Uncertainties and Probabilistic Methods 128
4.4.3 Example of Area/Depth Calculation 130

Chapter 5
ONE-PHASE FLUID MECHANICS AND WELL TEST INTERPRETATION

5.1 General. 133

5.2 Oil Flow around Wells. 134

5.2.1 Diffusivity Equation 134
5.2.2 Solutions of the Diffusivity Equation 136
5.2.3 Flow Equations 138
5.2.3.1 Constant Flow Rate 138
5.2.3.2 Variable Flow Rate 141
5.2.4 Equations of Pressure Build-Up after Shutting in the Well 142
5.2.5 Skin Effect or Damage 144

5.2.6 Productivity Index ... 148
5.2.7 Various Problems of Fluid Flow ... 148
5.2.8 Two-Phase Flows ... 149
5.2.9 Non-Flowing Wells ... 150

5.3 Gas Flow around Wells. ... 151

5.4 Different Well Tests. ... 154

5.4.1 Initial Tests ... 154
5.4.2 Tests Specific to Gas Wells ... 156
5.4.3 Periodical Tests ... 157
5.4.4 Interference Tests ... 157
5.4.5 Water Injection Wells ... 158

5.5 Tests during Drilling (Drill Stem Tests - DST). ... 159

5.6 Main Equations Used with Practical Drillsite Units. ... 161

5.6.1 Equations for Oil ... 161
5.6.2 Equations for Gas ... 162

5.7 Principle of Type Curves. ... 163

5.7.1 Presentation ... 163
5.7.2 Use ... 164
5.7.3 Semi-Automatic Test Analysis Programs: Well Models ... 167

5.8 Typical Test Interpretation. ... 168

Chapter 6
MULTIPHASE FLOW

6.1 General. ... 177

6.2 Review of Capillary Mechanisms Concept of Relative Permeability. ... 179

6.2.1 Capillary Doublet, Genesis of Drops and Jamin Effect ... 179
6.2.2 Concept of Relative Permeability ... 182
6.2.3 Variation in Relative Permeability as a Function of Saturation ... 183

6.2.3.1 Oil/Water (or Gas/Water) Pair ... 183

6.2.3.2 Oil/Gas Pair 184
6.2.4 Comments on Relative Permeabilities 185
6.2.5 Determination of Relative Permeabilities 185
6.2.5.1 Laboratory Measurements 185
6.2.5.2 Empirical Equations 187
6.2.6 Capillary Imbibition 187

6.3 Theory of Frontal Displacement. 189
6.3.1 Front Concept 189
6.3.2 Unidirectional Displacement (Capillarity Ignored) Buckley-Leverett Theory 190
6.3.2.1 Water-Cut f_w. 191
6.3.2.2 Calculation of the Displacement Speed of a Section S_w. 193
6.3.3 Effect of Capillarity Forces and Influence of Flow Rate 196
6.3.4 Practical Application 197

6.4 Two- and Three-Dimensional Two-Phase Flow. 200
6.4.1 Encroachment, Instability Mechanisms, Definition of the Mobility Ratio 201
6.4.2 Concept of Critical Speed, Formation of a Tongue 203
6.4.3 Fingering 204
6.4.4 Coning 205
6.4.4.1 Infra-Critical Flow 206
6.4.4.2 Supercritical Flow 207
6.4.4.3 Value of the Critical Flow Rate 207
6.4.4.4 Production Aspect, Coning Parameters 208

6.5 Conclusions. 210

Chapter 7
PRIMARY RECOVERY, ESTIMATION OF RESERVES

7.1 Production Mechanisms. 211
7.1.1 General 211

7.1.2 Reserves 212
7.1.3 Production and Recovery Mechanisms 212
7.1.4 Influence of the Production Rate 214
7.1.4.1 Oil Reservoir without Aquifer 214
7.1.4.2 Oil Reservoir Associated with an Aquifer with Mediocre Petrophysical Characteristics 214
7.1.4.3 Fractured Oil Reservoir Associated with a Large Aquifer with Good Characteristics 214
7.1.5 Compressibility Factors, Fluid Expansion 215
7.1.6 Multiphase Flow, Reservoir Heterogeneities 216
7.1.7 Simplified Calculation Methods 218

7.2 Recovery Statistics. 219
7.2.1 Recovery as a Function of the Type of Reservoir 219
7.2.2 World Reserves 219

7.3 Material Balance. 223
7.3.1 The Material Balance and its Use 223
7.3.2 Undersaturated Oil Reservoir 225
7.3.3 Expansion of Dissolved Gases 226
7.3.4 Oil Reservoir Associated with an Aquifer 228
7.3.4.1 Aquifer Extension, Bottom Aquifer and Edge Aquifer 228
7.3.4.2 Aquifer Functions 230
7.3.4.3 Calculation of Water Inflows 232
7.3.5 Generalized Material Balance for a Saturated Oil Reservoir with Gas-Cap and Aquifer 234
7.3.5.1 Expansion of Gas-Cap 234
7.3.5.2 Generalized Material Balance 235
7.3.5.3 Efficiency of Various Production Mechanisms 235
7.3.5.4 Segregation 236
7.3.5.5 Variation in Production Data According to Production Mechanisms 236
7.3.6 Specific Case of Volatile Oil 236
7.3.6.1 Jacoby and Berry Method ("Volumetric" Method) 238
7.3.6.2 Compositional Material Balance Program 238

7.3.7 Dry (or Wet) Gas Reservoir: not Giving Rise to Retrograde Condensation in the Reservoir 238

7.3.7.1 Material Balance (without Water Influx) 238

7.3.7.2 Calculation of Gas in Place 239

7.3.7.3 Recovery with Water Influx 239

7.3.7.4 Example of Dry Gas Material Balance (without Water Influx) 240

7.3.8 Gas Condensate Reservoirs 241

7.3.9 General Remarks 242

7.4 Statistical Correlations and Decline Laws 243

7.4.1 Statistical Correlations 243

7.4.2 Decline Laws 244

7.4.2.1 Exponential Decline of Flow Rate 244

7.4.2.2 Hyperbolic (and Harmonic) Decline 244

7.5 Production in Fractured Formations 245

7.5.1 General 245

7.5.2 Geological Aspect 247

7.5.2.1 Description of Cores 247

7.5.2.2 Observation of Outcrops 249

7.5.2.3 Rock Mechanics Model 249

7.5.2.4 Visual Analysis in the Borehole 249

7.5.2.5 Reservoir Seismic Prospecting 250

7.5.3 Data Obtained from Logs and Production Tests 250

7.5.3.1 Logs 250

7.5.3.2 Well Testing 250

7.5.4 Production Mechanisms 251

7.5.4.1 One-Phase Flow, Expansion 251

7.5.4.2 Two-Phase Flow 251

7.5.4.3 Transfer Function (Exudation) 254

7.6 A Specific Production Technique, the Horizontal Drain Hole 254

7.6.1 What Are the Cases for Application of the Horizontal Drain Hole? What Are their Advantages? 254

7.6.2 Ideal Situations 256
7.6.2.1 Fractured Reservoirs 256
7.6.2.2 Karst Reservoirs (Non-Porous Fractured) 258
7.6.2.3 Reservoirs with Aquifers 258
7.6.2.4 Enhanced Recovery 258
7.6.2.5 Value for Reservoir Characterization 259
7.6.2.6 Unfavorable Case 259

Chapter 8
SECONDARY AND ENHANCED OIL RECOVERY

8.1 General. 261

8.2 Factors Influencing Recovery. 264
8.2.1 Reservoir and Fluid Characteristics 264
8.2.1.1 Reservoir Geology 264
8.2.1.2 Permeability 264
8.2.1.3 Viscosity of Fluids and Mobility Ratio 266
8.2.2 Injection Characteristics 267
8.2.2.1 Volume of Injected Fluid 267
8.2.2.2 Type of Fluid 267
8.2.2.3 Flood Patterns 267

8.3 Efficiency Analysis. 271
8.3.1 Injection Efficiency and Definition 272
8.3.2 Areal Sweep Efficiency E_A. 272
8.3.3 Vertical (or Invasion) Efficiency E_V. 274
8.3.4 Displacement Efficiency E_D. 275
8.3.5 Conclusion 276

8.4 Waterflood. 277
8.4.1 Technical and Economic Aspects 278
8.4.1.1 Technical Aspect 278
8.4.1.2 Economic Aspect 278

8.4.2 Time and Start of Flooding 278

8.4.3 Implementation 279

8.5 Gas Injection (Non-Miscible). 279

8.5.1 Technical Aspect 279

8.5.2 Economic Aspect 280

8.5.3 Implementation 280

8.5.4 Comparison of Waterflood and Gas Injection 281

8.6 Gas Cycling in Retrograde Gas Condensate Reservoirs. 281

8.7 Enhanced Oil Recovery. 282

8.7.1 General 282

8.7.2 Miscible Methods 284

8.7.2.1 Miscible Displacements 284

8.7.2.2 Two Types of Miscibility 285

8.7.2.3 CO_2 Injection 288

8.7.3 Chemical Methods 289

8.7.3.1 Micro-Emulsions 289

8.7.3.2 Solutions of Polymers in Water 290

8.7.3.3 Micro-Emulsions + Polymers 290

8.7.4 Thermal Methods 291

8.7.4.1 Heavy Oils, and Principle of Thermal Methods 291

8.7.4.2 Steam Injection 293

8.7.4.3 *In Situ* Combustion (Fig. 8.17) 294

8.7.5 Enhanced Oil Recovery Pilot Flood 296

8.8 Conclusions. 296

Chapter 9
RESERVOIR SIMULATION MODELS

9.1 Role of Models. 299

9.2 Different Types of Models. 300

9.3 Numerical Models. 300

9.3.1 Principles 301

9.3.2 Modeling and Use 301

9.3.3 Specialized Models 305

9.3.3.1 Compositional and Miscible Models (Fig. 9.5) 305

9.3.3.2 Chemical Models 306

9.3.3.3 Thermal Models 306

9.3.3.4 Fractured Models 306

9.3.4 Recent Model 307

Chapter 10
DEVELOPMENT OF A FIELD

10.1 Development Project. 309

10.1.1 Analytical Phase 310

10.1.2 Modeling Phase 313

10.1.3 Forecasting Phase 313

10.2 Economic Concepts Used in the Development of a Field. 313

10.2.1 General 313

10.2.2 Different Cases Examined and Decision 316

10.2.2.1 Uncertainty in the Size of the Accumulation 317

10.2.2.2 Search for the Best Recovery Method 317

10.2.2.3 Comparison of Different Development Schedules 317

10.2.2.4 Influence of Inflation and Borrowing 317

10.2.2.5 Influence of Taxation 317

10.3 Research Directions in Reservoir Engineering. 318

Chapter 11
TYPICAL FIELDS

11.1 Oil and Gas Fields in France. 319

11.2 Typical Fields. 321

11.2.1 The Lacq Gas Field 321
11.2.2 The Frigg Field 326
11.2.3 The Vic Bilh Field 332
11.2.4 The Villeperdue Field 335

REFERENCES 339

INDEX 343

MAIN SYMBOLS AND INDUSTRY UNITS USED

Symbol	Definition	Industry units	Equivalence between industry and SI units
A	Flow area	ft^2	$9.3 \;.\; 10^{-2} m^2$
B	Formation volume factor (volume at reservoir conditions divided by volume at standard conditions)	Ratio	
bpd	Barrels per day	bpd	$1.84 \;.\; 10^{-6} m^3/s$
C	Deliverability curve coefficient (gas well test)		$\frac{1 \text{ cfd}}{(\text{psi})^2} = 6.9 \;.\; 10^{15} \frac{\dot{m}^3}{\text{s x pascal}^2}$
$\overline{C}$	Water-drive tabular function		
c	Isothermal compressibility $c = \frac{-1}{V} . \frac{dV}{dP}$	$\frac{V/V}{psi}$	$\frac{1V/V}{psi} = 14.5 \;.\; 10^{-5} \frac{V/V}{pascal}$
cfd	Cubic feet per day	cfd	$3.277 \;.\; 10^{-7} m^3/s$

Symbol	Definition	Industry units	Equivalence between industry and SI units
sp.gr.	Specific gravity	Ratio	–
E	Total efficiency	Ratio	–
E_A	Areal sweep efficiency	Ratio	–
E_V	Vertical efficiency	Ratio	–
E_D	Displacement efficiency	Ratio	–
f	Fractional flow rate (volumetric fraction of 1 phase in a flow)	Fraction	–
G	Gas in place at standard conditions	scf	0.02832 m^3
g	Gravity acceleration	32.16 ft/s^2	9.81 m/s^2
h	Bed thickness	ft	0.305 m
h_o	Oil pay thickness	ft	0.305 m
K	Diffusivity $K = k / \phi \mu c$	$md/cP. \frac{1}{psi}$	$\frac{1 \text{ md x psi}}{cP} \approx 6.89 . 10^{-9} \frac{1 \text{ m}^2 \text{ x Pa}}{Pa . s}$
k	Permeability	md	1 md = 0.987 . 10^{-15} m^2 or = 0.987 . 10^{-3} $(\mu m)^2$
k_i	Effective permeability to fluid i	md	(see preceding note)
k_{ri}	Relative permeability to fluid i	Ratio	–

Symbol	Definition	Industry units	Equivalence between industry and SI units
l	Length	ft	0.305 m
l_n	Napierian logarithm		–
log	Decimal logarithm		
M	Mobility ratio (displacing phase mobility divided by displaced phase mobility)	Ratio	–
M	Molecular weight	g	$1\ g = 10^{-3}\ kg$
m	Cementation factor	–	–
m_e, m_{10}	Pressure line slope in semi-log coordinates	$\frac{psi}{cycle\ log}$	$\frac{psi}{cycle\ log} = 3\ .\ 10^3\ \frac{pascal}{cycle\ l_n}$
N	Oil in place expressed in standard conditions (or weight)	Barrel (t)	$0.159\ m^3$ (10^3 kg)
Np	Cumulative production	bbl	$0.159\ m^3$
n	Deliverability curve slope in a gas well test (log x log)		
P	Pressure	psi	$1\ psi = 6.89\ 10^3$ pascal
PI	Oil well productivity index	bpd/psi	$\frac{1\ bpd}{psi} = 4.769\ .\ 10^{-11}\ \frac{m^3}{s \times pascal}$
Q	Flow rate at standard conditions	bpd	$1\ bpd = 1.84\ .\ 10^{-6}\ m^3/s$
Q_c	Critical flow rate	bpd	$1\ bpd = 1.84\ .\ 10^{-6}\ m^3/s$

Symbol	Definition	Industry units	Equivalence between industry and SI units
R	Resistivity	ohm-foot	0.305 ohm . m
R	Production GOR (standard gas volume divided by stock tank oil volume)	Ratio	
R	Universal ideal gas constant	10.7 mol/lb	1.287 mol/g
R	Drainage radius in circular radial flow	ft	0.305 m
R_S	Solution GOR		
r	Distance to well (variable) in radial flow	ft	0.305 m
r_w	Well radius	ft	0.305 m
S_i	Saturation with fluid i	Fraction	
S	Skin effect		
T	Flowing time (well testing)	h	1 h = 3,600 s
T	Absolute temperature	°R = °F + 460	°R = 1.8 K (Kelvin)
t°	Temperature	°F	32 + 1.8°C
t	Time	d (or h if specified)	1 d = 86,400 s

Symbol	Definition	Industry units	Equivalence between industry and SI units
u	Shape parameter characterizing porous medium	cm	1 cm = 10^{-2} m
U	Speed of injection (or filtration)	ft/d	1 ft/d = 3.54 10^{-6} m/s
V	Volume (oil) Volume (gas)	bbl cu.ft	0.159 m^3 0.02832 m^3
V_c	Critical velocity	ft/d	1 ft/d = 3.54 10^{-6} m/s
W_e	Water encroachment in oil zone	bbl	0.159 m^3
W_p	Water produced, cumulative (surface conditions)	bbl	0.159 m^3
Z	Compressibility factor of a gas		
α	Bed dip	Degree	
Δp	Pressure drop	psi	1 psi = 6.89 10^3 pascal
Δt	Time interval	d (or h if specified)	1 d = 86,400 s
θ	Angle in circular radial system	Radian	
θ	Angle of contact between wetting fluid and solid	Radian	
θ	Time of pressure build-up (after shut-in)	h	1 h = 3,600 s

Symbol	Definition	Industry units	Equivalence between industry and SI units
μ	Viscosity	cP	1 cP = 10^{-3} Pa.s
ρ	Density	g/cm^3	1 g/cm^3 = 10^3 kg/m^3
σ	Interfacial or surface tension for two fluids	dyne/cm	1 dyne/cm = 10^{-3} N/m
ϕ	Porosity	Fraction	

Subscripts:

g	Value relative to gas phase.
I	Value relative to injection.
i	Value relative to initial time.
j, n, t	Value relative to steps j, n, t.
o	Value relative to oil phase.
p	Value relative to production.
std	Value relative to standard conditions.
w	Value relative to water phase.

EQUIVALENT OF PRINCIPAL BRISTISH/US UNITS IN METRIC UNITS

1 Inch	=	2.54 cm
1 Foot	=	0.305 m
1 Yard	=	0.915 m
1 Square inch (sq. in.)	=	6.45 cm^2
1 Acre	=	4047 m^2
1 Cubic foot (cu. ft.)	=	28.32 dm^3
1 Gallon (US)	=	3.785 dm^3
1 Barrel	=	0.159 m^3
1 Cubic foot per barrel (GOR)	=	0.178 m^3/m^3
1 psi (pound-force per square inch)	=	0.06895 bar
1 psi (pound-force per square inch)	=	$6.89 \cdot 10^3$ pascal

Degrees API (°API)
60/60°F = weight of oil volume (at 60°F divided by the weight of the same volume of water at 60°F).

$$= \frac{141.5}{sp . gr . 60 / 60°F} - 131.5$$

Degrees Farenheit (°F) $= °C \times \frac{9}{5} + 32$

1 barrel per day (bpd) ≈ ± 50 tons/year

1 cubic foot per day (cfd) ≈ ± 10 m^3/year

INTRODUCTION

Reservoir engineering is a twentieth century discipline: the concepts of saturation and relative permeability, for example, emerged some fifty years ago. It is also a highly specific discipline, nourished by the earth sciences, by thermodynamics and fluid mechanics in particular.

Following a period in which reservoirs produced by their natural energy alone, and on exclusively empirical bases, the need for a more rational development was naturally imposed.

The science of reservoir engineering, which developed in recent decades, now draws on many advanced techniques in data acquisition and reservoir simulation.

Note also that producing and developing a reservoir implies a close interdependence of reservoir techniques, of those pertaining to drilling and well completion, and the surface techniques required for hydrocarbon gathering and processing before transport. Hence it naturally entails a multidisciplinary effort.

I.1 WHAT IS A RESERVOIR?

A reservoir is formed of one (or more) subsurface rock formations containing liquid and/or gaseous hydrocarbons, of sedimentary origin with very few exceptions. The reservoir rock is porous and permeable, and the structure is bounded by impermeable barriers which trap the hydrocarbons.

The vertical arrangement of the fluids in the structure is governed by gravitational forces. Figure 1 shows a cross-section of a typical hydrocarbon reservoir (classic anticline).

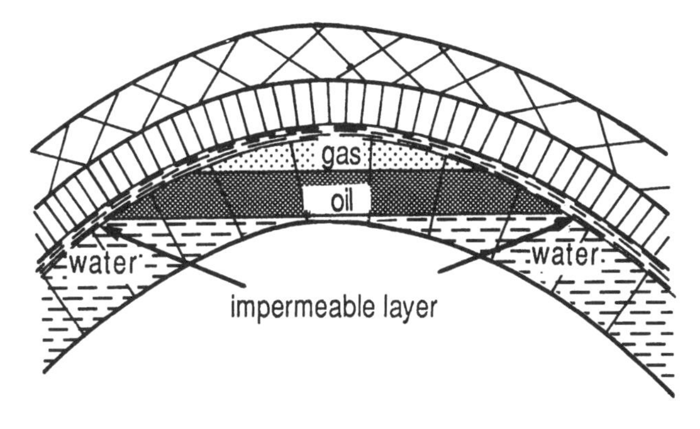

Fig. 1

A reservoir is also an **invisible** and **complex physical system** (porous medium) which must be analyzed as thoroughly as possible, knowing that our understanding of it will nevertheless be limited by the dearth of information.

I.2 WHAT IS RESERVOIR ENGINEERING?

The goal of reservoir engineering, starting with the discovery of a productive reservoir, is to set up a **development project** that attempts to **optimize** the hydrocarbon recovery as part of an overall economic policy. Reservoir specialists thus continue to study the reservoir throughout the life of the field to derive the information required for **optimal production from the reservoir**.

Sempé.

The following must accordingly be **estimated**, with the aim of optimal profitability for a given project:

(a) Volumes of hydrocarbons (oil and/or gas) in place.
(b) Recoverable reserves (estimated on the basis of several alternative production methods).
(c) Well production potential (initial productivity, changes).

It must be emphasized that the reservoir specialist works on a system to which he enjoys virtually no material access. He has to be satisfied with partial data, furnished by the wells, and therefore incomplete and insufficient. Consequently, he must extrapolate these local data over kilometers in order to compile a synthetic image of the reservoir, enabling him to make production forecasts that are fairly reliable for the near future, and much less so for a more distant future. Yet these forecasts are indispensable to determine an optimum development scheme.

The figure below helps to illustrate the different steps in reservoir engineering (Fig. 2).

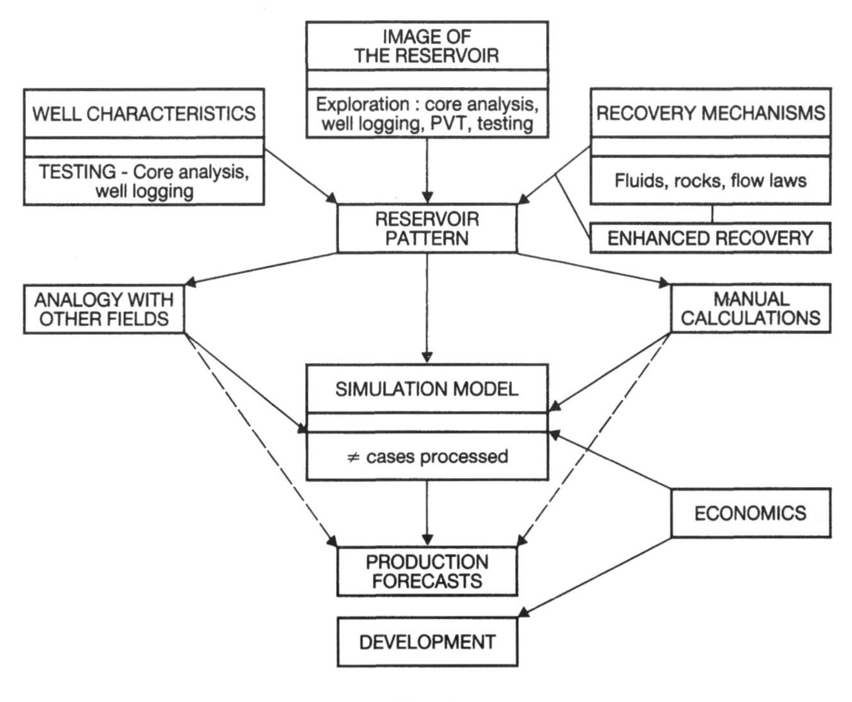

Fig. 2

I.2.1 Reservoir Image

The image of the reservoir is defined when the forms, boundaries, internal architecture (heterogeneities), and distribution and volumes of the fluids contained in the reservoir are known — or at least, to start with, approximated.

The methods used are partly included under the **term of reservoir (and production) geology**, and are based on petroleum geology and geophysics.

Figure 3 shows the different basic aspects involved in drawing up an image of the reservoir.

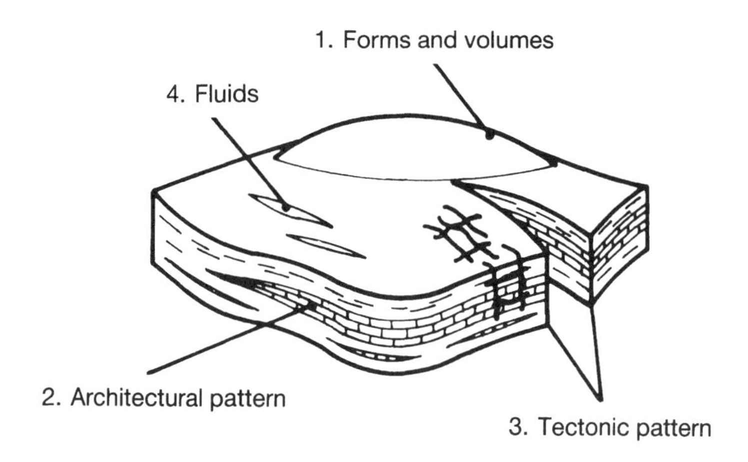

Fig. 3

These techniques also draw heavily on the direct and indirect analysis of the information obtained in the wells:

(a) Direct analysis: core analysis, and PVT analysis of fluids (pressure, volume, temperature). These measurements are performed in the laboratory.

(b) Indirect analysis: well logs, which are recorded during drilling and production. Recordings of physical parameters obtained by running instruments into the borehole at the end of a wire help to obtain vital data on the lithology, porosity and fluid saturations. **The range of investigation, a few decimeters, represents a survey area that is much larger than that investigated by core analysis.**

I.2.2 Well Characteristics

The production potential of the wells is another fundamental aspect in assessing the "value" of the reservoir discovered. To do this, **well tests** are performed, consisting in the measurement of the surface flow rates and surface and **downhole** fluid **pressures**. Well tests also help to obtain very important information about the reservoir, including the average permeability, but, in this case, in a radius of up to several hundred meters around the well.

In some situations, different layers are identified, a fractured formation is recognized, or a barrier may be detected. Before bringing in a well, these tests also help to select the appropriate well completion procedures for production.

I.2.3 Recovery Mechanisms

The withdrawal of fluids by the wells lowers the pressure of the fluids remaining in the reservoir.

In one-phase conditions, the laws and interpretation are relatively simple. By contrast, the physical laws governing multiphase flows in porous media are rather poorly known, because of their complexity and difficulty of analysis. In practice, the introduction of relative permeabilities breaking down the flow capacities between fluids (gas/oil/water) as a function of saturation helps to generalize the simple laws of one-phase flow.

The study of the mechanisms causing the displacement of fluids towards the wells by **natural drive** (primary recovery) places emphasis on the knowledge of the fluids in the reservoir and of its heterogeneities.

Depending on the type of reservoir and fluids, the **recovery rate** may not exceed a few percent of the volumes in place, with an average of 25% for the oil, or, on the contrary, it could be as high as 75% or more for gas (Fig. 4).

The term **reserves** is applied to the recoverable volumes that appear producible:

Reserves = volumes in place x recovery rate

If the natural forces of the reservoir do not allow good recovery of the **oil**, or even merely to accelerate production, **secondary or standard artificial recovery** is resorted to. This generally consists of the injection of water or associated gas. Since the recovery is rarely above 40%, other **improved me-**

thods (enhanced oil recovery) are designed and implemented whenever profitable. These include the injection of CO_2 or of chemicals added to water, intended to boost the oil recovery further.

RECOVERY STATISTICS	
Type of reservoir	Recovery
One-phase oil (1) (a few %)	< 10%
Dissolved gas drive (2)	5 to 25%
Gas-cap drive (3)	10 to 40%
Water drive (4)	10 to 60%
Gas (figure similar to (1))	60 to 95%

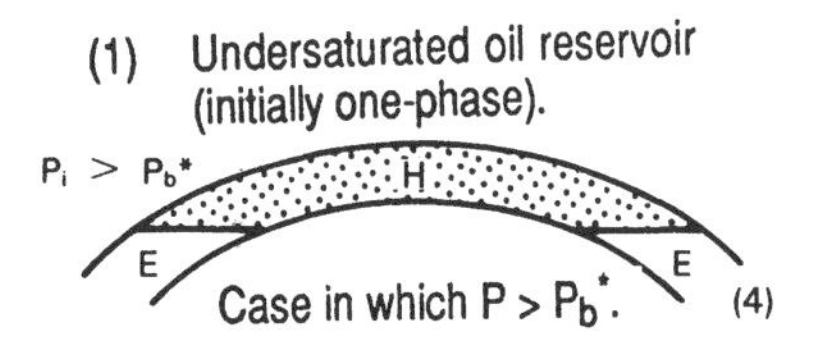

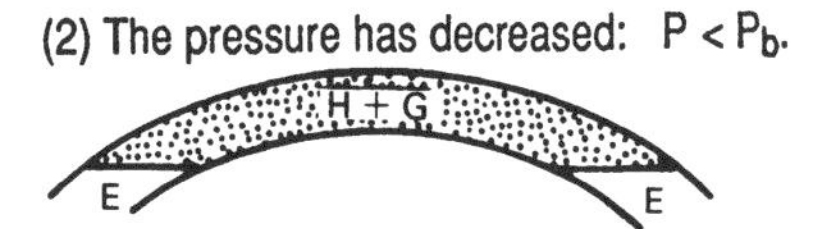

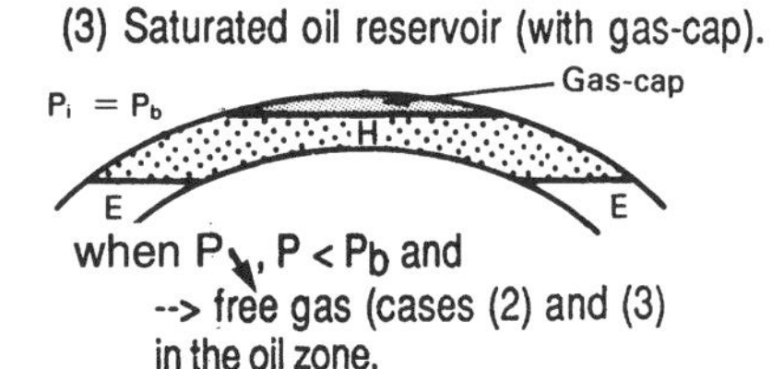

* Pb = bubble point.

Fig. 4

Mechanisms of primary recovery.

For **heavy oil**, the use of **thermal methods**, by raising the reservoir temperature, helps to improve production and recovery, by making the oil more fluid (**steam** injection, in situ combustion).

These reservoir studies characterize:

(a) The image of the reservoir → volumes in place.
(b) The well potential → productivity.
(c) The recovery mechanisms → (recoverable) reserves.

They serve to obtain the schematic **reservoir model**, which represents the synthesis of the data and of our knowledge about each reservoir.

I.2.4 Uncertainties

A large body of data is thus compiled, but it represents only an infinitesimal portion of the actual space of a reservoir. Thus the concept of uncertainty must be kept constantly in mind, and this means the conception of new approaches in the analysis and use of the data. What are the possible margins of error in the volumes in place of oil N, gas G, and also of water (aquifer) W? And in the recovery mechanisms? What are their consequences on our production forecasts over time of oil Np (t), of gas Gp (t) and of water Wp (t)? Better control of the probabilistic factors is necessary in this area (Fig. 5).

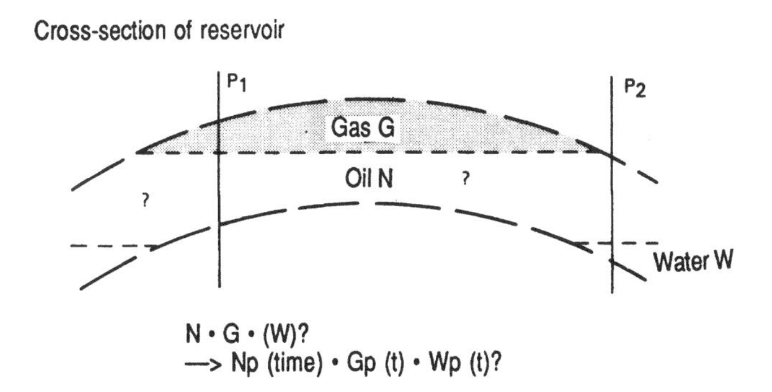

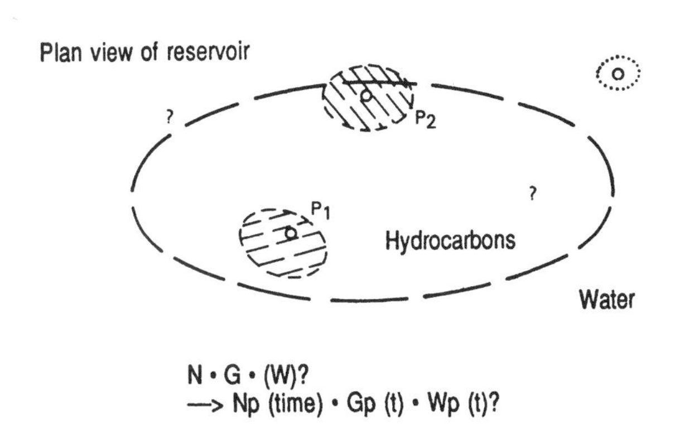

Fig. 5

Knowledge required and uncertainties.

I.2.5 Simulation of Reservoir Production

It is necessary to emphasize the fact that, while the physical laws governing reservoirs are rather general, the complexity of the porous media and the variety of the fluids are so great that the production of each reservoir represents a special case, and must give rise to a series of specific forecasting studies.

For many years, these forecasting studies were conducted by manual calculations, and by analogy with the results of other fields. The advent of the computer allowed a much more accurate simulation by the discretization of the reservoir in space and time, which is thus represented by a complex **numerical model** integrating the specific reservoir data and the laws of flows in porous media (Fig. 6). However, it is indispensable to make simplified calculations (manual or on a microcomputer) before going on to a large model, especially if the information is not extensive at the outset of production.

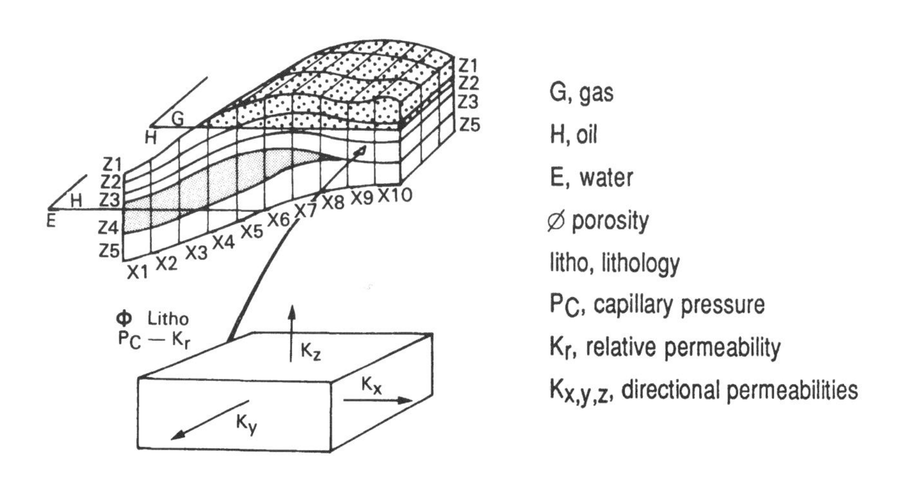

Fig. 6

Mathematical reservoir simulation model.

I.2.6 Development Project, Optimization

Several development plans are compared (number of wells, locations, injection, etc.) on the simulation model(s).

The initial choices naturally take account of related techniques, including well and surface production, and depend on the economic data. In each case, these models help to assess **probable future production** (or at least an order of magnitude), leading to an anticipated income, which is compared with the capital investment (drilling, completion, crude oil processing and transport), with the corresponding operating costs and the tax and market conditions. The configuration yielding the best return is adopted.

In conclusion, the working methods of the reservoir specialist involve three classic phases, analytical, then synthetic, and finally forecasting. These different phases can be repeated as new boreholes are drilled, and can be supplemented during the development phase and even throughout the life of the reservoir. The table on page 12 summarizes the main phases of reservoir engineering.

We must also point out that, in this speciality, which demands both rigor and imagination, the research effort must be sustained to pursue the challenge made to Nature.

I.3 THE PLACE OF RESERVOIR ENGINEERING IN PRODUCTION

After the discovery well(s), operations go from an "exploration" function to a "production" function. The successive study and operational phases culminating in the field development and production program, described in detail in Chapter 10, can be classed systematically in two groups:

(a) Assessment studies, which are chiefly methodological, essentially with reservoir engineering.
(b) Material operations, of which the main aspect is technological: drilling, well testing, completion, gathering lines, crude oil processing and transport.

The table on page 13 shows these different functions.

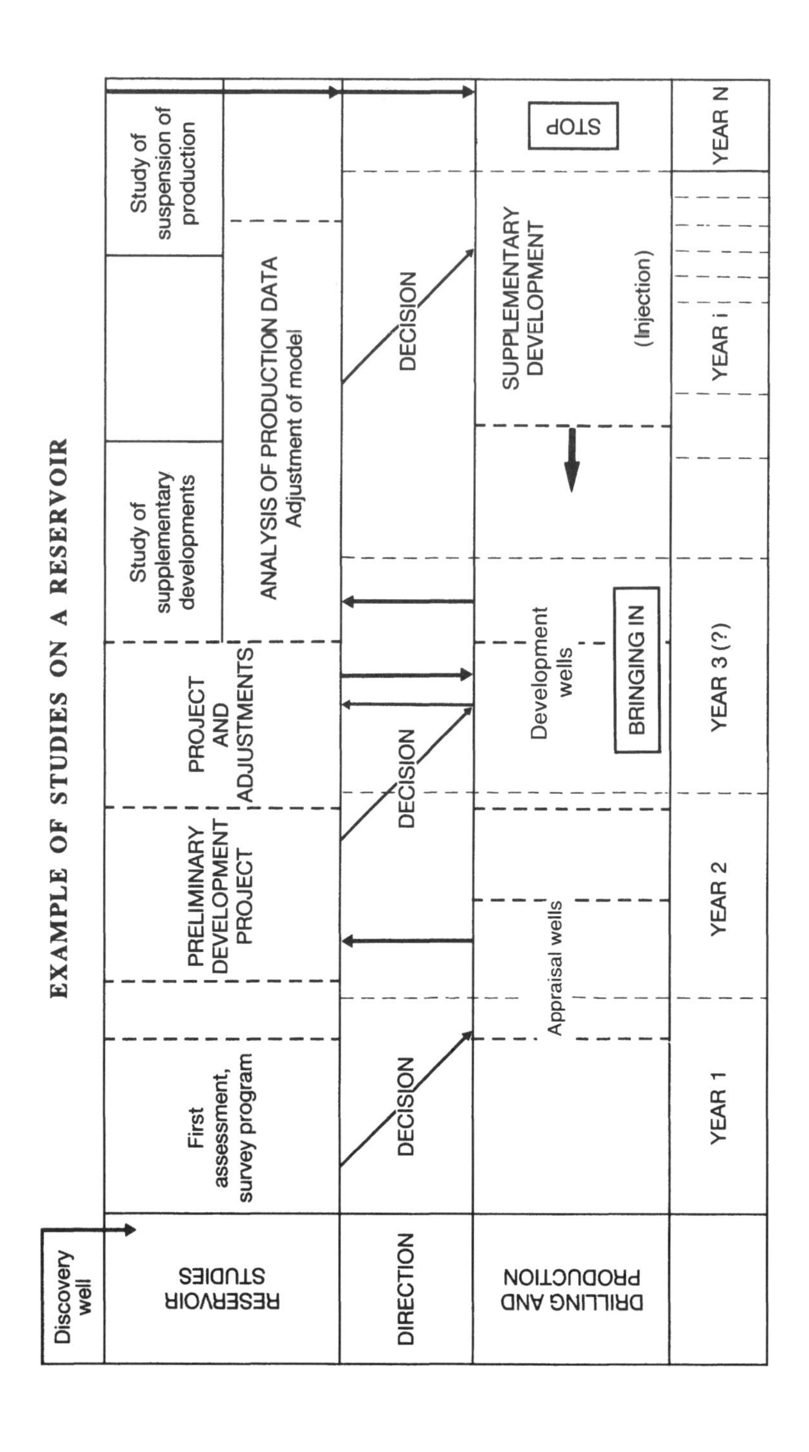
EXAMPLE OF STUDIES ON A RESERVOIR
Discovery well
RESERVOIR STUDIES
DIRECTION
DRILLING AND PRODUCTION
First assessment, survey program
PRELIMINARY DEVELOPMENT PROJECT
PROJECT AND ADJUSTMENTS
Study of supplementary developments
ANALYSIS OF PRODUCTION DATA
Adjustment of model
Study of suspension of production
DECISION
DECISION
DECISION
Appraisal wells
Development wells
BRINGING IN
SUPPLEMENTARY DEVELOPMENT
(Injection)
STOP
YEAR 1
YEAR 2
YEAR 3 (?)
YEAR i
YEAR N

PLACE OF RESERVOIR ENGINEERING

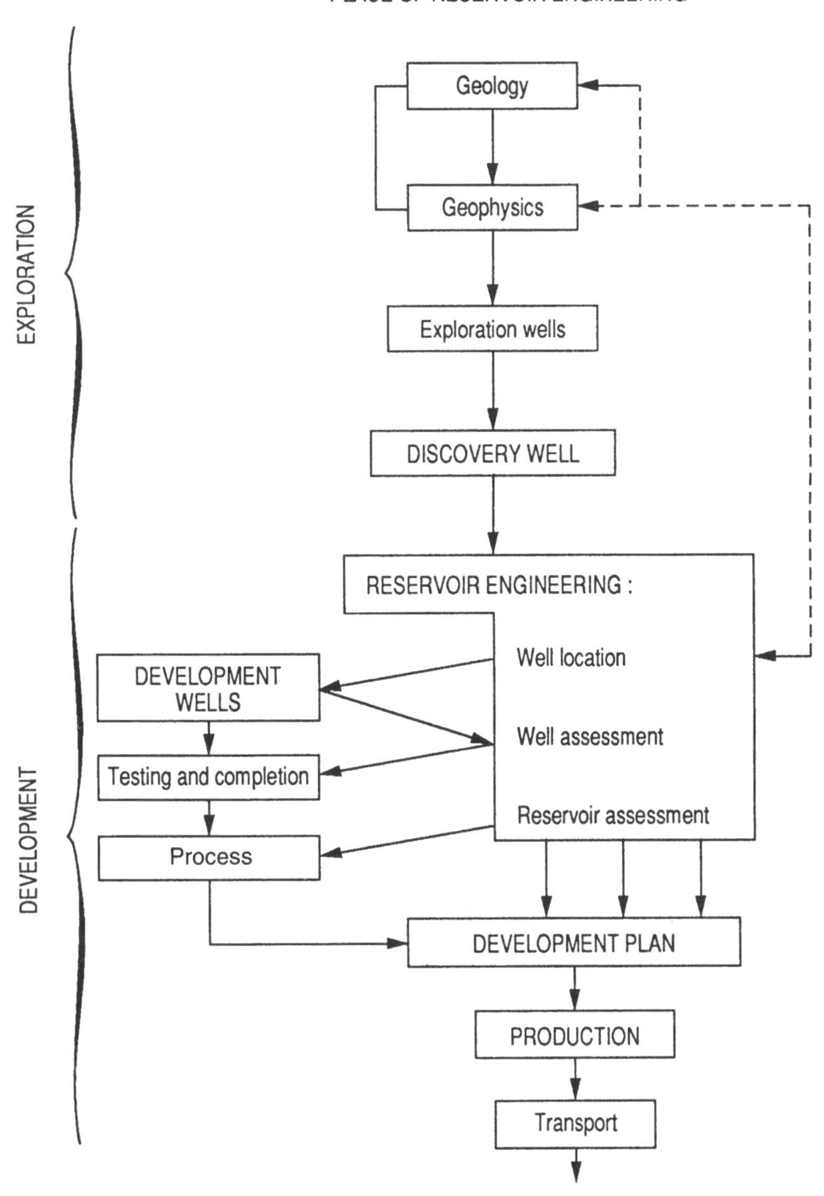

Chapter 1

RESERVOIR GEOLOGY AND GEOPHYSICS

1.1 GOALS
Image of the Reservoir (Geological Model)

The image of the reservoir develops when the forms, boundaries, internal architecture (heterogeneities), distribution and volumes of the fluids contained in the reservoir are known — or at least, to start with, approximated.

The methods used are partly included under the term of reservoir (and production) geology, and are based on **petroleum geology** and **geophysics**.

Figure 1.1 shows the different basic parameters involved in drawing up an image of the reservoir.

These techniques also draw heavily on the direct and indirect analysis of the information obtained in the wells:

(a) Direct analysis: **core analysis**, and **PVT analysis of fluids** (pressure, volume, temperature). These measurements are performed in the laboratory.

(b) Indirect analysis: well logging, in which some logs are recorded during drilling and some during production. Recordings of physical parameters obtained by running instruments into the borehole at the end of a wire, logs help to obtain vital data on the lithology, porosity and fluid saturations. **The range of investigation, a few decimeters, represents a survey area that is much larger than that investigated by core analysis.**

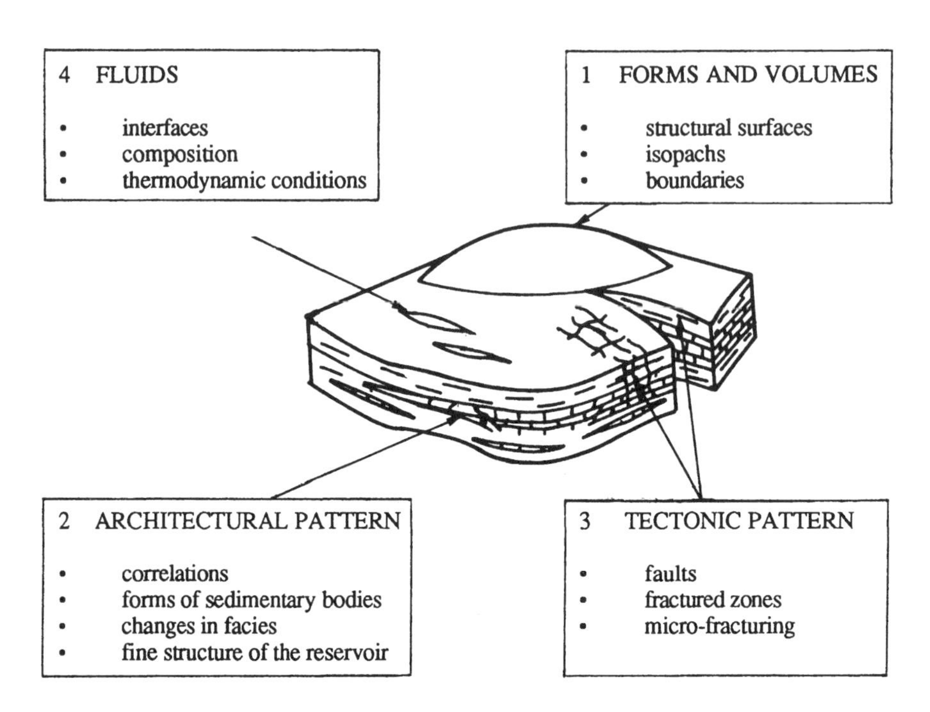

Fig. 1.1

Other methods are also indispensable to characterize this reservoir image:

(a) **Seismic surveys**, which give the form of the reservoir, the faults, and sometimes the variations in facies and fluid boundaries.
(b) **Sedimentology** which, from the analysis of cores, cuttings and logs, defines the nature of deposition, its extension and its probable heterogeneities.
(c) **Chemical measurements** for the mineralogical composition, the percentage of organic matter, and the hydrocarbon family.
(d) **Tectonics or microtectonics** for the detailed description of the fractures from the core analyses, overall surface studies, and aerial and satellite photographs.
(e) **Production data** from tests to determine the flow rates, interferences between wells, calculation of transmissivities by flow rates or pressure build-ups, temperatures, type and specific gravities of fluids in bottom-hole conditions, pressure distribution, heterogeneities (fractures, boundaries by faults or pinchouts).

This list is obviously incomplete.

This image, in other words the **geological model**, is used to:

(a) Enhance the value of a discovery by calculating the **volumes in place** (Chapter 4).
(b) Decide on the location of the wells, which is very important for development.
(c) Provide "static" details which are introduced into the simulation model, which is designed to supply production forecasts and to identify the ideal development method (Chapter 9).

The **results** are illustrated by:

(a) Vertical profiles such as composite well logs.
(b) Correlation cross-sections and facies cross-sections (Fig. 1.2).

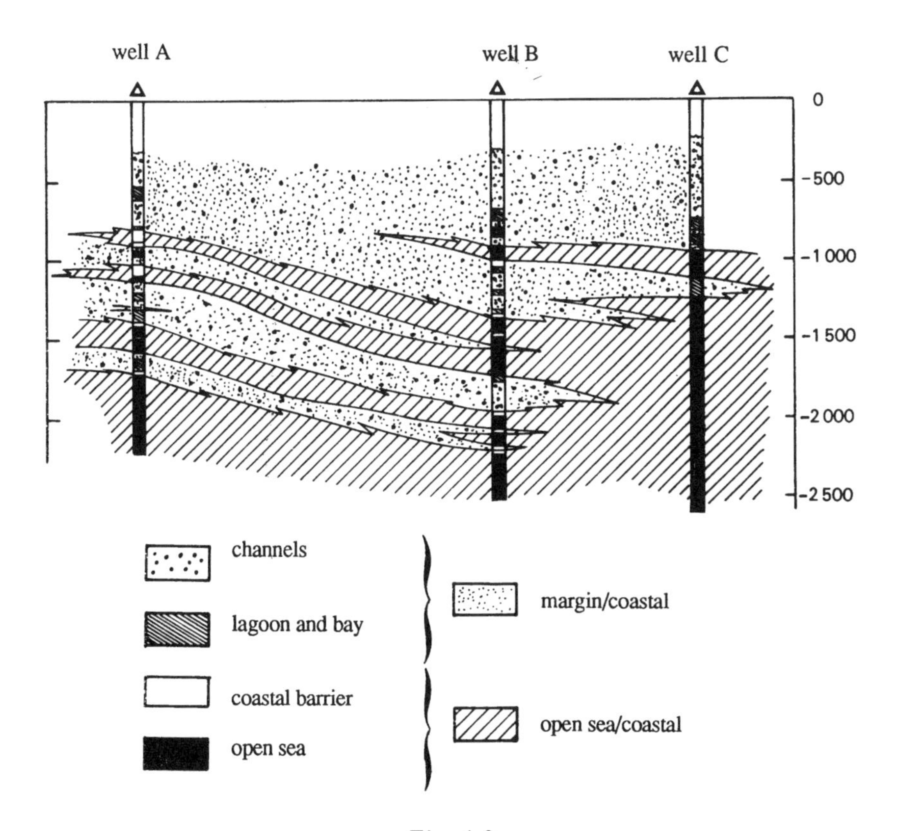

Fig. 1.2

Correlations between wells.

(c) Isobath, isopach (same thickness), isofacies and isopercentage maps, especially "h x ϕ" and "h x k" maps, which characterize the value of the reservoir.

This geological model is not fixed. It must be reviewed continually based on the data gathered throughout the lifetime of the field.

Before reservoirs and oil and gas traps are dealt with some details concerning hydrocarbon generation and migration are provided in Section 2 below.

1.2 HYDROCARBON GENERATION AND MIGRATION

1.2.1 Generation

Hydrocarbons originate in the **organic matter** contained in sediments.

This organic matter very often consists of microalgae and microorganisms which are deposited in aquatic environments, especially on the sea bed.

During deposition and at the onset of the burial of these sediments, most of the materials are broken down, particularly by oxidation.

The remaining part contains **kerogen** which, due to the gradual burial of the sediments causing high compression and a sharp rise in temperatures — during very long time intervals — is transformed into **hydrocarbons** by thermal cracking.

The kerogen, which is initially immature, is converted into **oil** above 50 to 70°C. Around 120 to 150°C, the oil is cracked in turn, yielding first wet and then dry **gas**. The **"oil window"** lies between these two temperature levels. It usually corresponds to burial depths between 1000 and 3500 m (Fig. 1.3).

Source rocks are rocks containing this organic matter in sufficient quantities: at least 0.5% organic carbon and 100 ppm (parts per million, 1/1,000,000) of organic matter extractable by solvents.

These source rocks are chiefly **shales**, and sometimes carbonates. They occur in concentrated form in certain horizons, or in dispersed form.

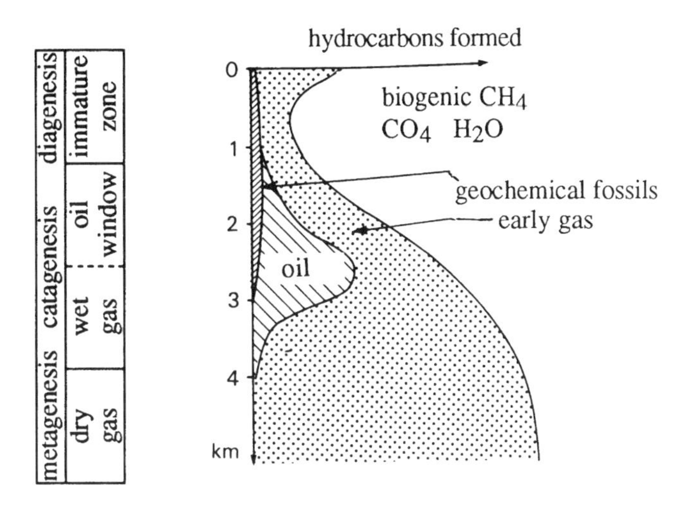

Fig. 1.3

General diagram of hydrocarbon generation
as a function of depth for an average geothermal degree.

The oil potential of these source rocks can be analyzed by the pyrolysis at increasing temperature of a sample of ground rock (**Rock Eval** method). In this way, the "source rock" formations can be identified during drilling.

1.2.2 Migration

The hydrocarbons formed in the source rocks are generally expelled from them towards lower pressure zones, which means that, especially with respect to shales, the source rock was still permeable at the time of migration. Two successive migrations can be distinguished.

1.2.2.1 Primary Migration

The hydrocarbons are expelled from the source rock to a more porous adjacent environment, where the fluids can move. At the start of the process, it is essentially the forces associated with burial and compaction that cause this expulsion, which may be lateral, *per descensum* or *per ascensum*.

1.2.2.2 Secondary Migration

This takes place from the neighborhood of the source rock to the reservoir, where the hydrocarbons are **trapped**. This upward movement takes place in one (or more) reservoirs through faults, fracture zones, etc. **Why is the flow upward?** Simply because the densities of gas and oil are lower than that of water. Under the effect of gravity, the hydrocarbons therefore rise naturally towards the earth's surface (a. in Fig. 1.4).

They sometimes travel very long distances — hundreds of kilometers, it has been said, in the Middle East. The forces acting are gravity and capillarity (Chapter 2). The distances travelled may also be very short, as in lenses (b. in Fig. 1.4).

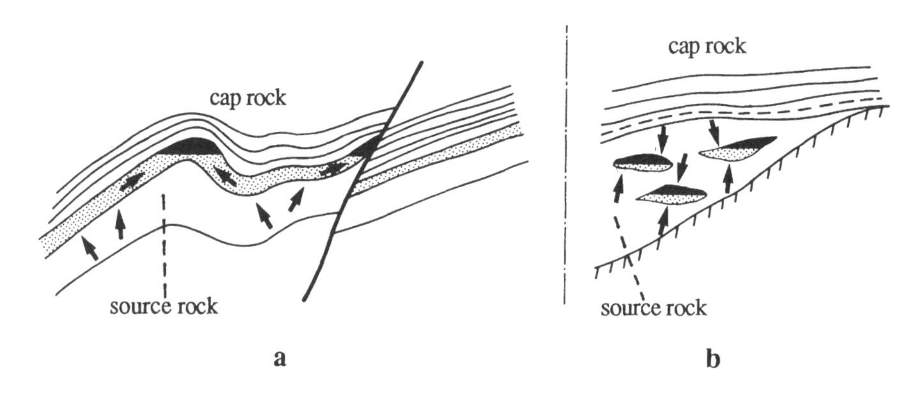

Fig. 1.4

Different concepts of hydrocarbon migration.

However, migration mechanisms are still very poorly understood. Gas moves by seepage or dissolved in water. Oil, which is insoluble in water, flows in a continuous phase, this being caused first by compaction and then by gravity.

1.3 RESERVOIRS

1.3.1 Definitions

A **reservoir** is a porous and permeable subsoil formation containing a natural, individual and separate accumulation of hydrocarbons (oil and/or gas). It is bounded by a barrier of impermeable rock and often by an aquifer barrier, and characterized by only one natural pressure system.

This definition incorporates the concepts of:

(a) Petrophysical properties: porosity/permeability.
(b) Natural accumulation of hydrocarbons.
(c) Cap rock/closure: rock or aquifer barrier.
(d) Natural pressure state before production.

The reservoir, "impregnated with hydrocarbons", is normally subdivided into **layers**, or levels, or units (which are individualized lithologically, by the analysis of cores and well logs).

A reservoir consists of one or more superimposed or laterally nearby pools. Some reservoirs are made up of some dozen or even hundreds of pools. These are called "multi-layer" reservoirs.

The essential feature of these pools is that they are **porous media**. The fluids are stored and move in pores of minute size (about 1 μm), leading to significant capillarity forces which enter into action (Chapters 2 and 6).

A pool may contain oil, gas, or both fluids superimposed.

1.3.2 Reservoir Rocks

The main reservoir rocks are made up of sandstones and/or carbonates (99% of the total). These are sedimentary rocks, in other words rocks made up of sediments formed at the earth's surface by debris (mineral, animal and vegetable) or chemical precipitations. They are stratified in successive beds.

1.3.2.1 Sandstone Reservoirs

These are by far the most widespread, accounting for 80% of all reservoirs and 60% of oil reserves.

The rock is formed of grains of quartz (silica SiO_2). If the grains are free, they form **sand**. If the grains are cemented together, they form **sandstone**. Shaly sandstone, carbonate sandstone, etc. also exist (Fig. 1.5a).

Sandstones are very often stratified in a simply superimposed pattern, or with intersecting beds. This results from successive depositions at the shoreline or in the form of fluvial or deltaic alluvia.

A vertical cross-section generally shows alternating deposits of sands, shaly sands, silts and clays or shales, forming a "shaly/sandstonelike" whole.

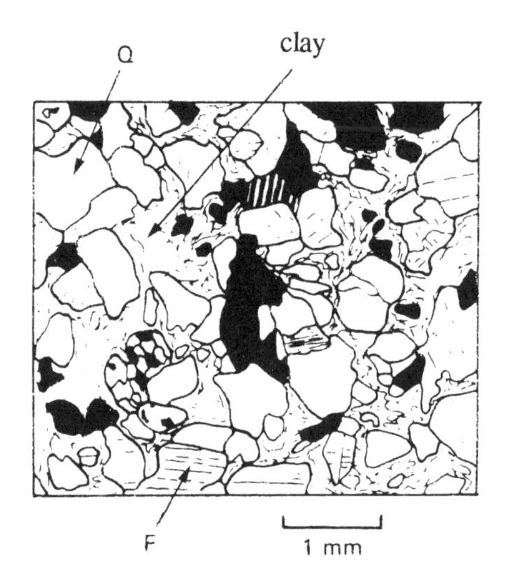

Quartz and feldspars with a shaly cement.

Fig. 1.5a

Shaly cement sandstone.

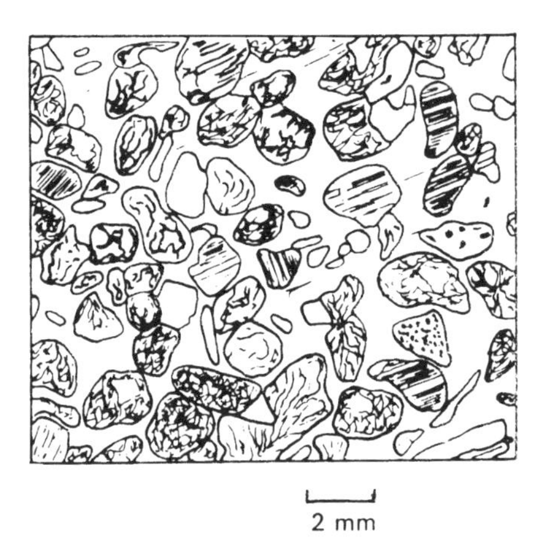

Debris of various types (~clasts) buried in a calcite cement.

Fig. 1.5b

Skeletal limestone.

1.3.2.2 Carbonate Reservoirs

Carbonate rocks are of varied origins:

(a) **Detrital:** formed of debris (grains of limestone, shells, etc.).
(b) **Constructed**: of the **reef** type.

(c) **Chemical:** formed by the precipitation of bicarbonate and originating in marine muds.

They consist of **limestone** ($CaCO_3$) and/or **dolomite** ($CaCO_3$, $MgCO_3$), and often display reservoir characteristics. Shaly carbonates also exist. But "marls", which contain between 35 and 65% shale, are not reservoirs. This is because a small proportion of shale, binding the grains together, considerably decreases the permeability (Fig. 1.5b).

Chalk and karst are two special cases:

(a) **Chalk** is formed by the stacking of small single-cell algae (coccoliths). The porosity is high, but the permeability is low or very low (about 1 millidarcy), the pores being very small (0.2 to 2 μm).
(b) **Karst** originates in an emerged limestone mass subjected to erosion. Rainfall causes dissolution, forming a very discontinuous network of "pockets". Very porous and permeable, but highly variable by zones.

1.4 TRAPS

The existence of reservoirs impregnated with hydrocarbons indicates the presence of a trap capable of stopping the hydrocarbons from migrating. A trap is an area bounded by a barrier lying upwards from the flow. The reservoir's upward seal is provided by a layer of impermeable rocks called cap rock (usually shale, salt or anhydrite).

1.4.1 Classification of Traps

There may be structural, stratigraphic or combination traps (Fig. 1.6):

(a) **Structural traps:** due to deformation of the rock, simple anticlines or faults. Round traps are called domes.
(b) **Stratigraphic traps:** trapping is due to variations in facies, the rock becoming impermeable laterally. Examples are sandstone lenses in a shaly/sandstonelike whole, depositional or erosional pinch-outs, and carbonate reefs.
(c) **Combination traps:** eroded anticlines, traps associated with salt domes.

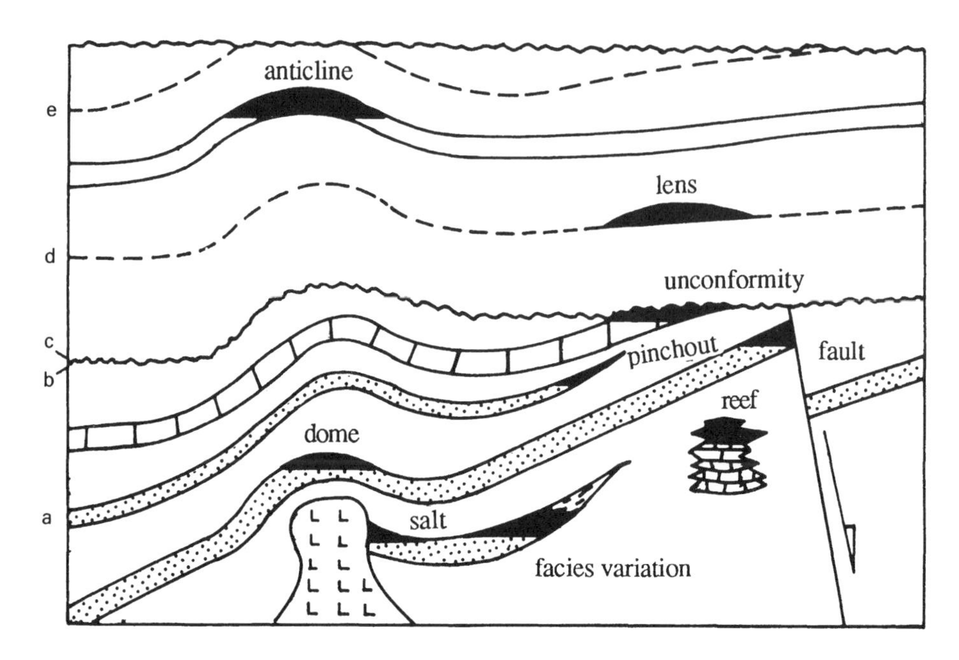

Fig. 1.6
Different types of trap.

1.4.2 Characteristics

Irrespective of the trap, its size is determined from the following criteria:

(a) Closed area: bounded by the deepest closed isobath (some dozen or hundred square kilometers).
(b) Closure: height between the uppermost and lowest points (spillpoint) generally a few dozen or hundred meters.
(c) Impregnated zone closure: thickness impregnated with hydrocarbons.
(d) Filling ratio: practical closure/closure.

The numerical values of these trap characteristics are extremely variable (Chapter 1, Section 6).

1.5 RESERVOIR SEISMIC SURVEY

1.5.1 Principles

The continuous progress achieved in seismic techniques and interpretation make them invaluable in understanding reservoirs. Given the relatively small size and thickness of the reservoirs compared with the basin being explored, "conventional" seismic prospecting must be adapted to **reservoir survey**, which is in full development at the present time.

Let us briefly review the principle of seismic reflection shooting: earth tremors created at ground level (dynamite blasts, air gun, vibro-seismic method, etc.) generate waves which are reflected in the subsoil against the boundaries of geological layers (markers). They return to the surface where they are detected. The primary aim of seismic reflection shooting is thus to obtain a **structural image** of the geological layers (Fig. 1.7).

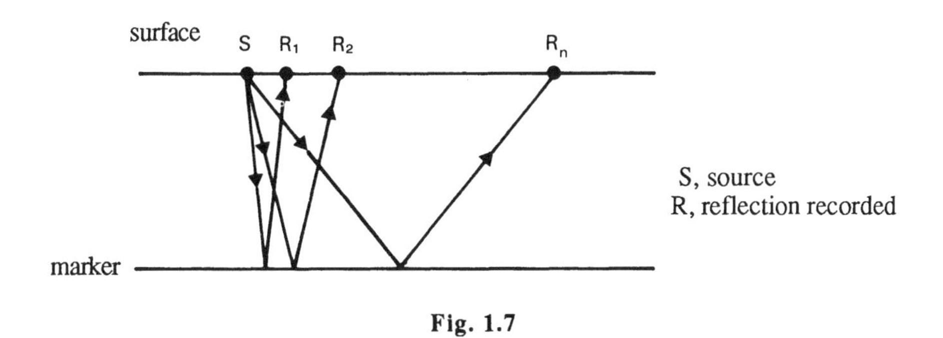

Fig. 1.7

In actual fact, these reflections, "picked" on seismic profiles (which are vertical time sections), help to plot **"isochrone"** (equal time) curves. They are then converted into **isobaths** (equal depth), using the acoustic logs recorded in the wells, which give us the wave propagation velocity in each layer (Chapter 2).

As we pointed out, exploration seismic survey is inappropriate for the structural analysis of a reservoir. A tighter recording grid spacing is necessary, due to the size of the reservoirs, and the source and recording array must also be adapted to their depth.

As to the grid pattern in exploration seismic survey, the density of the seismic profiles rarely exceeds one profile per kilometer. In a reservoir survey, profiles are required at 500, 250 or even 100 m intervals.

One of the basic problems concerns the **accurate identification of the different markers**. Emphasis is also placed on the means employed to identify them with respect to the top and base of the reservoir(s). Vertical seismic profiles, offset vertical seismic profiles, sonic logs and synthetic seismic films are some of the tools that serve to obtain and to refine the depth/time calibration of the seismic sections.

Conventional 2-D (two-dimensional) shooting presents the drawback of bringing all the events directly underneath the profile, distorting the time image of the subsoil.

1.5.2 3-D Seismic Survey

The interpretation must also be very thorough, in order to achieve the finest possible detection. This is done by **3-D shooting**, which places the reflections back in their true position.

Using this technique, the profiles are very close together (at 50 m intervals, for example) and they are interpreted with **three-dimensional** migration. What is migration?

The principle of **migration** is the following:

A sloping element M of a reflecting horizon, struck by a seismic raypath from O, gives a response on the seismogram corrected back to O, which does not appear on seismic trace 2 passing through M, but on trace 1 passing through O in M'. In other words, on a seismic section, sloping events are not in their true place, and the migration of the section is designed to reposition them (M' → M) (Fig. 1.8).

These 3-D methods have the disadvantage of being very costly (grid pattern + interpretation), especially for onshore surveys, but they are increasingly used offshore.

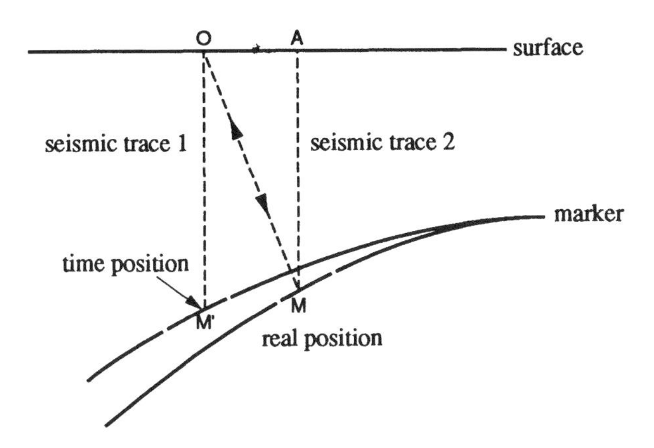

Fig. 1.8 Principle of migration.

1.5.3 Vertical Seismic Profile (VSP) and Offset Vertical Seismic Profile (OSP)

The depth/time calibration is carried out from recordings made in wells by VSP and OSP. These methods also help to obtain an accurate image of the reservoirs in the neighborhood of the well. The recordings in the wells are very close together, generally 10 to 20 m apart (Figs 1.9a and 1.9b).

1.5.4 Detection of Fluids

The presence of gas (and, to a lesser degree, oil) causes a decrease in the apparent density "d" of the reservoir and in the acoustic propagation velocity "v". **The contrast in acoustic impedance** "z" ($z = v \times d$) between the gas zone and the cap rock, or between the gas (or oil) zone and the aquifer, gives rise to high reflection coefficients that lead to different marked reflections called **bright spots, flat spots** and **pull-downs**, depending on each specific case (Fig. 1.10). These particular figures are not generally obtained, and only appear in favorable and limited conditions.

Thus, the recording laboratories and seismic devices are not "sniffers", but, in certain limited cases, they can help to detect **probable** impregnated reservoirs, and subsequently, after discovery, they can determine the extension of the reservoir. This leads to the possible detection of hydrocarbons directly from the earth's surface. Long live technical progress!!

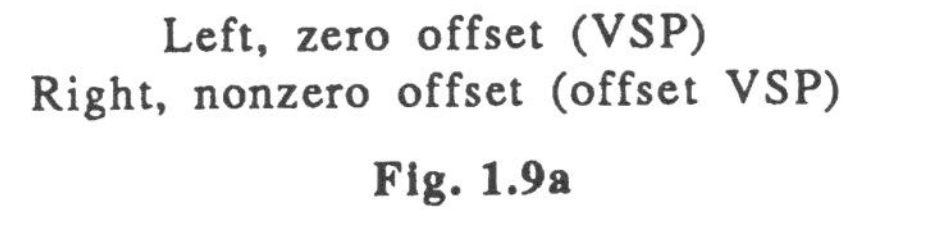

Left, zero offset (VSP)
Right, nonzero offset (offset VSP)

Fig. 1.9a

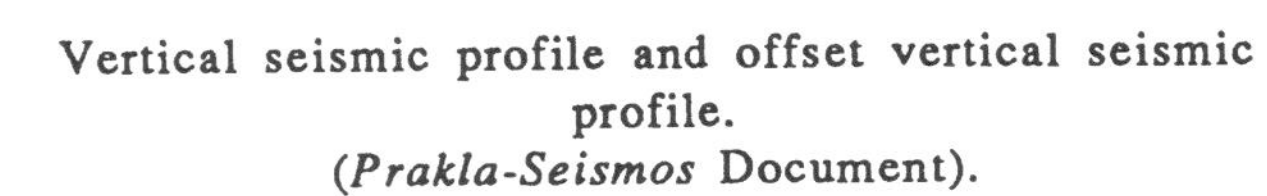

Vertical seismic profile and offset vertical seismic profile.
(*Prakla-Seismos* Document).

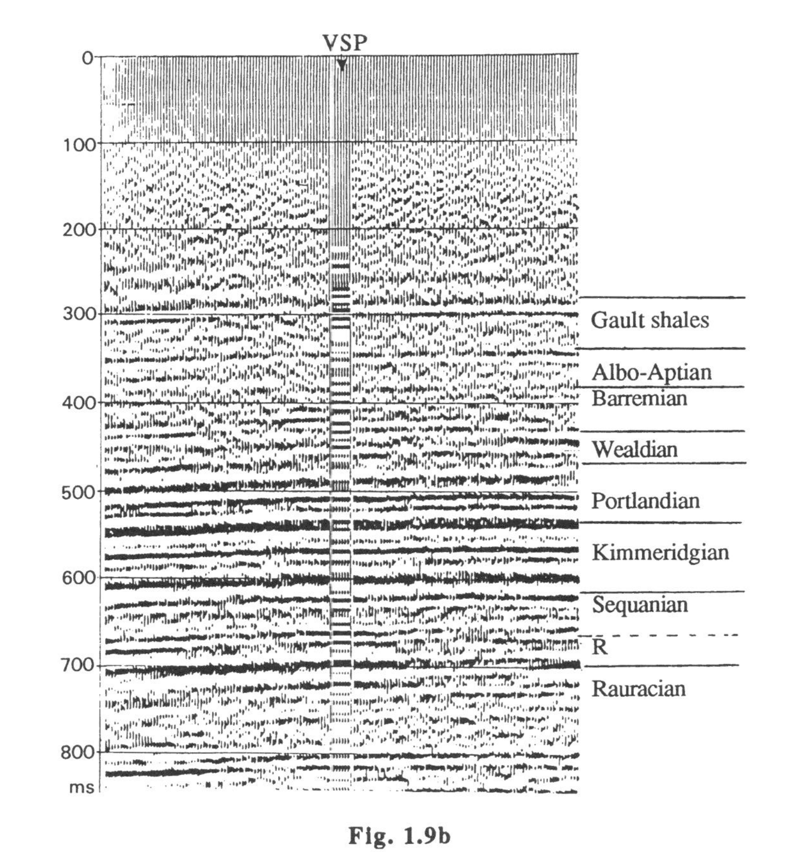

Fig. 1.9b

Calibration of a VSP against a reflection shooting section.
(*IFP* Document).

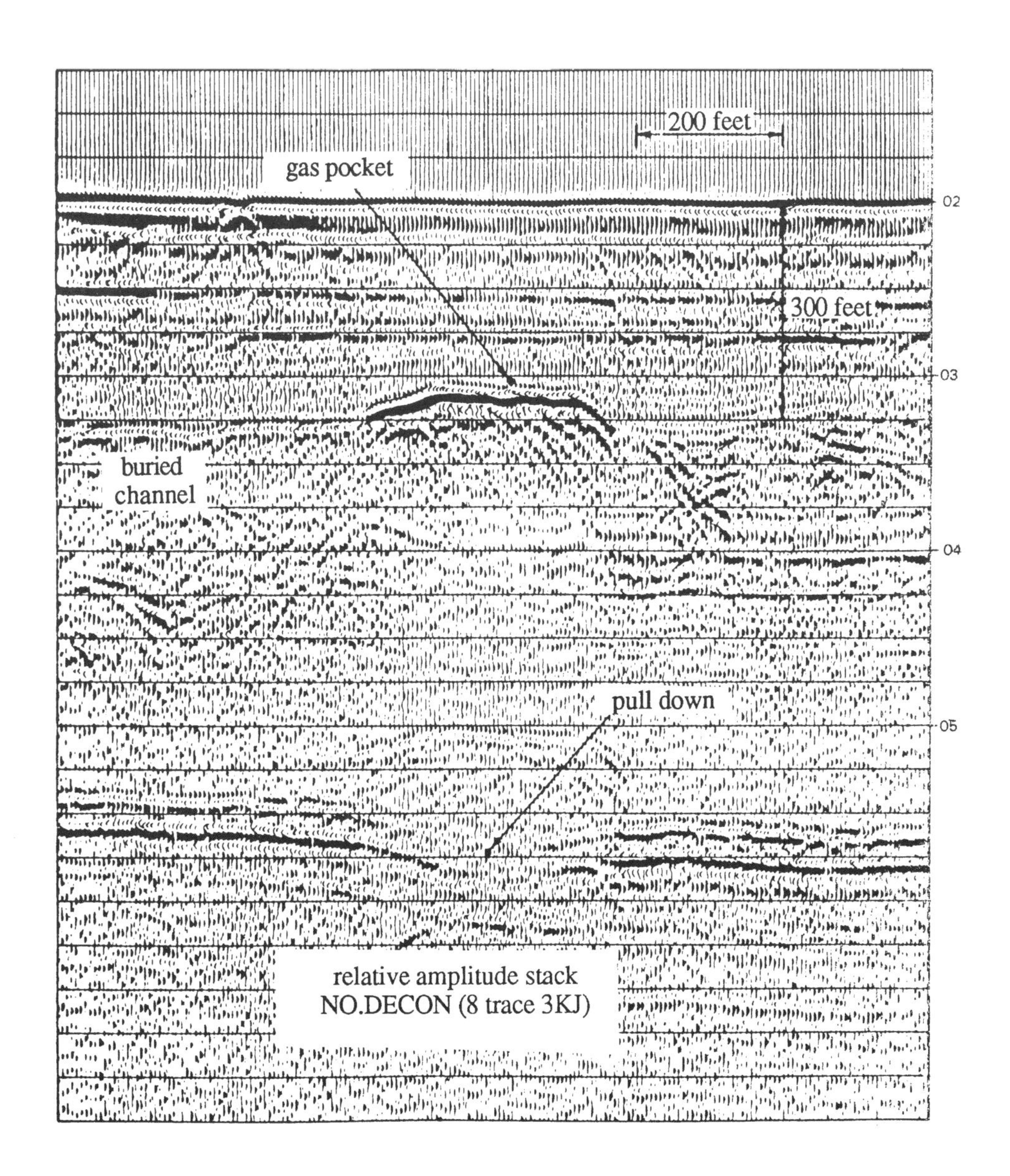

Fig. 1.10

Gas pocket bright spot.

1.6 EXAMPLES OF RESERVOIRS

A number of examples of fields have been selected as being representative of the type of trap, fluids, and petroleum zones known, especially in France.

There are approximately 30,000 fields recognized in the world today. Their size, characteristics and depth are extremely varied. **Ghawar** in Saudi Arabia (the world's biggest) has a closed area of more than 8000 km^2, and estimated reserves of 10 Gt (10 billion tons).

At the other extreme, some developed fields are smaller than 1 km^2, and have reserves of less than 100,000 t.

Among the large fields, **Perrodon** distinguishes the following:

(a) Major fields: reserves > 14 Mt oil (100 Mb)[1].
(b) Giant fields: reserves > 70 Mt oil (500 Mb).
(c) Supergiant fields: reserves > 700 Mt oil (5 Gb).

For gas, the same figures are multiplied by 1000, in cubic meters.

Note that 90% of the **giant** fields are anticlines. They only account for 1% of the total number of fields, but represent two-thirds of world production.

Fields are found at depths of a few dozen meters and others at depths of over 6000 m.

The age of the fields varies considerably. There are three highly favorable periods, that can be correlated with periods of tectonic activity: the Tertiary (Alpine period), the Upper Cretaceous (Laramian period), and the Permo-Carboniferous (Hercynian period).

These examples of fields are described at the end of the book.

1. b, barrel: 1 barrel = 0.159 m^3, or approximately 0.14 t.

STRATIGRAPHIC SCALE

Era	System	Stage
Quaternary (Anthro-pozoic)	Holocene (Neolithic) Pleistocene (Paleo-lithic)	Flandrian Tyrrhenian Sicilian
Tertiary (Cenozoic)	Pliocene	Calabrian (Villafranchian) Astian Plaisancian
	Miocene	Sahelian (Pontian) Vindobonian Burdigalian
	Oligocene	Aquitanian Chattian Stampian Sannoisian
	Eocene	Ludian Bartonian Lutetian Ypresian Sparnacian Thanetian Montian

Era	System	Stage
Secondary (Mesozoic)	Upper Cretaceous (Neo-cretaceous)	Danian Senonian Turonian Cenomanian
	Lower Cretaceous (Eo-cretaceous)	Albian Aptian Barremian (Urginian) Hauterivian Valanginian
	Upper Jurassic (Malm)	(Purbeckian) Portlandian Tithonic Kimmeridgian Sequanian Rauracian Argovian Oxfordian Callovian
	Middle Jurassic (Dogger)	Bathonian Bajocian
	Lower Jurassic (Lias)	Aalenian Toarcian Charmouthian Sinemurian Hettangian Rhetian
	Triassic	Keuper Muschelkalk Variegated sandstone

Era	System	Stage
Primary (Paleo-zoic)	Permian	Zechstein or Thuringian Saxonian Autunian
	Carboni-ferous	Coal measures (Stephanian) (Westphalian) Dinantian (Culm)
	Devonian	Famennian Frasnian Givetian Eifelian Coblenzian Gedinnian Downtonian
	Silurian	Gothlandian Ordovician
	Cambrian	Potsdamian Acadian Georgian
	Precambrian (Algonkian)	
	Archean	

Chapter 2

CHARACTERIZATION OF RESERVOIR ROCKS

Petrophysics is the study of the physical properties of rocks. For a rock to form a reservoir:

(a) It must have a certain storage capacity: this property is characterized by the **porosity**.
(b) The fluids must be able to flow in the rock: this property is characterized by the **permeability**.
(c) It must contain a sufficient quantity of hydrocarbons, with a sufficient concentration: the impregnated volume is a factor here, as well as the **saturations**.

The methods used to characterize reservoir rocks are essentially core analysis and well logging.

2.1 POROSITY

2.1.1 Definition

Let us consider a rock sample. Its apparent volume, or total volume V_T, consists of a solid volume V_S and a pore volume V_p. The porosity ϕ is:

$$\phi = \frac{V_{pores}}{V_{total}} \qquad \text{expressed in } \%$$

The porosity of interest to the reservoir specialist, that which allows the fluids in the pores to circulate, is the effective porosity ϕ_u, which corresponds to the pores connected to each other and to other formations.

Also defined is the total porosity ϕ_t, corresponding to all the pores, whether interconnected or not, and the residual porosity ϕ_r, which only takes account of isolated pores (Fig. 2.1):

$$\phi_t = \phi_u + \phi_r$$

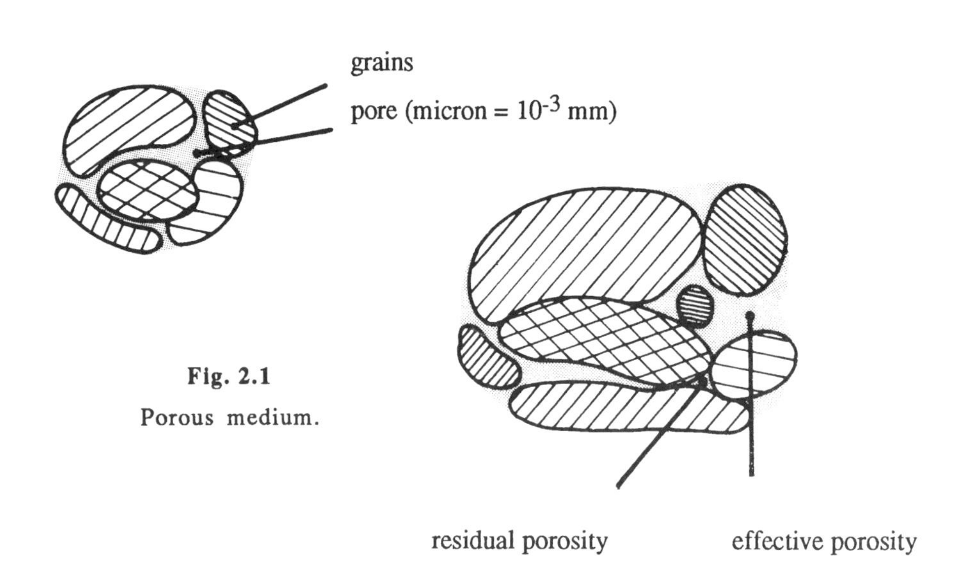

Fig. 2.1
Porous medium.

The effective porosity of rocks varies between less than 1% and over 40%.

It is often stated that the porosity is:

(a) Low if $\phi < 5\%$.
(b) Mediocre if $5\% < \phi < 10\%$.
(c) Average if $10\% < \phi < 20\%$.
(d) Good if $20\% < \phi < 30\%$.
(e) Excellent if $\phi > 30\%$.

A distinction is made between intergranular porosity, dissolution porosity (as in limestones, for example), and fracture porosity. For fractured rocks, the fracture porosity related to the rock volume is often **much less than 1%**.

As a rule, porosity decreases with increasing depth.

2.1.2 Determination of Porosity

The porosities are determined by core analysis or by well logging (see Section 4).

2.1.2.1 Core Analysis

The following equation applies: $\phi = \frac{V_P}{V_T} = \frac{V_T - V_S}{V_T} = 1 - \frac{V_S}{V_T}$

On a sample of generally simple geometric form, two of the three values V_p, V_S and V_T are therefore determined.

The standard sample (plug) is cylindrical. Its cross-section measures about 4 to 12 cm^2 and its length varies between 2 and 5 cm.

The plugs are first washed and dried. The measuring instruments are coupled to microcomputers to process the results rapidly.

A. Measurement of V_T

a. Measurement of the buoyancy exerted by mercury on the sample immersed in it (IFP apparatus) (Fig. 2.2)

The apparatus has a frame C connected by a rod to a float F immersed in a beaker containing mercury. A reference index R is fixed to the rod. A plate B is suspended from the frame.

(a) First measurement: the sample is placed on plate B with a weight P_1 to bring R into contact with the mercury.

(b) Second measurement: the sample is placed under the hooks of the float, and weight P_2 is placed on B to bring R into contact with the mercury,

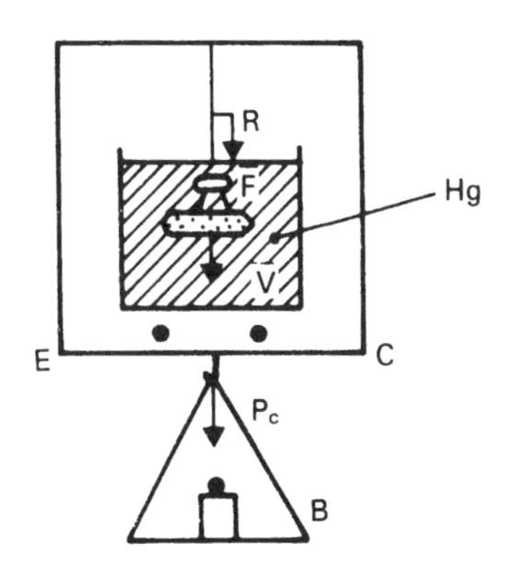

Fig. 2.2
IFP apparatus.

$$V_T = \frac{P_2 - P_1}{\rho Hg}$$

where ρHg is the density of mercury at the measurement temperature.

For the measurement of V_T to be valid, the mercury must not penetrate into the sample.

b. Use of a mercury positive displacement pump (Fig. 2.3)

Without a sample, using the piston, the mercury is pushed to the mark indicated on the reference valve. The vernier of the pump is set to zero.

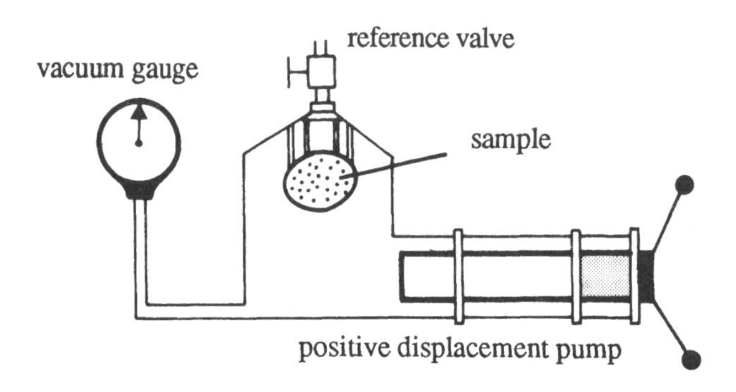

Fig. 2.3

Measurement of V_T with mercury pump.

With the sample in place, the mercury is again pushed to the same mark. The vernier is read, and volume V_T is obtained.

The accuracy is $\pm 0.01\ cm^3$. Here also, the measurement is only valid if the mercury does not penetrate into the pores.

c. Measurement

The foregoing methods are unsuitable if the rock contains fissures or macropores, because the mercury would penetrate into them.

Here a piece of cylindrical core's diameter "d" and height "h" can be measured using a sliding caliper:

$$V_T = \frac{\pi d^2 h}{4}$$

B. Determination of V_S

a. Measurement of the buoyancy exerted on the sample by a solvent with which it is saturated (Fig. 2.4)

This method is the most accurate, but it is difficult and time-consuming to achieve complete saturation. The operations are standardized.

The difference between the weights of the dry sample in air (P_{air}) and in the solvent in which it is immersed ($P_{immersed}$) gives V_S:

$$V_S = \frac{P_{air} - P_{immersed}}{\rho_{solvent}}$$

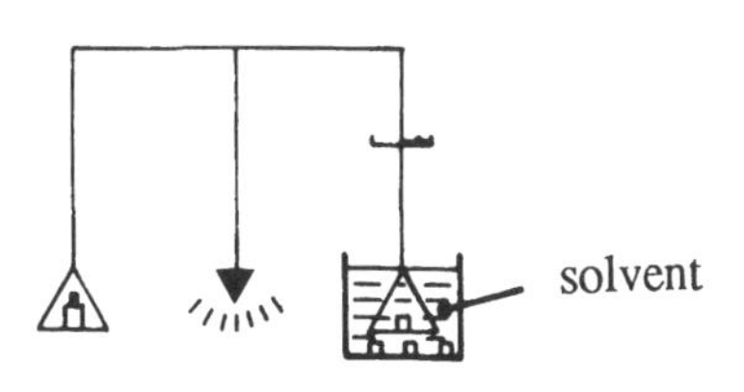

Fig. 2.4

Determination of V_S by immersion weighing.

b. Use of a compression chamber and Mariotte-Boyle's Law

Regardless of the specific apparatus used, it always has a chamber in which the sample is placed at a given initial pressure. The pressure is then changed by varying the volume of the chamber, and Boyle's Law is used to determine the volume not concerned by decompression, or V_S.

C. Determination of V_p, volume of effective pores

The pore volume can be measured directly, by measuring the volume of air in the pores, by weighing a liquid filling the pores, or by mercury injection.

a. Measurement of air in the pores

The mercury positive displacement pump is used for the purpose. After measuring V_T, the valve of the sample-holder is closed and the air in the interconnected pores is expanded. The variations in volume and pressure are measured and Boyle's Law used.

b. Measurement by weighing a liquid filling the effective pores

This liquid is often brine.

c. Measurement by mercury injection

In this case, the mercury never totally invades the interconnected pores. Hence the value obtained for V_p is under par.

D. A special method: fluid summation

This method involves the analysis of a "fresh" sample containing water, oil and gas. The distribution of these fluids is not the same as in the reservoir, because the core has been invaded by the mud filtrate, and then decompressed when pulled out. But the sum of the volumes of these three fluids, for a unit total volume of rock, gives the effective porosity of the sample (the total volume is determined by a mercury positive displacement pump).

2.1.3 Effect of Pressure

Rocks are compressible. In the reservoir, they are subjected to the geostatic pressure and the pressure of the fluids present in the pores. If this pressure falls due to production, the rock is compressed until a new equilibrium is reached. This is reflected by a reduction in porosity.

We can define:

$$c_r = \frac{1}{V_T}\left(\frac{\partial V_T}{\partial p}\right)_T \qquad \text{and} \qquad c_p = \frac{1}{V_p}\left(\frac{\partial V_p}{\partial p}\right)_T$$

Since the compressibility of the solid volume is negligible, we have:

$$c_p = \frac{1}{\phi}\, c_r$$

The value of c_p varies with the pressure, the type of rock, and the porosity. It varies from $0.3 \cdot 10^{-4}$ to $3 \cdot 10^{-4}$ $(\text{bar})^{-1}$.

In general, the porosity obtained by the methods described above is not corrected to account for the differences between reservoir and laboratory stress, because porosity variations are low, and a core cannot be representative of the entire reservoir.

2.2 PERMEABILITY

During production, the fluids flow in the rock pores with greater or lesser difficulty, depending on the characteristics of the porous medium.

2.2.1 Definition, Darcy's Law

The **specific** or **absolute permeability** of a rock is the ability of the rock to allow a **fluid** with which it is saturated to flow through its pores. Permeability can be determined by Darcy's Law, an experimental law.

Let us consider a sample of length dx and cross-section A, saturated with a fluid of dynamic viscosity μ, and crossed horizontally by a flow rate Q (**measured in the conditions of section dx**). In steady-state conditions, the upstream pressure is P, and the downstream pressure is P – dP. The lateral sides are impervious to fluids. If the fluid does not react with the rock, which is the general case:

$$Q = A \cdot \frac{k}{\mu} \cdot \frac{dP}{dx} \qquad \text{Darcy's Law}$$

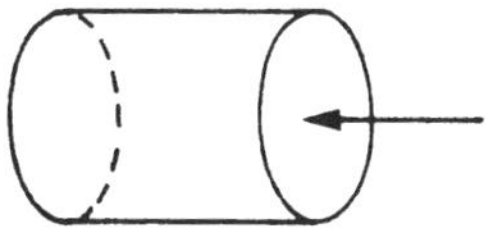

k is called the permeability coefficient, and is independent of the fluid considered as a first approximation (Section 2.2.5). It is the absolute or specific permeability of the sample in the direction considered. Permeability is expressed like an **area**.

In the SI International System, k is expressed in square meters, and this is an enormous unit!

(a) **SI** :

$$Q_{(m^3/s)} = k_{(m^2)} \cdot \frac{A_{(m^2)}}{\mu_{(pascals\,.\,s)}} \cdot \frac{dP_{(pascals)}}{dx_{(m)}}$$

(b) **Practical system** (in standard use in the profession):

$$Q_{(cm^3/s)} = k_{(darcys)} \cdot \frac{A_{(cm^2)}}{\mu_{(cP)}} \cdot \frac{dP_{(atmospheres)}}{dx_{(cm)}}$$

The customary unit is the **millidarcy**:

$$1\ mD = 0.987 \times 10^{-15}\ m^2$$

In practice:

$$1\ mD \approx 10^{-15}\ m^2 \qquad 1\ darcy = 1\ (\mu m)^2$$

The range of permeabilities found is very wide. It varies from 0.1 mD to more than 10 D. The following terms can be employed to specify the value of the permeability:

< 1 mD :	Very low
1 to 10 mD :	Low
10 to 50 mD :	Mediocre
50 to 200 mD :	Average
200 to 500 mD :	Good
> 500 mD :	Excellent

N.B.: in a porous medium, the permeability generally varies with the flow direction.

2.2.2 Laws of Horizontal Flow in Steady-State Conditions

The elementary pressure drop law in a porous medium is not Darcy's Law, but a more general law of the form:

$$dp = \frac{\mu Q_m}{A k \rho}\left(1 + \frac{\bar{u} Q_m}{\mu A}\right) dx$$

where

Q_m = mass flow rate,
ρ = fluid density in the pressure and temperature conditions of the section dx considered,
$\bar{u}$ = shape parameter characterizing the shape of the pores (about 10^{-5} to 10^{-4} m).

For **liquids**, whether in reservoirs or in the laboratory, the term $\bar{u}\, Q_m / \mu A$ is generally negligible compared with 1 (except for flows at high velocity).

Since liquids are also relatively incompressible and since the usual pressure variations are limited, Darcy's Law can be used:

$$\rho \approx \text{constant} \qquad \text{and} \qquad \frac{Q_m}{\rho} = Q \approx \text{constant}$$

i.e.:

$$dp = \frac{\mu Q}{A k} \, . \, dx$$

The term $\bar{u}\, Q_m / \mu A$ can be ignored for **gases**, since the viscosity of gases is low and the mass flow rates generally high, except for flows at low velocity (as in the laboratory).

In brief:

— Oil, water — Gas at low velocity	Darcy's Law $\left(\frac{\bar{u}\,Q_m}{\mu A} << 1\right)$
— Gas — Oil, water at high velocity	Elementary pressure drop law.

The laws of horizontal flow in steady-state conditions are the following:

2.2.2.1 Liquids

Parallel Flow

$$Q = k \frac{A}{\mu} \frac{P_1 - P_2}{l}$$

where l is the distance between the upstream point where pressure P_1 prevails and the downstream point where pressure P_2 prevails.

Cylindrical Steady State Flow (Fig. 2.5)

$$Q = \frac{2\pi h k}{\mu} \frac{P_1 - P_2}{\ln \frac{r_1}{r_2}}$$

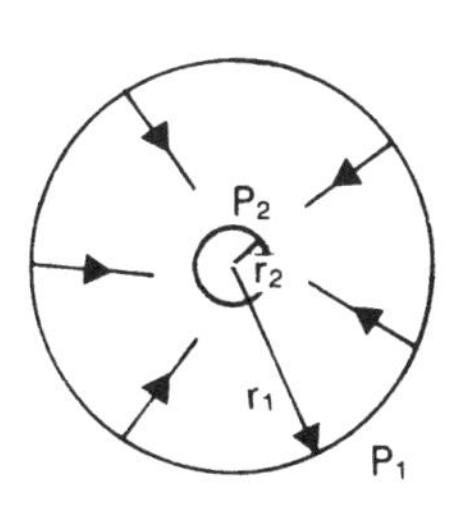

Fig. 2.5

where h is the thickness of rock assumed to be constant in which the fluid flows, and r_1 and r_2 are the distances from the cylinder axis where P_1 and P_2 prevail.

For a well drilled far from the boundaries of a layer, we have in particular:

$$Q = \frac{2\pi h k}{\mu} \frac{P_i - P_{wf}}{\ln \frac{R}{r_w}}$$

where

r_w = borehole radius (at pressure P_{wf}),
R = drainage radius,
P_i = initial reservoir pressure.

In steady-state conditions, the flow rate is the same in any annulus centered on the well. The velocity of the fluid decreases at increasing distance from the well, as does as the pressure drop gradient. From distance R, there is practically no more pressure drop, and the reservoir pressure P_i prevails. Matters proceed as if the well drained a portion of a cylinder of height h, bounded by the radii r_w and R.

2.2.2.2 Gases

a) If Darcy's Law applies:

Parallel Flow

$$Q = A \frac{k}{\mu} \frac{P_1^2 - P_2^2}{2 l P}$$

Cylindrical Steady State Flow

$$Q = \frac{2 \pi h k}{\mu} \frac{P_1^2 - P_2^2}{2 P . \ln \frac{r_1}{r_2}}$$

where Q is the volume flow rate measured at pressure P and at the flow temperature.

b) If Darcy's Law does not apply:

$$P_1^2 - P_2^2 = A' Q_m + B' Q_m^2$$

or

$$P_1^2 - P_2^2 = A Q_{std} + B Q_{std}^2$$

where Q_{std} is the volume flow rate of gas expressed in standard conditions:

$$Q_m = Q_{std} \times \rho_{std}$$

2.2.3 Associations of Formations of Different Permeabilities

Average Permeability k_m

Association in Parallel, Layers without Vertical Communication (Fig. 2.6)

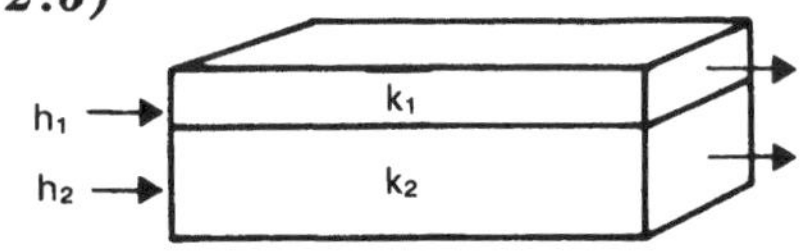

$$k_m = \frac{h_1 k_1 + h_2 k_2}{h_1 + h_2}$$

The upstream and downstream pressures are the same. The flow rates are added together.

Association in Series

The flow rate remains the same. The pressure drops are added together.

Parallel Flow (Fig. 2.7)

$$\frac{l_1 + l_2}{k_m} = \frac{l_1}{k_1} + \frac{l_2}{k_2}$$

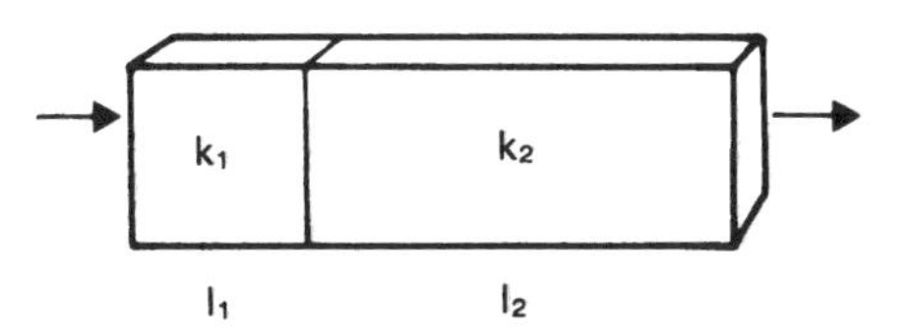

Fig. 2.7

Cylindrical Steady State Flow (Fig. 2.8)

$$\frac{\log \frac{r_1}{r_2}}{k_m} = \frac{\log \frac{r_1}{r}}{k_1} + \frac{\log \frac{r}{r_2}}{k_2}$$

This shows that the deterioration of the zone close to the well may result in a significant decrease in the average permeability of the layer around the borehole, and hence in the capacity of the well (the highest pressure drops occur in the zone close to the hole).

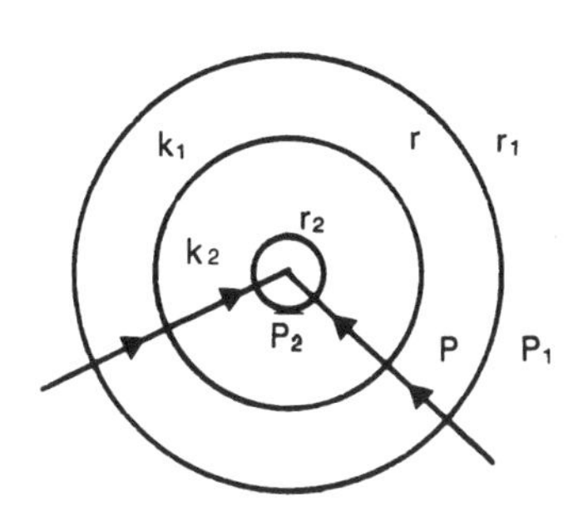

Fig. 2.8

2.2.4 Absolute Permeability, Effective and Relative Permeabilities

The absolute permeability of a rock depends on the direction considered (rocks are not isotropic). In particular, the horizontal permeability k_h (parallel flow towards the well) and vertical permeability k_v (segregation problems of fluids of different densities) are defined. Due to **stratification, the k_v are generally much lower than k_h** (ratio of 1 to 10, for example).

In practice, in gas and oil reservoirs, at least two fluids are always present (water + hydrocarbons). Darcy's Law serves to determine an effective permeability for each of the fluids. For two fluids 1 and 2:

$$Q_1 = A \frac{k_1}{\mu_1} \cdot \frac{dP_1}{dx}$$

$$Q_2 = A \frac{k_2}{\mu_2} \cdot \frac{dP_2}{dx}$$

Since the pressures in fluids 1 and 2 differ due to capillary mechanisms (Section 3), the concept of relative permeability is mainly employed, for example:

$$\text{Relative permeability to oil} = \frac{\text{Effective permeability to oil}}{\text{Permeability of the rock}}$$

These relative permeabilities depend on the rock sample concerned and the proportions of the fluids present (Chapter 6).

2.2.5 Measurement of Absolute Permeability

The absolute permeability can be determined on a sample by circulating a liquid that does not react with the rock. The problem arises of saturation of the sample with the liquid beforehand.

So absolute permeability is usually determined by circulating air (constant or variable head air permeameters).

These measurements are performed with air under pressures close to atmospheric pressure. In these conditions, given the narrowness of the pores, the number of collisions of the molecules against the pore walls is high in comparison with the number of collisions of the molecules between each other. The viscosity is poorly defined and the permeability obtained is higher than that with a **liquid**, considered as being the **correct** permeability (**Klinkenberg** effect):

$$k_{air} = k_{liq} \left(1 + \frac{b}{P_a}\right)$$

P_a = average pressure of flow.

For normal requirements, the k_{air} is sufficient, except for low values of k, for which the correction is systematically applied.

It should also be observed that, since the stress on the sample is not the same in the laboratory as in the reservoir, the permeability is not the same either. Permeability measurements under stress are useful for low permeabilities, which vary considerably with stress.

The permeability to hydrocarbons is obtained from well tests. It represents the average of a large volume of formation.

2.2.5.1 Measurement Using the Variable Head Permeameter (Fig. 2.9), Initial Apparatus, Measurement Principle

The cylindrical sample, washed and dried, is placed in a rubber plug 1. The plug is placed in a metal body 2 containing a hollow space with the same taper as the plug. A clamping system 3 compresses the plug so that tightness is achieved on the perimeter of the sample, and also between the base of the plug and the seat. The measurement is taken as follows.

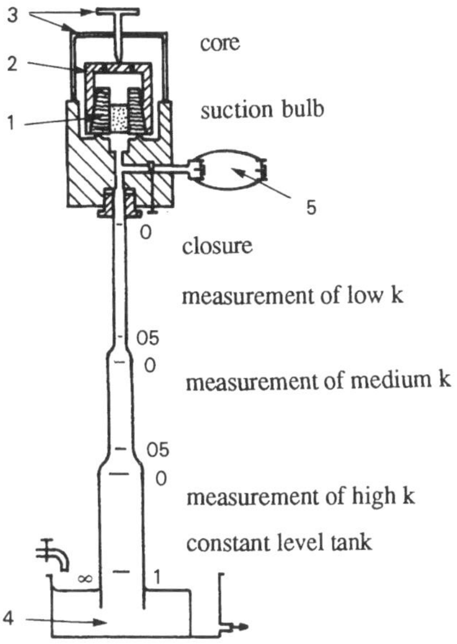

Water flows into the constant level tank 4; its level in the glass tube is caused to rise using a suction bulb 5. Using a chronometer, a measurement is taken of the time t required for the water to flow between the two marks of one of the calibrated tubes when air passes through the sample, with the suction bulb isolated from the tube.

The permeability is given by:

$$k = \frac{B\,l\,\mu}{A\,t}$$

The constant B, given by a table, depends on the diameter of tube selected.

A recent *automatic* instrument allows continuous measurements of 36 samples placed on a tray in 3 to 4 h. It is naturally coupled with a microcomputer (Laboratoires *Beicip*).

Fig. 2.9

2.2.5.2 Constant Head Air Permeameter

After obtaining steady-state conditions, Q is measured (at pressure P), the upstream pressure P_1 and downstream pressure P_2 are measured, and the following equation applied:

$$Q = \frac{A\,k}{\mu} \cdot \frac{P_1^2 - P_2^2}{2\,l\,P}$$

The experimental conditions are selected so that this formula is applicable, in other words with a high flow rate.

2.2.6 Porosity/Permeability Relationship

In some cases, a correlation has been established between porosity and permeability for a given sediment (Fig. 2.10). An attempt is made to write an equation of the type:

$$\log k = a \phi + b$$

Tube models can be used to develop an equation between the porosity and permeability.

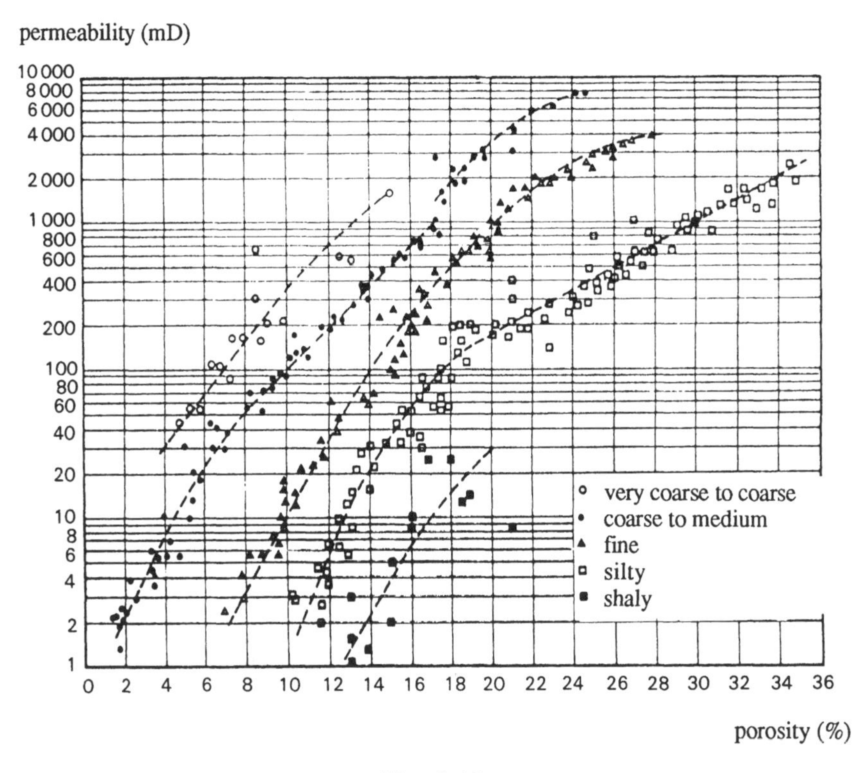

Fig. 2.10

Statistics on 610 sandstone samples.
(*After* Chilingar).

Let us consider two samples of the same porosity, but differing in pore size. It is easy to see that the permeability of the sample with the larger pores is greater than that of the other. This is shown by the equation obtained with a model consisting of a stack of capillary tubes of the same length and the same radius as the sample considered, and which would also have the same porosity and permeability as the sample. We obtain:

$$k = \phi \frac{r^2}{8}$$

A more accurate model can be obtained taking account of the distribution of the pore radii, determined from a capillary pressure curve, or from the observation of thin sections.

2.2.7 Porosity/Permeability Exercise

1. A sample has a diameter of 24 mm and a length of 32 mm.

The dry sample weighs 41.250 g. It is saturated with water, of specific gravity 1. A second weighing of the sample gives a weight of 43.165 g.

What is the porosity of the sample?

2. Its permeability to air is measured, at a temperature of 20°C.

It is found that 100 cm^3 of air passes through, measured at atmospheric pressure and at 20°C, in 2 min 20 s, the difference in pressure between the two sides having been kept constant and equal to 12 cmHg and the upstream pressure being 76 cmHg.

What is the permeability of this sample in millidarcys?

Viscosity of air at 20°C, $\mu = 180.8 \cdot 10^{-6}$ P $(180.8 \cdot 10^{-7}$ Pa/s).

Density of Hg = 13.6 g/cm^3 (13,600 kg/m^3).

Answers

1. Porosity

Weight of water contained in the sample:

$$43.165 - 41.250 = 1.915 \text{ g}$$

The volume of the pores is hence:

$$V_p = 1.915 \text{ cm}^3$$

Total volume:

$$V_T = \frac{\pi}{4} \times 2.4^2 \times 3.2 = 14.496 \text{ cm}^3$$

Porosity:

$$\phi = \frac{V_P}{V_T} = \frac{1.915}{14.496} = 13.2\%$$

2. Permeability

The elementary pressure drop law is written:

$$dp = \frac{Q_m}{A k \rho}\left(1 + \frac{\bar{u} Q_m}{\mu A}\right) dx$$

We shall first calculate the term $\frac{\bar{u} Q_m}{\mu A}$. We have:

$$Q_m = \left(\frac{100}{140} \times 10^{-6} \times \frac{273}{293} \times 1.293\right) \text{ kg / s}$$

(flow rate in cubic meters per second expressed at 0 °C and 1 atmosphere, at which the density of air is 1.293 kg/m³) = $0.86 \cdot 10^{-6}$ m³/s,

μ = 180.8 x 10^{-7} Pa/s,

A = 4.53 x 10^{-4} m²,

$\frac{\bar{u} Q_m}{\mu A} \approx 10^{-3}$ or 10^{-2} depending on whether $\bar{u}$ is taken as 10^{-5} or 10^{-4} m, a term which is negligible compared with 1.

Darcy's Law is therefore applicable taking account of the fact that, for gases in steady-state flow, only the mass flow rate is used. The volume flow rate, measured at pressure P across a sample of length l, is given by:

$$Q_v = \frac{k A}{\mu} \frac{P_1^2 - P_2^2}{2 l P}$$

where P_1 and P_2 are the upstream and downstream pressures. Hence:

$$k = \frac{2 \mu l P Q_v}{A\left(P_1^2 - P_2^2\right)}$$

We therefore obtain the following:

	International System	Practical system
μ	180.8×10^{-7} Pa/s	180.8×10^{-4} cP
l	3.2×10^{-2} m	1.26 inch
$P = P_1$	$0.76 \times 13{,}600 \times 9.81$ Pa	1 atm.
P_2	$(0.76 - 0.12)\ 13{,}600 \times 9.81$ Pa	$\frac{76 - 12}{76}$ atm.
Q_v	$\frac{100 \times 10^{-6}}{140}$ m^3/s	$\frac{100}{140}$ cm^3/s
A	4.53×10^{-4} m^2	0.7 $inch^2$
k	62.0×10^{-15} m^2	$62.8 \, . \, 10^{-3}$ D = 62.8 mD

2.3 SATURATIONS

2.3.1 Definition

In the pore volume V_p are found a volume V_W of water, a volume V_O of oil, and a volume V_G of gas ($V_W + V_O + V_G = V_p$).

The oil, water and gas saturations are:

$S_W = \frac{V_W}{V_p}$	$S_O = \frac{V_O}{V_p}$	$S_G = \frac{V_G}{V_p}$

expressed in percent, with $S_W + S_O + S_G = 100\%$.

Knowing the volumes of oil and gas in place in a reservoir requires knowing the saturations at every point, or at least a satisfactory approximation.

2.3.2 Distribution of the Different Fluids in an Initial-State Reservoir, Capillary Mechanisms

Reservoirs initially contained only water. The hydrocarbons displaced part of this water to accumulate therein. Since the transverse dimensions of the pores are very small (from about 1 μm to a few dozen μm), the equilibrium of an initial-state reservoir is thus governed by capillary mechanisms.

2.3.2.1 Capillary Properties of Rocks

A. *Wettability*

Let us consider a solid surface in the presence of two fluids. It can be seen that one of these fluids tends to spread on the solid. The angle θ of connection of the interface with the solid, measured in this fluid, is less than $\pi/2$. Measured in the other fluid, it is greater than $\pi/2$. The fluid that tends to spread out is said to wet the surface better than the other fluid. It is also said that it is wetting, and the other is nonwetting.

Examples:

If a drop of water is allowed to fall on a clean plate of glass, the water spreads. The water is wetting (Fig. 2.11a).

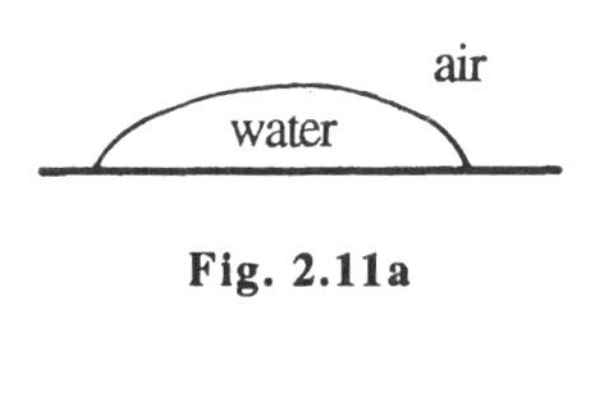

Fig. 2.11a

If a droplet of mercury is allowed to fall on a clean plate of glass, the mercury remains in spherical form. Air is the wetting fluid (Fig. 2.11b).

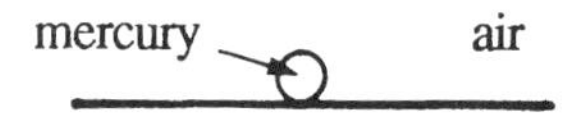

Fig. 2.11b

In reservoirs, it is estimated that, in most situations, the **wetting fluid is water** (water/oil and water/gas pairs). In some cases, however, the oil may be wetting preferentially to the water, especially for a number of limestones.

These mechanisms correspond to molecular attractions and repulsions exerted between each fluid and the solid present.

Figure 2.12 shows the oil distribution (nonwetting fluid) in the pores of a rock filled with water (wetting fluid).

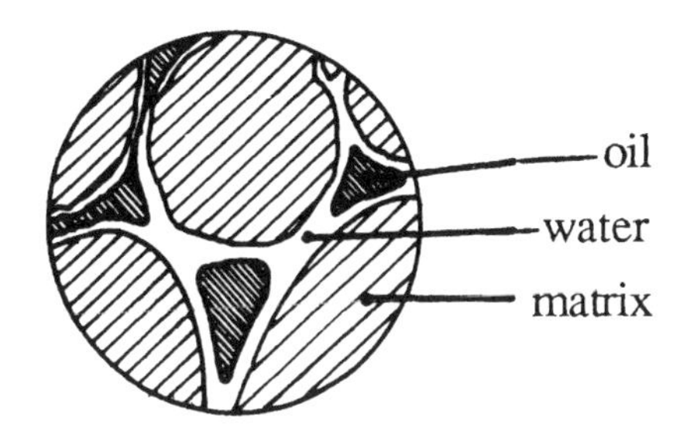

Fig. 2.12

Water/oil:

Water is often the wetting fluid.

Water/gas:

Water is always the wetting fluid.

Oil/gas:

Oil is the wetting fluid.

Wettability Test (IFP/Beicip)

The wettability of a sample is estimated in standard temperature and pressure conditions, using a combination of brine and Soltrol 130 refined oil (isoparaffinic oil). The reconstituted reservoir brine is used for each sample. The ratio of the viscosities of the oil and the water is always close to 1.2.

The test selected is derived from the one proposed by Amott.

With the sample saturated with oil and brine and the brine at residual saturation, the test consists in performing the following four experiments in succession: imbibition in brine, displacement by brine, imbibition in Soltrol oil, displacement by Soltrol oil.

A water **wettability index** r_w is evaluated, such that:

$$r_w = \frac{\text{Quantity of oil displaced by brine by imbibition}}{\text{Quantity of oil displaced by brine by imbibition and by displacement}}$$

and an oil wettability index r_o:

$$r_o = \frac{\text{Quantity of brine displaced by oil by imbibition}}{\text{Quantity of brine displaced by oil by imbibition and by displacement}}$$

If a rock is preferentially water-wet, r_o is zero and the greater the affinity for water, the closer r_w is to 1.

If the rock is preferentially oil-wet, the values of the two ratios are reversed.

For a rock of intermediate or neutral wettability, the two ratios are zero or close to zero.

The difference $r_w - r_o$ is also used; it ranges from +1 to −1 for a strongly water-wet rock and a strongly oil-wet rock.

B. Interfacial Tension

At the interface of two immiscible fluids, the forces acting on the molecules of each of these fluids are not the same as within each phase, and the system behaves as though the two phases were separated by a membrane.

The **interfacial tension** σ can be defined as the force per unit length necessary to maintain contact between the two lips of an incision assumed in the interface.

Orders of magnitude:

Oil/gas	:	0 to 15 dyn/cm (0 to 15 mN/m)	(reservoir)
Water/oil	:	15 to 35 dyn/cm	(reservoir)
Water/gas	:	35 to 55 dyn/cm	(reservoir)
Air/mercury	:	480 dyn/cm	(laboratory)
Air/water	:	72 dyn/cm	(laboratory)

C. Capillary Pressure, Jurin's Law (Fig. 2.13)

Let us consider a cylindrical capillary tube of radius r, immersed in a receptacle containing water. The water rises in the tube to a height h above the interface in the container. It is also found that the water/air interface is spherical, with the center of curvature in the air. This simple fact implies that, in the neighborhood of the interface, the pressure of the air is higher than that of the water (which is the wetting fluid since it tends to spread).

The capillary pressure P_C is the pressure difference existing between two points A and B infinitesimally close together and situated on either side of the interface:

$$P_C = P_A - P_B$$

where A is the non-wetting fluid.

At equilibrium, the vertical resultant of the surface tension forces is offset by the action of the capillary pressure on the tube cross-section:

$$2\pi r \,.\, \sigma \,.\, \cos\theta = P_C \,.\, \pi r^2$$

$$P_{Ch} = \frac{2\sigma\cos\theta}{r}$$

For a fissure of thickness e, we obtain:

$$P_C = 2\,\sigma \cos\theta / e$$

(these equations can also be obtained from the law of Laplace).

Apart from points A and B in the tube, let us now consider two points A' and B' on either side of the interface in the container. Since we can go from A to A' while remaining in air and from B to B' while remaining in water, we have:

$$P_A = P_{A'} - h\,\rho_{air}\,g$$

$$P_B = P_{B'} - h\,\rho_W\,g$$

where ρ_W and ρ_{air} are the densities of the water and air.

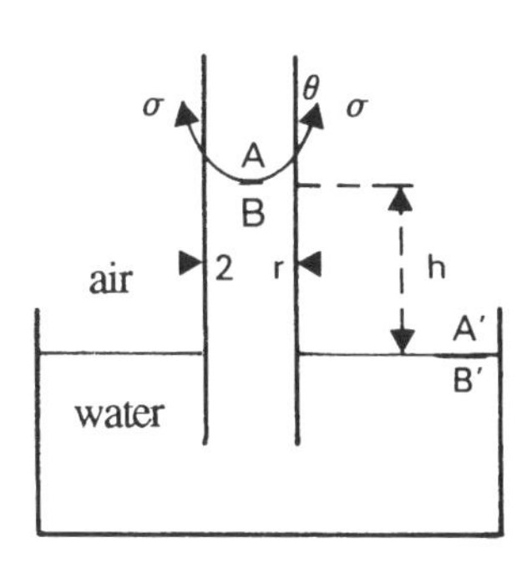

Fig. 2.13

Hence:

$$P_A - P_B = P_{A'} - P_{B'} + h\,(\rho W - \rho_{air})\,g$$

$$P_{C\,high} = P_{C\,low} + h\,(\rho_W - \rho_{air})\,g$$

If the container has a large diameter:

$$P_{C\,high} = h\,(\rho_W - \rho_{air})\,g$$

The case examined above is similar to that of water-wet reservoirs with two fluid phases, in which the hydrocarbon phase plays the role of the air. If ρ_{HC} and ρ_W are the densities of the hydrocarbon phase and of the water respectively, we obtain:

$$P_{C\,high} = P_{C\,low} + h\,(\rho_W - \rho_{HC})\,g$$

If the wetting phase is less dense than the non-wetting phase, the level of the interface in the capillary tubes is lower than that in the container, and we obtain:

$$P_{C\,low} = P_{C\,high} + h\,(\rho_{fnw} - \rho_{fw})\,g$$

where ρ_{fnw} and ρ_{fw} are the densities of the non-wetting and wetting phases respectively.

D. Conclusions

The foregoing discussion shows that, for a rock sample saturated with a fluid and surrounded by another fluid:

(a) **If the saturating fluid is wetting, it is displaced by the surrounding fluid only if the excess pressure applied to the surrounding fluid is at least equal to the capillary pressure for the largest pores.**
(b) **If the saturating fluid is non-wetting, it is displaced spontaneously by the surrounding fluid.**

2.3.2.2 Equilibrium of an Initial-State Reservoir

Let us consider a homogeneous reservoir that has just been discovered by a borehole, contains single-phase oil (or gas) and water, and is water-wet. Hydrodynamism is absent.

This first well has been drilled into the aquifer zone, cased and perforated over the entire height of the layer. After clearing, the well is shut in at the wellhead.

By analogy with the above discussion on the air/water combination, the following can be observed (Fig. 2.14):

(a) The oil-bearing and water-bearing zones are separated by a transition zone of a certain thickness. This can be seen by considering that the reservoir consists of several juxtaposed capillary tubes of variable dimensions. It can also be stated that the higher the point above the base of the oil accumulation, the larger the term "$h (\rho_w - \rho_o) g$" (where ρ_o is the density of the oil), and hence the larger the difference (oil pressure - water pressure) enabling the oil to penetrate into increasingly small pores because:

$$P_c = 2 \sigma \cos \theta / \Gamma$$

(b) The water/oil interface is lower in the well than in the reservoir, or at the same level if the layer contains macropores or large fissures. In fact, since the well has a large diameter, capillary mechanisms are practically absent.

The level in the well materializes the depth of the **"zero capillarity plane"**. This well is a reference (or observation) well. It serves to monitor changes in the aquifer during the production of the reservoir, but not necessarily the rise of the water in the pores (high friction).

The variation in water saturation with depth can be shown by Fig. 2.14.

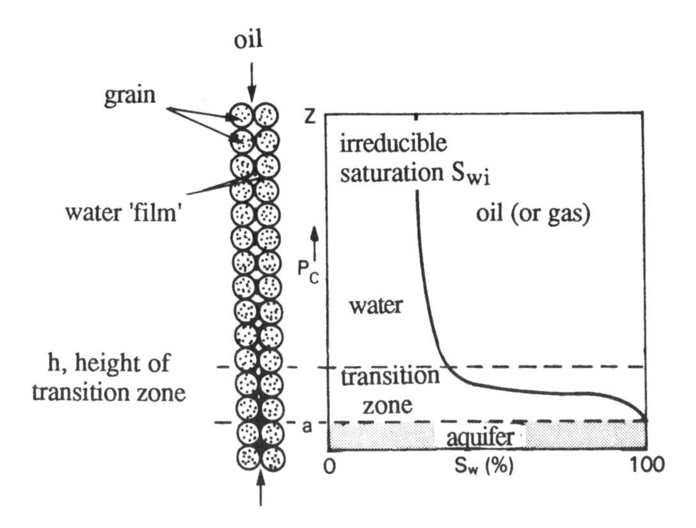

Fig. 2.14

At a level below the base of the continuous oil accumulation zone, the water saturation is 100% (except on the oil migration path from the source rock). This is the aquifer zone.

Above the transition zone, some **pore water saturation** S_{wi} remains, corresponding to the water along the pore walls and in the small pores or **interstices**. This water cannot flow, and is **irreducible** (interstitial or connate water).

In the case of **hydrodynamism**, the plane corresponding to the base of the continuous hydrocarbon accumulation (water level) is inclined in the flow direction (Fig. 2.15).

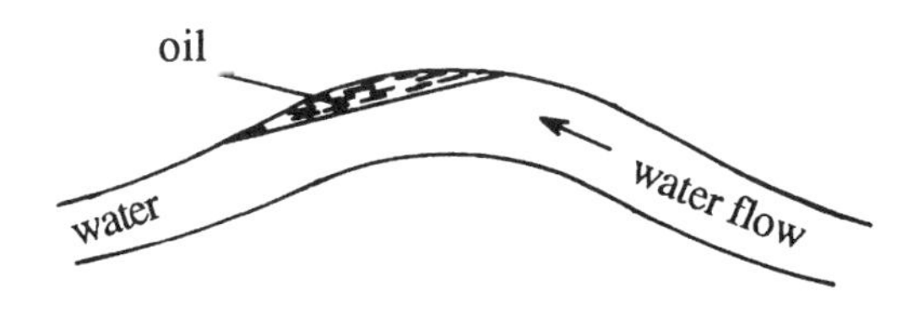

Fig. 2.15

If the reservoir is not homogeneous, the change in saturation with depth is not uniform as indicated in Fig. 3.21 (see Section 2.3.5).

2.3.2.3 Capillary Migration

When a drop of oil moves from the source rock to the reservoir rock, it passes through a succession of narrowing and widenings pores. Let us consider a drop that has stopped at a constriction (Fig. 2.16). The following relationship exists between the capillary pressures above and below the drop:

$$P_{Ch} = P_{Cl} + h\,(\rho_W - \rho_O)\,g$$

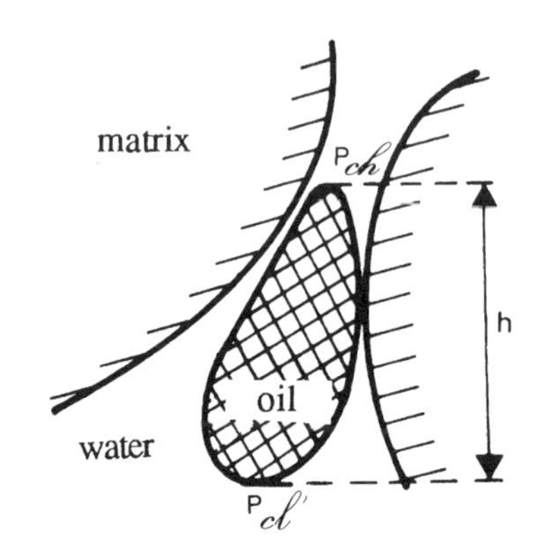

Fig. 2.16

For the drop to be able to pass through a constriction of radius r, it is necessary for:

$$P_{Ch} = \frac{2\,\sigma \cos\theta}{r}$$

and hence for h to be sufficiently large.

This means that the drop will pass through when it has been increased by another drop. At this time, two alternatives are available:

(a) Sudden widening of the pore after the constriction: the drop may break, with part remaining stopped by the constriction.
(b) Progressive widening: the whole drop can pass through.

Thus, along the hydrocarbon migration path from the source rock to the reservoir rock, some oil saturation may remain.

2.3.3 Capillary Pressure Curves, Curves Representing the Average Capillary Properties of a Reservoir

2.3.3.1 Capillary Pressure Curve of a Sample

This curve is obtained by drainage, i.e. by the displacement of a wetting fluid, which saturates the sample, by a nonwetting fluid. Two pairs of fluids are used.

A. *Mercury/air (Purcell Method) (Fig. 2.17)*

The sample, washed and dried, is first placed in a vacuum. Mercury is injected into it at increasing pressure stages. At each stage, the volume of mercury injected (positive displacement pump) is recorded. The pore volume is determined beforehand. The capillary pressure is the absolute pressure of the mercury. The injection pressure reaches 250 bar.

B. *Air/Brine (Desorber, Restored State Method) (Fig. 2.18)*

The sample, saturated with brine, is placed under a bell jar, on a porous plate saturated with brine and covered with diatom powder (good contact). Below the plate, a capillary tube is used to measure the volumes of water leaving the bell. Air is injected at increasing pressure stages. The pore volume is also determined. The capillary pressure is the relative pressure of the air.

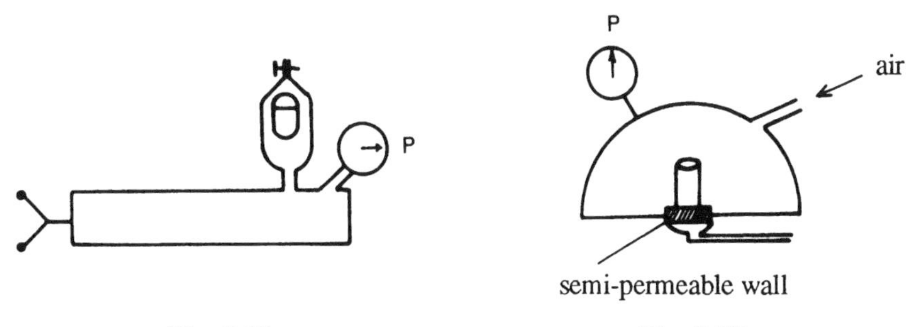

Fig. 2.17 **Fig. 2.18**

C. *Another Method: Centrifugation*

This method can be used to supplement the capillary pressure curve obtained with the desorber, which is limited to P_C = 8 bar (the porous plate allows air to pass through at higher pressures).

The curves obtained have the following shape:

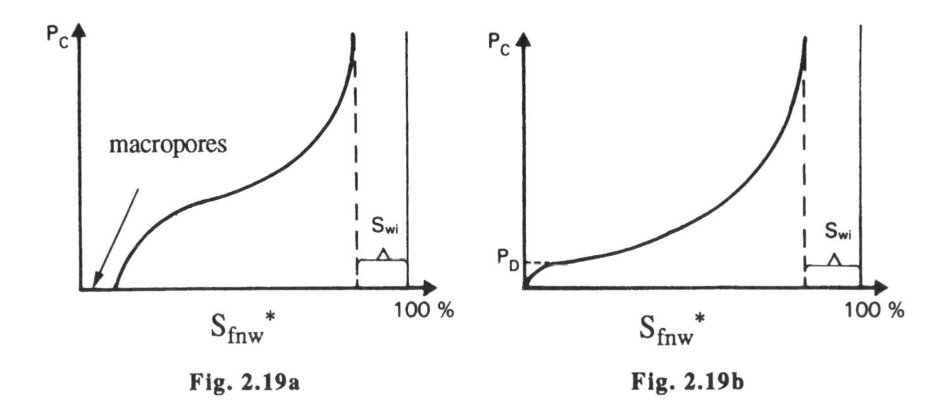

Fig. 2.19a **Fig. 2.19b**

The first curve (see Fig. 2.19a) shows the presence of macropores, invaded when the pressures of both fluids are equal. The second shows a homogeneous medium, with a displacement pressure P_D corresponding to the onset of invasion of the medium.

In a water-wet reservoir, the displacement pressure exists at the base of the transition zone.

The use of the air/brine combination achieves the irreducible pore water saturation S_{wi} (see Figs 19a and 19b).

The mercury/air combination does not ensure reaching this state.

The curves also help to derive an idea of the morphology of the pores, because the P_C can be used to determine their radius:

$$P_C = \frac{2\,\sigma \cos\theta}{r}$$

2.3.3.2 From a $(P_C - S)_{lab}$ Curve to a $(P_C - S)_{reservoir}$ Curve

The processes in the laboratory and in the reservoir (when the hydrocarbons accumulate) are drainage processes at increasing capillary pressure.

---

* S_{fnw}, saturation with nonwetting fluid.

Saturation with a nonwetting fluid S_{fnw} corresponds to the invasion by this fluid of the pores of radius ≥ r regardless of the pair of fluids concerned. For this saturation, we therefore have:

$$P_{C\,lab} = \frac{2(\sigma \cos \theta)_{lab}}{r}$$

$$P_{C\,reservoir} = \frac{2(\sigma \cos \theta)_{reservoir}}{r}$$

hence:

$$\frac{P_{C\,lab}}{(\sigma \cos \theta)_{lab}} = \frac{P_{C\,reservoir}}{(\sigma \cos \theta)_{reservoir}}$$

$\frac{P_C}{\sigma \cos \theta}$ is invariable for a given saturation.

2.3.3.3 Curves Showing the Average Capillary Properties of a Reservoir

The capillary pressure/saturation curves are plotted for only a limited number of samples.

A curve representing the average capillary pressures of a geological formation can be obtained from the P_C curves available, by two methods:

(a) Determination of an average capillary pressure curve from the permeability distribution results on the field.

(b) Plot of the capillary pressure function:

$$J(S_{fw}) = \frac{P_C}{\sigma \cos \theta} \sqrt{\frac{k}{\phi}}$$

S_{fw} = saturation by wetting fluid.

Each P_C curve available gives a J curve. The average J curve is plotted (Fig. 2.20). Using the J curve, a $P_C - S$ curve can be plotted for a given sediment sample, if its k and ϕ are known.

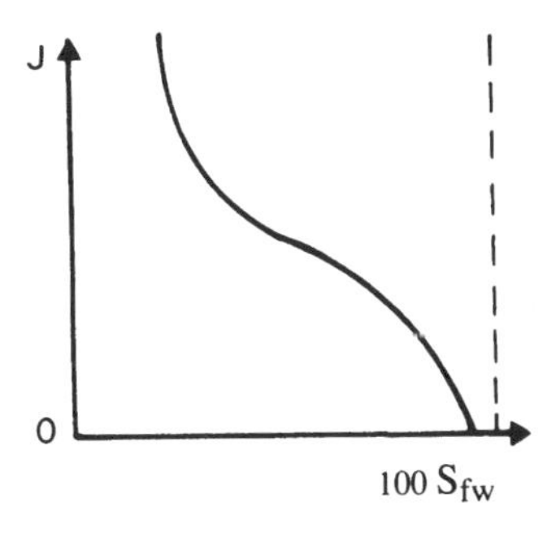

Fig. 2.20

A curve must be plotted for each sediment.

2.3.4 Determination of *in Situ* Saturation

2.3.4.1 Direct Method (Core Analysis)

This is virtually never possible, even if precautions are observed in core drilling. This is because, when the core is pulled up, the pressure and temperature fall. The fluids contained expand, and their distribution at the earth's surface is no longer the distribution that prevailed *in situ*.

In some cases (appropriate drilling mud), the saturation with irreducible water can be determined on samples taken from the center of the core.

2.3.4.2 Indirect Method by Analysis of Capillary Mechanisms

Let us consider a sample taken at a depth h above the plane $P_C = 0$. The pressure difference between the hydrocarbons (single-phase) and the water is:

$$P_{C\ reservoir} = h\ (\rho_W - \rho_{HC})\ g$$

As we showed in Section 2.3.3.2, this corresponds to a given hydrocarbon saturation S_{HC}.

If the $J(S_{fw})$ curve is available for the sediment concerned, the value of J corresponding to $P_{C\ reservoir}$ is calculated:

$$J = \frac{h\,(\rho_W - \rho_{HC})\,g}{(\sigma \cos\theta)_{reservoir}} \sqrt{\frac{k}{\phi}}$$

where k and ϕ of the sample have been determined.

This value is plotted on the curve. Hence:

$$S_{fw} = 1 - S_{HC}$$

If the curve P_C lab $\leftrightarrow$ S_{fnw} is available for the sample concerned, the P_C lab is calculated:

$$P_{C\ lab} = \frac{h\,(\rho_W - \rho_{HC})\,g}{(\sigma \cos\theta)_{reservoir}}\,(\sigma \cos\theta)_{lab}$$

The curve accordingly gives:

$$S_{fnw} = S_{HC}$$

2.3.4.3 Well Log Method

This is the **basic method** for determining saturations (Section 2.4).

2.3.5 Typical Cross-Section of Saturations in a Reservoir

If a cross-section of a reservoir is made, the saturations shown below may actually be obtained, for example, as a function of depth, and according to the heterogeneity of the rock (Fig. 2.21).

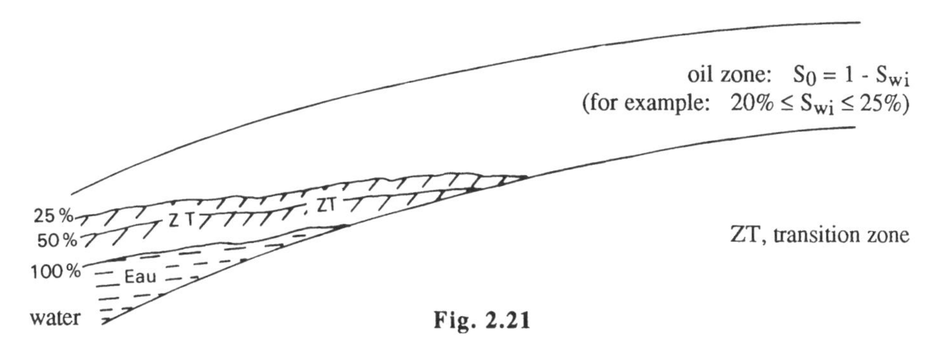

Fig. 2.21

Large reservoirs can be characterized zone by zone (ϕ, k, S_w). For small reservoirs, it is often adequate to calculate a mean S_{wi}, the height of the transition zone h (TZ), and the depth of the water/oil interface. This interface corresponds to a plane surface only in fairly permeable reservoirs. In impermeable reservoirs, the depths of S_w = 100% may vary laterally by several meters, often 10 to 20 m or more.

2.3.6 Surveys with a Scanner
Porosity, Pore Morphology, Sweeping by another Fluid

X-rays were used in petroleum laboratories for many years. But the scanner, that only entered use recently, allows a much more complete analysis, thanks to interpretation on a computer (Fig. 2.22).

On a core sample (*SNEA (P)* experiments), the scanner serves to obtain cross-sections perpendicular, parallel or oblique to the axis of the object. Their thicknesses can be adjusted at the operator's discretion, to 1, 5 or 10 mm. Cross-sections may be separate, adjacent or overlapping.

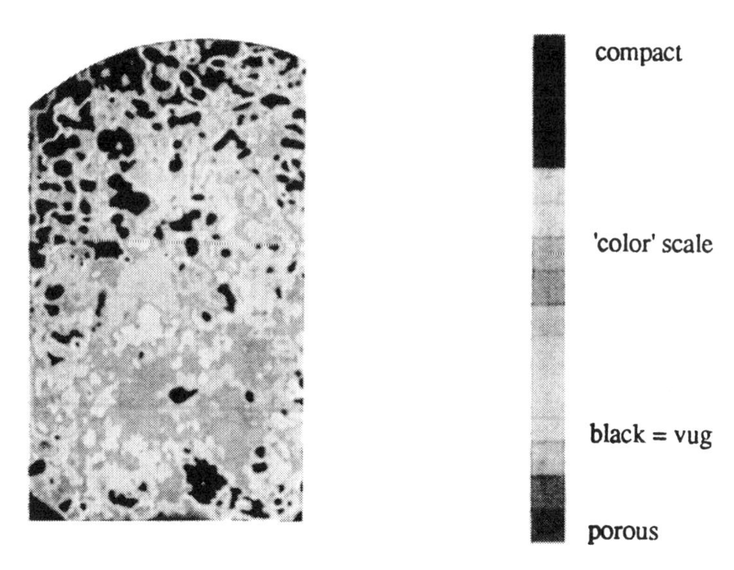

Fig. 2.22

For a given cross-section, the data acquisition time may be as short as 3.4 s, and the 2-D image is available on the screen for display and processing after about 30 s.

These studies allow detailed visualization of the porous medium, the morphology of the pores, the heterogeneities and the fractures. They also allow calculation of the porosity, the saturations and the saturation variations (sweeping). These methods will undoubtedly continue to develop. *IFP* has had a scanner since late 1988.

2.4 WELL LOGGING

2.4.1 General Introduction

A well log is the recording (usually continuous) of a characteristic of the formations intersected by a borehole, as a function of depth.

Electric well logs are recorded when drilling is interrupted and are the subject of this section (mud logging is during drilling). The data are recorded and transmitted to the surface instantaneously.

Well logs are essential instruments for reservoir assessments.

Purpose of Electric Well Logs:

(a) Identification of the reservoirs: the **lithology, porosity, saturations** (water/oil/gas) as a function of depth. Permeability values are not obtained (but research is under way in this direction).
(b) The dip of the beds.
(c) Survey of the well: diameter, inclination, casing cementing, formation/hole connection (perforations).
(d) Comparison among several wells, by "electric" correlations which highlight variations in depth, thickness, facies, etc.

Means employed:

A recording truck (or cab in offshore operations) + computerized interpretation + motor-driven winch + electric cable + sonde (Fig. 2.23). Leading company: *Schlumberger*.

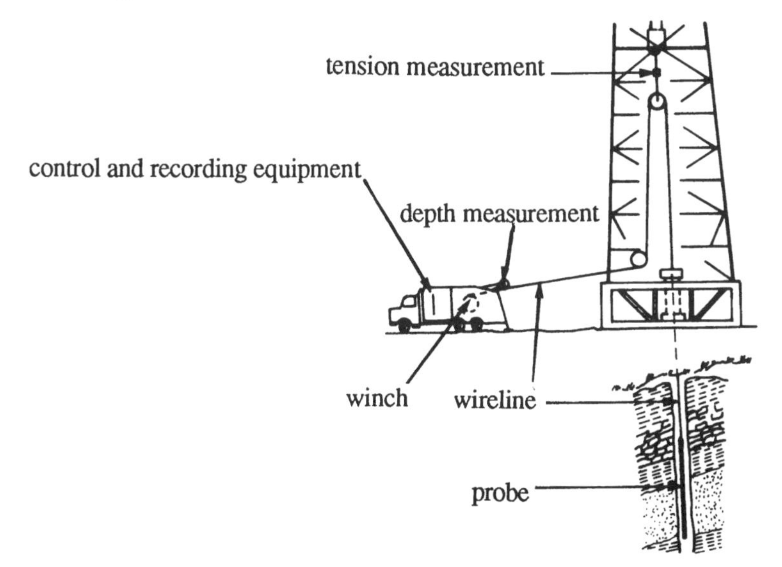

Fig. 2.23

Electric well logging.

The sonde: electronic unit with transmitters (electric, nuclear or acoustic), receivers and amplifiers. The investigation radius is normally about 1 m. The instruments are often combined (simultaneous recording).

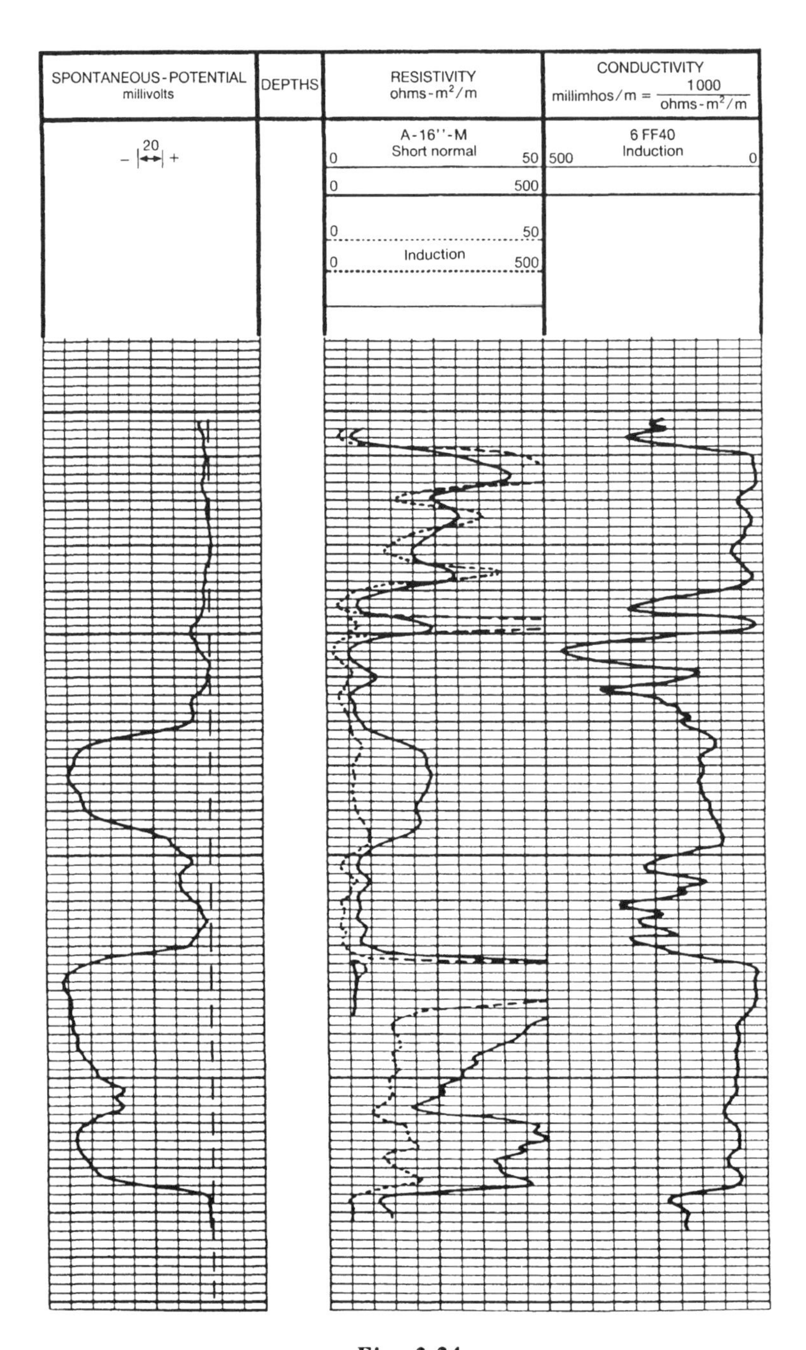
SPONTANEOUS-POTENTIAL
millivolts
DEPTHS
RESISTIVITY
ohms-m²/m
CONDUCTIVITY
millimhos/m = 1000 / ohms-m²/m
20
− +
A-16''-M
Short normal
0 50
0 500
0 50
Induction
0 500
6 FF40
Induction
500 0

Fig. 2.24

Recording: the receiver signal is monitored, calibrated and recorded on film and magnetic tape (digitized signals). Depth scale: 1/200 (and 1/500). The recording is called a **log**.

Main characteristics recorded:

(a) Resistivity and spontaneous potential.
(b) Natural radioactivity (gamma rays) and induced radioactivity (neutron/gamma-gamma).
(c) Speed of sound, attenuation of acoustic waves.
(d) Borehole diameter and deviation. Dip of beds.

2.4.2 Electric Logs

2.4.2.1 Spontaneous Potential SP

This log is not based on a transmitter but involves only natural currents. The variations in electrical potential are measured directly between a surface electrode and the sonde (second electrode).

A deflection is observed opposite the reservoir rocks compared with a "base line" of shales (or marls) (Fig. 2.24). Causes: different salinity of the reservoir water and drilling mud.

The characteristics obtained are the **boundary of the reservoir beds** and the resistivity of the pore water $\mathbf{R_w}$.

2.4.2.2 Resistivity Log

A system of electrodes sends an electric current into the formation. The apparent resistivity of the reservoir is measured, in ohms per meter.

The most widely used instruments are the **laterolog LL** (focusing) and **induction IL** (solenoid) (Fig. 2.24). The characteristics obtained are a function of the **porosity and saturation** (water/hydrocarbons). The rock matrices are insulating and the hydrocarbons have high resistivity, whereas the resistivity of the water decreases with increasing salinity. **The resistivity can thus differentiate the water from the hydrocarbons.**

Empirical equations:

$$F = \frac{R_o}{R_w} \approx \frac{a}{\phi^m}$$

with

$a \approx 1$ and $m \approx 2$ in general,

F = formation factor (constant for a given sample), sometimes denoted F_R,

R_0 = resistivity of rocks 100% saturated with water of resistivity R_w, **hence ϕ.**

Archie's equation:

$$S_w \approx \frac{1}{\phi} \sqrt[n]{\frac{R_w}{R_t}}$$

with

n = 2 for formations without fractures or vugs,

R_t = calculated resistivity of the rock whose water saturation is S_w, **hence S_w.**

This equation is satisfied for **clean** reservoirs (with very little shale).

Note that Archie's equation can confirm the formula giving F:

$$100\% \approx \frac{1}{\phi} \sqrt{\frac{R_w}{R_o}}$$

hence:

$$\frac{R_o}{R_w} \approx \frac{1}{\phi^2}$$

2.4.2.3 Microresistivity Log

Concept of invaded zone: the mud filtrate invades the zone near the hole, partly expelling the liquids in place. The resistivity of this zone provides data on the ϕ and S_w of the reservoir. Instruments employed: microlog ML and **microlaterolog MLL** or **MSFL** (electrodes close together).

2.4.3 Radioactivity Logs

2.4.3.1 Gamma Ray Log (GR)

The natural radioactivity of the formations is measured (no transmitter). Shales and marls are generally more radioactive than sandstones and limestones, leading to the **identification of reservoir beds** (Fig. 2.25).

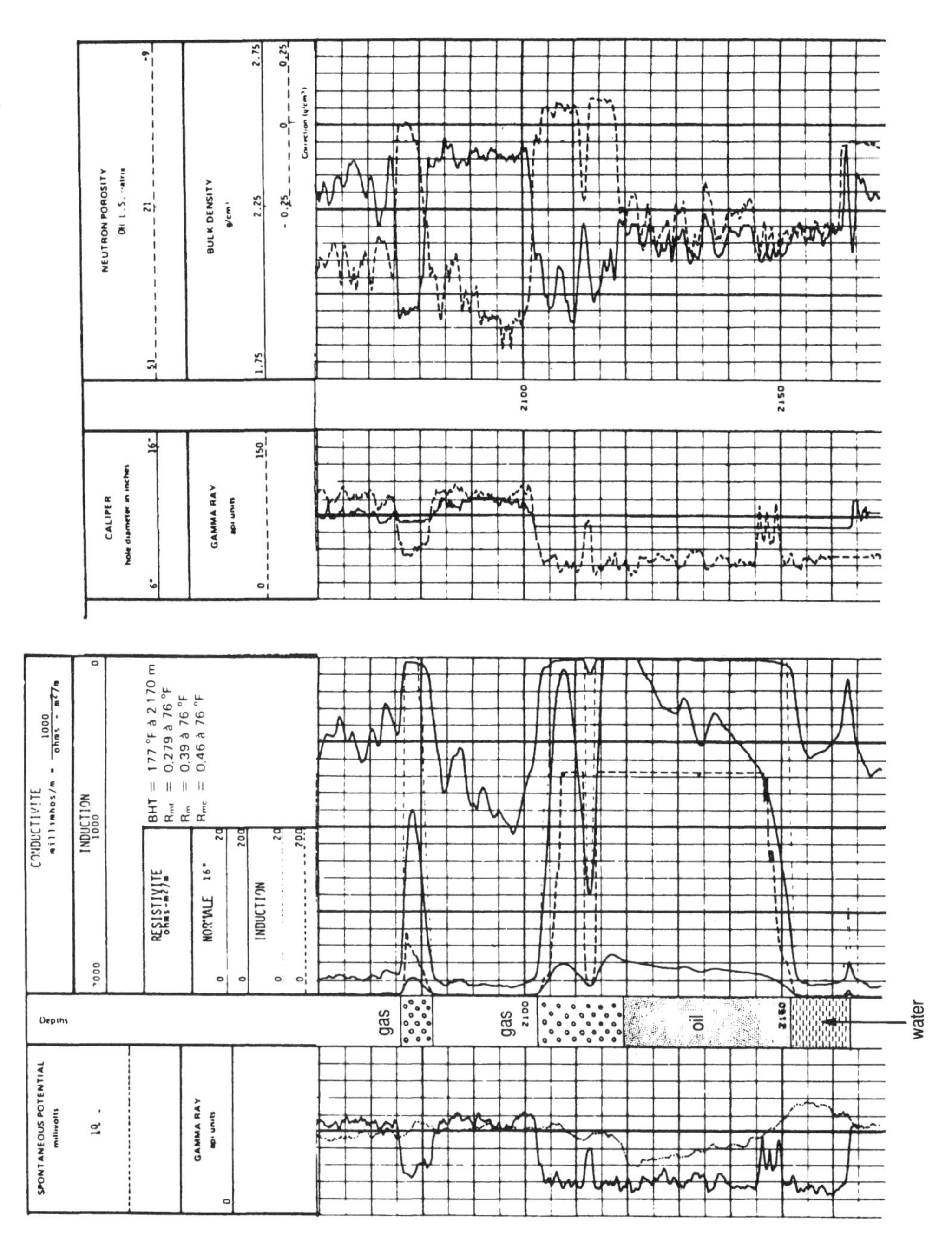

Fig. 2.25

Example of 1/200 composite log.

2.4.3.2 Neutron Log N

The sonde emits fast neutrons which bombard the formation giving rise to slow neutrons, which are detected, and of which the number N depends on the quantity of hydrogen, and accordingly on ϕ (Fig. 2.25).

Empirical equation:

$$N = a - b \log \phi \qquad \textbf{hence } \phi$$

2.4.3.3 Density Log (gamma/gamma) D

The formations are irradiated by gamma rays that are received as a function of the density of the formation (Fig. 2.25). This is written:

$$D = \phi \,.\, D_f + (1 - \phi)\, D_m$$

with

D = total density read on the log (also denoted ρ),
D_f = fluid density (filtrate),
D_m = density of rock matrix.

Values of D_m: sands/sandstones 2.65, shales 2.65 to 2.70, limestones 2.71, and dolomites 2.85, which leads to the calculation of the **porosity** ϕ.

2.4.3.4 Neutron Relaxation TDT

Based on a neutron emission, the parameter recorded is related to the quantity of chlorine, hence to the apparent salinity which leads to the saturation S_w. This log helps to locate the hydrocarbons **behind the casing** and to monitor changes in the interfaces during **production** (Fig. 2.26).

2.4.4 Sonic (or Acoustic) Log

These logs consist of the transmission and reception of sound waves.

Sonic Log

Empirical equation:

$$\frac{1}{V} = \frac{\phi}{V_f} + \frac{1 - \phi}{V_m} \qquad \textbf{hence } \phi$$

or

$$\Delta t = \phi \,.\, \Delta tf + (1 - \phi) \,.\, \Delta tm$$

with Δt = travel time in the transmitter/receiver interval.

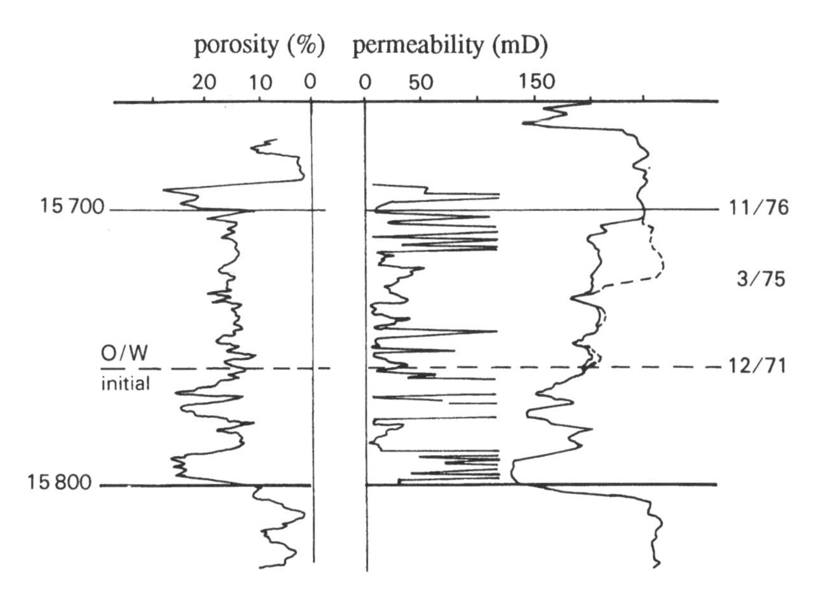

Fig. 2.26 TDT log.

Speeds: air ≈ (1100 ft/s), 335 m/s; oil (4250 ft/s), 1300 m/s; water (5000/to 6000 ft/s), 1500 to 1800 m/s; shales (5200 to 15,700 ft/s), 1600 to 4800 m/s; sandstones (18,000 ft/s), 5500 m/s; and limestones (< 23 000 ft/s), < 7000 m/s.

2.4.5 Auxiliary Logs

2.4.5.1 Caliper Log

This is a system with arms giving the borehole diameter (caving, constrictions → quantities of cement) (see Figs 2.24 and 2.25).

2.4.5.2 Dipmeter Log

This is the simultaneous recording of four **microlaterolog** curves along four 90 degree generating lines in a plane normal to the borehole axis: the difference in the four curves gives the value of the **dip** and its direction.

2.4.5.3 Cement Bond Log (CBL)

This involves the measurement of the **amplitude** of the acoustic signal received through the casing:

(a) In the presence of cement, the signal is weak because the cement attenuates the vibrations of the metal.
(b) In the absence of cement, the casing vibrates freely, generating a strong signal (Fig. 2.27).

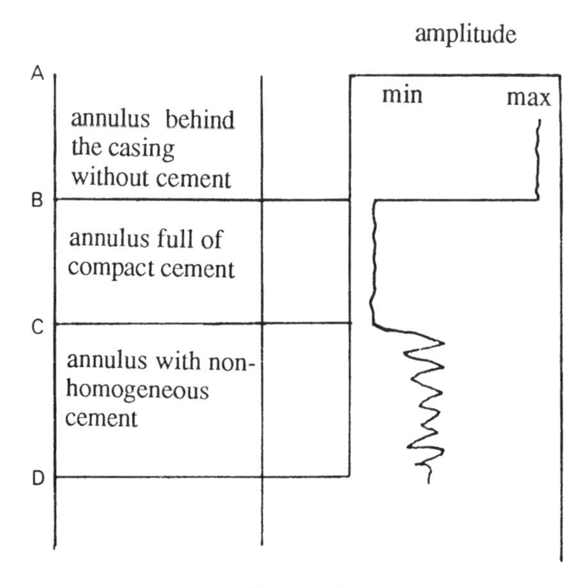

Fig. 2.27

Schematic response of CBL.

The analysis of the signals provides data on the **presence and bonding of the cement** to the casing and to the formation.

2.4.6 Determination of Lithology, Porosity and Saturations

2.4.6.1 Lithology

Gamma ray and/or SP + cuttings (and cores).

"Neutron + density + sonic" combination.

2.4.6.2 Porosity

Resistivity, neutron, density, sonic.

Case of shaly sands (sandstone): resistivity and gamma ray (SP). Diagrams with two porosity logs (N-D, N-S and S-D).

2.4.6.3 Saturations

(a) Resistivity (+ SP):

$$S_w \approx \frac{1}{\phi} \sqrt{\frac{R_w}{R_t}}$$

→ **water and oil (or gas).**

(b) "Neutron + density" combination → gas: lower density and higher neutron, from which D indicates too strong an apparent ϕ and N too weak an apparent ϕ (Fig. 2.25). **Hence**:

Resistivity (+SP) + N + D → water, oil and gas

For formations containing shale (shaly sandstones and carbonates), the interpretation is complex and less accurate.

2.4.6.4 Composite Log

The essential combination of the logs is interpreted giving rise to a "composite" log in which the different fluids are indicated (Fig. 2.25).

2.4.6.5 Overall Interpretations

Quick Look

Rapid interpretations, called **"quick looks"**, based on a **simple visual examination** of the logs, help to derive a preliminary idea about the formations and fluids. The method consists of the following:

(a) Identification of the reservoirs by eliminating the shale beds and compact beds.

(b) Comparison of the resistivity and porosity logs in the reservoirs:

- Comparison of the resistivity logs identifies the **water/hydrocarbon contact** and gives an approximate value of the water saturation S_w.
- Comparison of the "neutron" and "density" logs helps to determine the lithology in the water zone, distinguish the gas from the oil, and estimate the porosity of the formations.

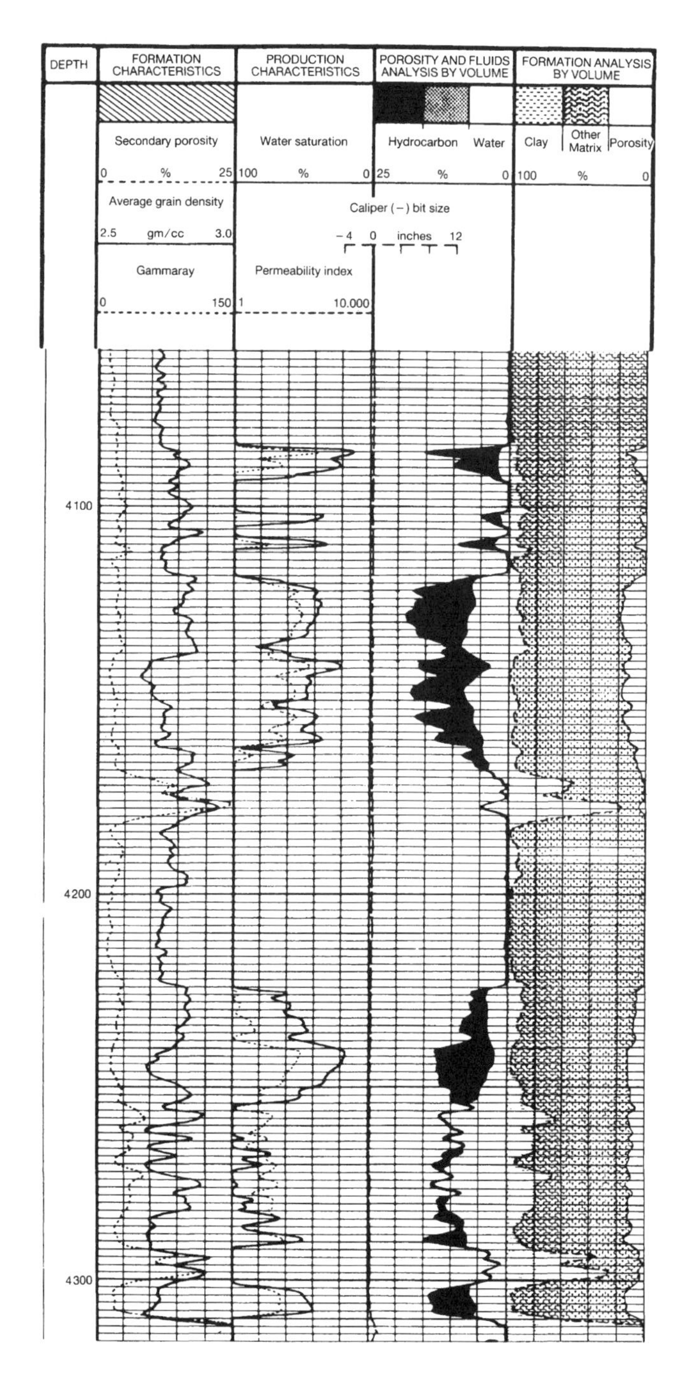

Fig. 2.28

Interpretation of logs.

Computerized Interpretation

The use of **computer programs** offers considerable flexibility and greater speed in obtaining the results. A preliminary interpretation is thus directly made on the site by the specialized company (microcomputers in truck or cab). For example, preliminary processing gives the volume of the pores filled by hydrocarbons $\phi \cdot S_{HC}$ (Fig. 2.28); it distinguishes the gas from the oil, and serves to make an initial calculation of the shale content and the density of the matrix rock.

More thorough analyses are then performed by the service companies and oil companies. These interpretations demand the use of more powerful computers.

2.4.6.6 Cost

Depending on the boreholes (exploration or development) and their location (onshore or offshore), the cost of well logging can generally be estimated at **about 5 to 15% of the total cost of the borehole.**

RFT

An instrument for repeated formation tests also exists, the RFT (Repeat Formation Tester), which is a log (data transmission by carrier cable) designed for a spot microtest (Chapter 5) giving the static pressure of the reservoir fluids, the type of fluid, and possibly an order of magnitude of the permeability. The principle of the RFT is to establish a connection between the bed and two sampling chambers through a pad applied to the wall (Fig. 2.29). An open-hole version and a new cased-hole version (RFTT-N) exist.

These tests can be performed at the desired depths to give valuable information on vertical thickness parameter values in the reservoirs.

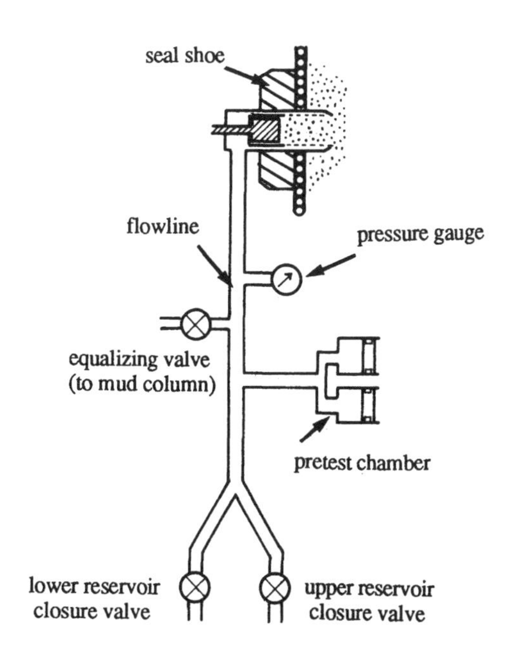

Fig. 2.29 RFT.

2.4.7 Production Logs

These logs analyze the production and characteristics of the fluids **level by level**. They must answer the following questions:

(a) What are the production (or injection) intervals?
(b) What are the fluids produced and the production for each level?
(c) What is the quality of a completion (well equipment), treatment (acidizing, etc.), or a cementing job seal?

Logs are accordingly run **during the production** of a well, and the small-diameter instruments can be run into the production tubing.

The instruments are either simple (flow meter, gradiomanometer, etc.) or combined, like *Schlumberger*'s **Production Logging Tool** (PLT) (Fig. 2.30).

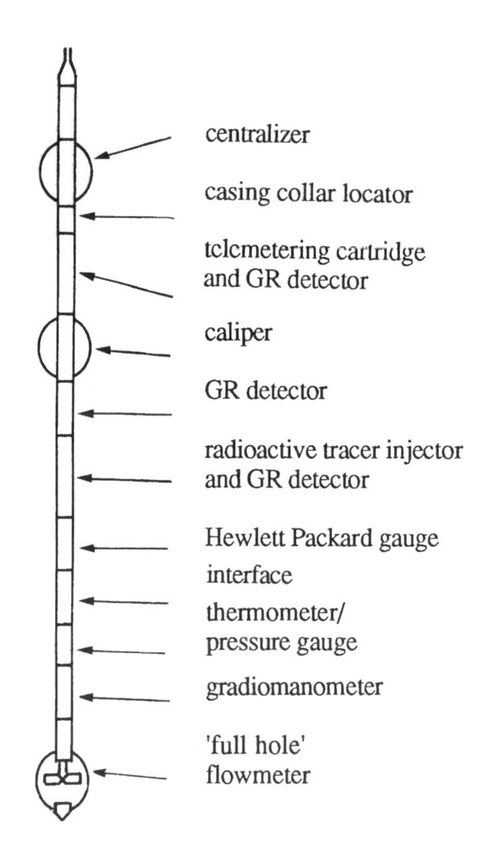

Fig. 2.30 PLT recording.
(*Schumberger* document).

Based on the recorded measurements, the interpretation (Fig. 2.31) gives:

(a) The flow rate, water cut, GOR, temperature, density, etc., for each level.
(b) The results in surface conditions, including the production for each interval.

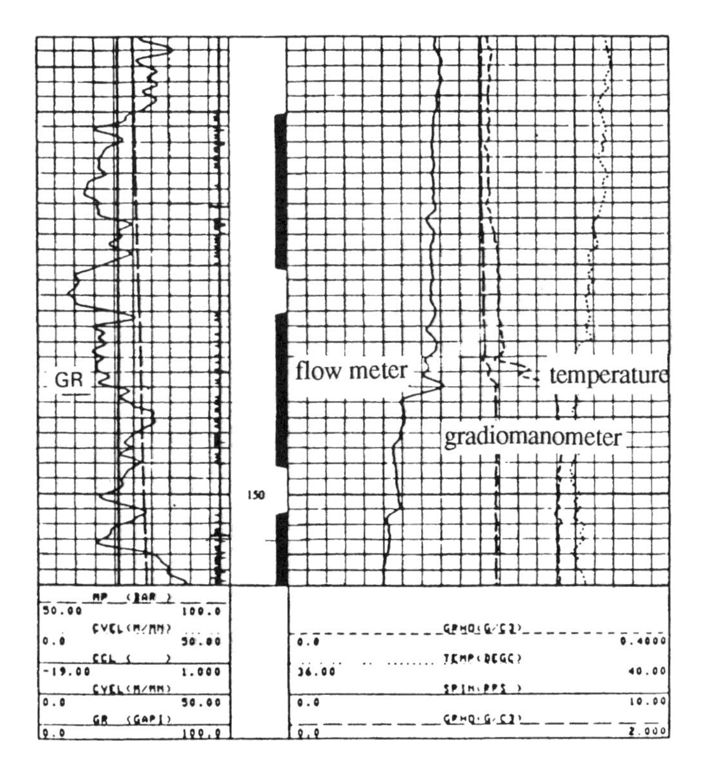

Fig. 2.31

(*Schlumberger* Document).

Chapter 3

FLUIDS AND PVT STUDIES

Oils and gases are **mixtures**. Before defining them, however, let us make a brief review of thermodynamics.

3.1 GENERAL BEHAVIOR

Fluids may be **pure substances**, in other words consisting of identical molecules, or **mixtures**, made up of different molecules.

Their behavior is different in each situation, because the molecular interactions are different.

The variance V indicates the number of parameters conditioning the behavior of a fluid, as a function of the number of independent components C and the number of phases φ. According to Gibbs' rule:

$$V = C + 2 - \varphi$$

For a **pure substance** (C = 1), if it is a one-phase substance ($\varphi = 1$), its behavior depends on two parameters: the volume occupied by a given mass depends on the pressure and the temperature (V = 2). If it is a two-phase substance ($\varphi = 2$), the behavior depends only on one parameter: the pressure is a function of the temperature alone and not of the volumes of the two phases (for a three-phase substance, only one temperature and one pressure are possible, V = 0).

For a **mixture** of N components, the number of parameters conditioning the behavior is the same as above plus (N – 1). This behavior is hence much more complex.

We shall rapidly show the behavior of pure substances on pressure/specific volume and pressure/temperature diagrams, and then examine the behavior of mixtures in greater detail, on the same diagrams.

3.1.1 Pure Substances

Pure substances consist of identical molecules.

3.1.1.1 Pressure/Specific Volume Diagram (Clapeyron Diagram) (Fig. 3.1)

Starting with a pressure and temperature such that the pure substance is in the liquid state (point A in the diagram), if the volume available to the substance is increased very slowly (at constant temperature), the following is observed in succession:

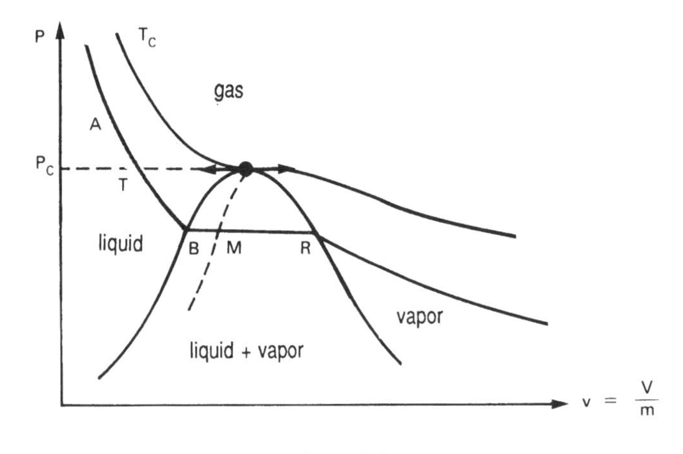

Fig. 3.1

(a) A rapid drop in pressure as long as the pure substance remains in the liquid phase: liquids are relatively incompressible.
(b) The appearance of a vapor phase (point B in the figure): bubble point.

(c) An increase in the vapor phase and a decrease in the liquid phase, with pressure remaining constant.
(d) The disappearance of the last drop of liquid (point R in the figure): dew point.
(e) A relatively slow decrease in pressure: vapors are compressible.

This experiment can be repeated at temperatures lower and higher than T, up to a maximum temperature T_C, the critical temperature, above which the pure substance is always one-phase irrespective of the pressure: the super-critical state called "gas". Its behavior is in fact represented by the equation of state of gases.

The different bubble points form the **bubble-point curve**, and the different dew points the **dew-point curve**. The combination of these two curves is the **saturation curve**.

The critical isotherm is tangent to the saturation curve at the point of inter-section of the bubble- and dew-point curves, the **critical point** (P_C, T_C) along a tangent parallel to axis v.

N.B.:

Any point in the two-phase zone represents a given distribution of the pure substance between the liquid and vapor phases:

$$\frac{BM}{BR} = \frac{\text{Mass of vapor}}{\text{Mass of pure substance}}$$

The series of points corresponding to the same distribution form an equal concentration curve, which terminates in the critical point.

3.1.1.2 Pressure/Temperature Diagram (Fig. 3.2)

If the bubble and dew points are plotted in P and T coordinates, the points merge for each temperature, because the bubble- and dew-point pressures are the same.

The two-phase state is hence also represented by the saturation curve which terminates in the critical point (P_C, T_C).

The liquid/vapor equilibrium pressure is called the vapor pressure curve. It depends only on the temperature and not on the quantities of the two phases present.

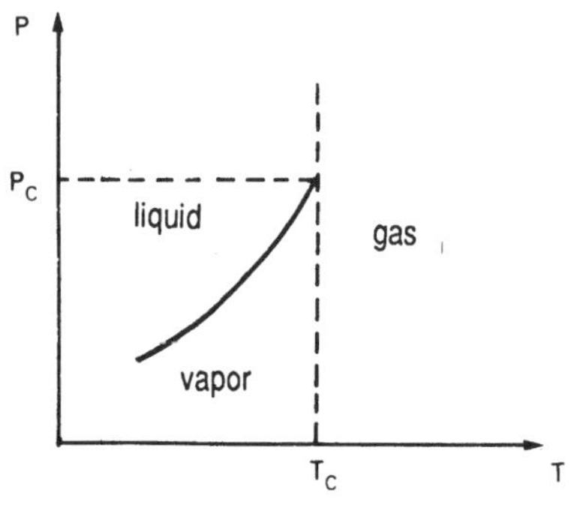

Fig. 3.2

3.1.2 Mixtures

Mixtures consist of several types of molecules.

3.1.2.1 Pressure/Specific Volume Diagram (Fig. 3.3)

$$T_1 < T_C < T_2 < T_{CC}$$

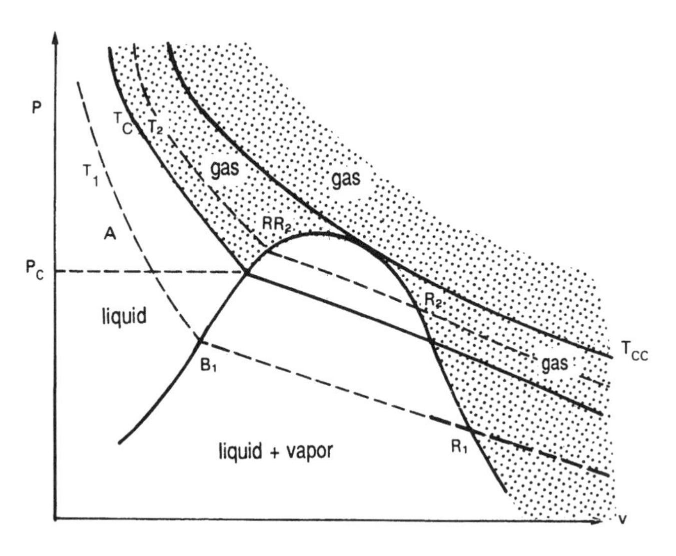

Fig. 3.3

As above, let us start with a pressure and a temperature T_1 (point A) such that the mixture is in the liquid state, and let us slowly increase the volume available to it. The following can be observed:

(a) A rapid decrease in pressure in the liquid phase.
(b) The appearance of a vapor phase (point B_1): bubble point.
(c) An increase in the vapor phase and a decrease in the liquid phase: the pressure decreases, but less rapidly than with the liquid phase alone.
(d) Disappearance of the last drop of liquid at the dew point R_1.
(e) The entire mixture is in the vapor state: the pressure decreases.

This behavior is observed up to the critical temperature T_C. Above this point, as long as the temperature remains below the critical condensation temperature T_{CC}, the following is observed at constant temperature T_2 (and slowly increasing volume):

(a) A "gas" phase (supercritical state): the pressure falls.
(b) The appearance of a liquid phase at the **retrograde dew point** R_2.
(c) An increase in the liquid phase to a peak, followed by its decrease: the pressure falls.
(d) Disappearance of the last drop of liquid at the dew point R_2.
(e) The entire mixture is in the "gas" state: the pressure falls.

Above the **critical condensation temperature** (cricondentherm), the mixture is always in the "gas" state.

N.B.:

Each point in the two-phase zone represents a certain distribution of the mixture between the liquid and vapor phases, as well as given compositions of these phases.

For a given temperature, the compositions of the liquid and vapor vary with the pressure.

3.1.2.2 Pressure/Temperature Diagram (Fig. 3.4)

As for pure substances, the bubble and dew points for a mixture are shown on the P/T diagram.

At a given temperature, the bubble- and dew-point pressures are not the same. The bubble- and dew-point curves are hence distinct. The figure shows equal-composition curves in the liquid phase (0% = dew-point curve, 100% = bubble-point curve).

The zone corresponding to retrograde phenomena is cross-hatched. It corresponds to a **condensation** occurring by isothermal pressure drop. Its boundary is the locus of the condensation maxima for each temperature.

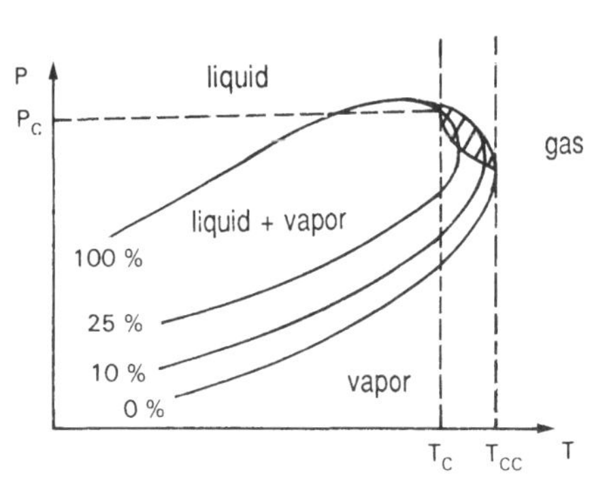

Fig. 3.4

3.1.2.3 Different Types of Reservoirs

Figure 3.5 gives a classification of hydrocarbon reservoirs on a P/T diagram.

A. Oil Reservoirs (Fig. 3.5)

The reservoir temperature is lower than the critical temperature of the mixture contained therein.

If the hydrocarbons are initially one-phase, i.e. **under-saturated oil** reservoirs, and if the same mixture exists throughout the reservoir, an important point is the one at which a gas phase appears (bubble-point, or saturation pressure). From this point on, the passage available to the oil decreases, and its viscosity also increases, decreasing the productivity of the wells. Moreover, the gas, which does not circulate initially (trapping of "bubbles"), begins to flow increasingly faster. The production of a unit volume of oil "costs" more and more in gas. This results in a fairly rapid drop in pressure and low recovery.

If a hydrocarbon liquid phase and a vapor phase are initially in equilibrium, i.e. **oil with gas cap**, part of the gas released by production can be added to the gas cap. If the gas cap is not produced, its "piston" effect maintains the pressure and recovery is better. In this situation, the initial bubble-point

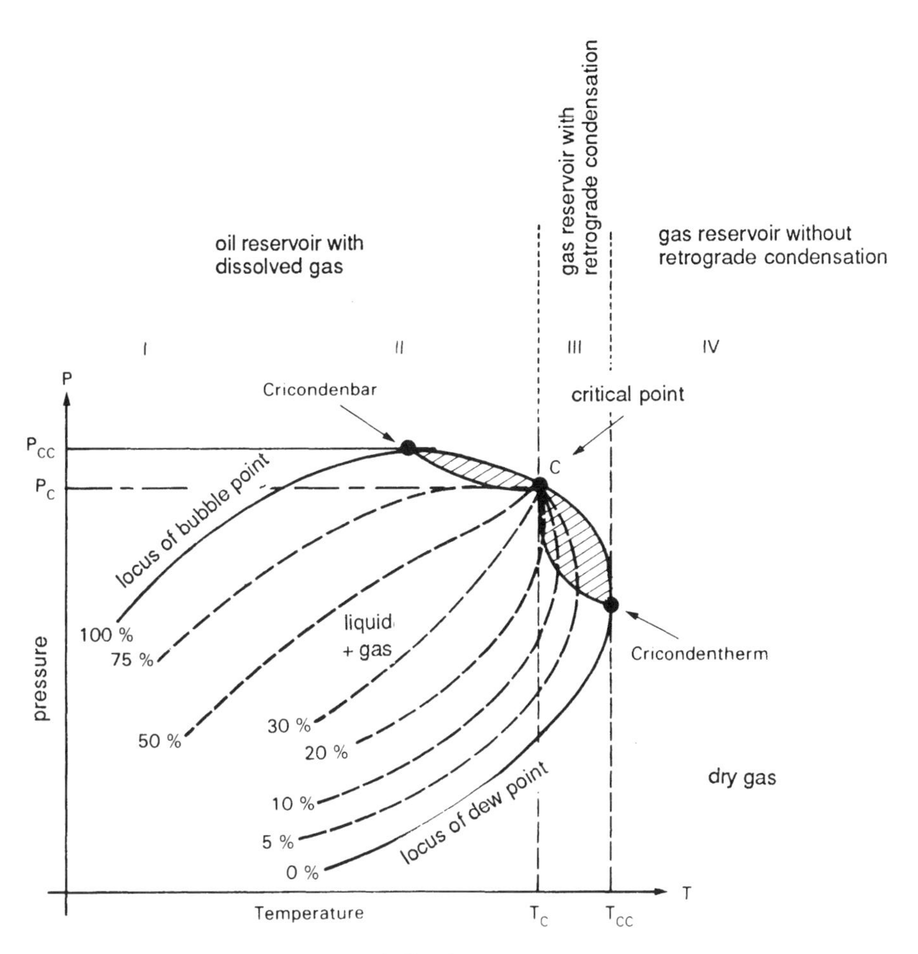

Zone I: Little or no influence of dissolved gases.
Zone II: Significant influence of dissolved gases (volatile oils).
Zone III: Retrograde with condensation of liquid in the reservoir.
Zone IV: Dry or wet gas.

Isothermal and isobaric retrograde condensation occurs for: $T_c \leq T \leq T_{cc}$
and also for: $P_{cc} \geq P \geq P_c$ (cross-hatched area).

Fig. 3.5

Classification of hydrocarbon reservoirs based on thermodynamic criteria: P/T diagram.
(*After* R. Monicard).

pressure of the hydrocarbon liquid phase is the pressure of the gas-cap gas. The oil is said to be saturated.

Whether or not the oil is initially one-phase, the **bubble-point pressure of the oil generally decreases** with increasing depth for a given reservoir. This means that the composition of the oil varies vertically (slightly lighter at the top of the oil zone). In some cases, the bubble point also varies laterally in the reservoir.

B. Retrograde Condensate Gas Reservoirs

Examples of this type of reservoir are Saint Marcet and Hassi R'Mel. The reservoir temperature ranges between the critical and "critical condensation" temperatures of the mixture in the reservoir.

In most cases, the initial pressure is the retrograde dew-point pressure (not at Saint Marcet). This means that production causes a very rapid condensation of hydrocarbons in the reservoir. As the condensed products are mainly heavy components, and since the condensed volume does not flow because of the low saturation, the gas produced is depleted of heavy products, i.e. condensables.

C. Gas Reservoirs One-Phase in Reservoir Conditions

Typical examples are Lacq and Frigg. The reservoir temperature is higher than the critical condensation temperature of the mixture in the reservoir. Most of the gases yield condensates at the earth's surface: they are called **wet** gases. Some of them, however, which practically consist of pure methane and ethane, yield only gas at the earth's surface: these are **dry** gases.

3.1.3 Behavior of Oil and Gas between the Reservoir and the Earth's Surface

The oil and gas in the reservoirs yield fluids which differ significantly in volume and in quality when they reach the surface.

Thus light oils (richer in C_1 to C_4 light and intermediate components) produce a great deal of gas at the surface. Conversely, heavy oils produce only very little gas, if at all (dead oils).

Dry gases yield only gas at the earth's surface and, conversely condensate gases may yield a great deal of condensates (or casinghead gasoline). This is summarized in the following table.

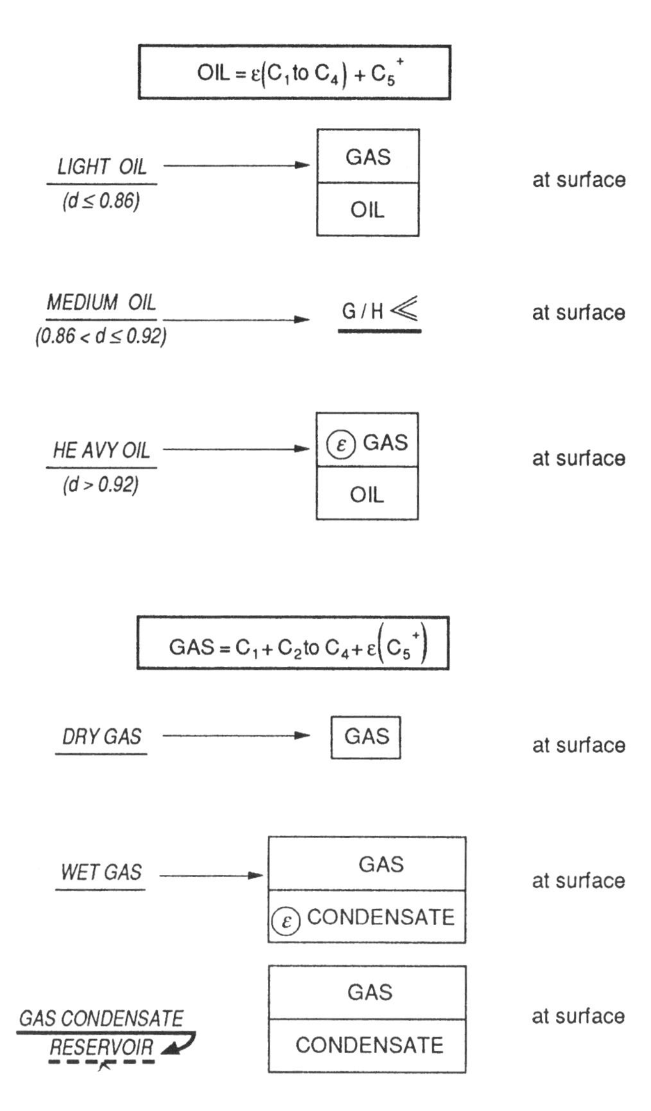

G / H ⩽ indicating that the production gas/oil ratio is lower than for light oils

(ε) = little gas (or condensate)

These significant transformations of fluids between the reservoir and the surface are analyzed in the following sections.

3.1.4 Physical Constants of Hydrocarbons and Other Components

See table below.

3.2 NATURAL GASES

3.2.1 Practical Equation of State

The equation employed is the ideal gas equation, with a factor Z indicating the difference in behavior between the real gas and an ideal gas:

$$PV = Zn\,RT = Z\,\frac{m}{M}\,RT$$

where
- P = absolute pressure (measured from vacuum),
- T = absolute temperature (measured from absolute zero):
  - TK = t°C + 273 (French)
  - T°R = t°F + 460 (British/US),
- V = volume occupied at P and T by n moles of gas,
- m = mass of gas considered,
- M = molecular weight of gas,
- R = universal ideal gas constant:
  - MKSA system: R = 8.32 J/K/mol.g
  - US system: R = 10.7 mol/lb (P in psia, T in °R, and V in cubic feet).

The factor Z, called the **compressibility factor**, depends on the gas composition, the pressure and the temperature.

As the pressure approaches 0, molecular interactions decrease. The behavior of the gas approaches that of an ideal gas and Z approaches 1 (Fig. 3.6).

PHYSICAL CONSTANTS OF HYDROCARBONS AND OTHER COMPONENTS

Symbol	Molecular weight	Boiling point at 1 atm. (K)	Critical constants: Absolute pressure (bar)	Critical constants: Absolute temperature (K)	Critical constants: Specific volume (cm^3/g)	Properties of liquid at 15°C and 1 atm.: Specific gravity (15°C/15°C)	Properties of gas at 15°C and 1 atm.: Gas gravity* (air = 1)	Properties of gas at 15°C and 1 atm.: Spec. vol. (cm^3/g)
C_1	16.042	111.8	46.41	191.2	6.186	0.3 (a)	0.554	1,472.8
C_2	30.068	184.7	48.94	305.7	4.919	0.377 (b)	1.038	785.8
C_3	44.094	231.3	42.57	370.1	4.545	0.508 (b)	1.522	535.9
IC_4	58.120	261.6	36.48	408.3	4.520	0.563 (b)	2.006	406.6
nC_4	58.120	272.8	37.97	425.3	4.382	0.584 (b)	2.006	406.6
IC_5	72.145	301.2	33.30	460.7	4.276	0.625	2.491	327.5
nC_5	72.146	309.4	33.75	469.7	4.307	0.631	2.491	327.5
C_6	86.172	342.1	30.34	507.8	4.276	0.664	2.975	274.2
C_7	100.198	371.8	27.36	540.3	4.257	0.688	3.459	235.8
C_8	114.224	399.0	24.97	568.7	4.257	0.707	3.943	206.9
C_9	128.250	424.1	22.89	594.7	4.239	0.722	4.428	184.2
C_{10}	142.276	447.4	20.96	617.7	4.239	0.734	4.913	166.1
C_{11}	156.302	469.2	20.00	639.4	4.183	0.744		
C_{12}	170.328	489.6	18.75	658.3	4.164	0.753		
C_{13}	184.354	508.8	17.72	676.1	4.158	0.760		
CO	28.010	81.3	34.96	133.3	3.321	0.801 (c)	0.967	843.5
CO_2	44.010	194.8	73.84	304.3	2.141	0.827 (b)	1.519	536.8
H_2S	34.076	213.7	90.05	373.7	2.878	0.79	1.176	693.4
SO_2	64.060	263.3	78.81	430.8	1.898	1.397	2.212	368.8
Air	28.966	79.1	37.72	132.6	2.227	0.856 (c)	1.000	815.7
H_2	2.016	20.6	12.97	33.4	32.205	0.07 (c)	0.070	11,719.2
O_2	32.000	90.3	50.75	154.9	2.341	1.14 (c)	1.105	738.4
N_2	28.016	77.5	33.92	126.3	3.209	0.808 (c)	0.967	843.3
H_2O	18.016	373.3	221.18	647.5	3.121	1.000	0.622	

(a) Bulk specific gravity in a medium crude oil.
(b) Property of the liquid at 15°C and at the bubble-point pressure.
(c) Specific gravity at 1 atm. and boiling point.
(*) To obtain the gas density, multiply by $1.226 \cdot 10^{-3}$, which is the density of air at 1 atm. and 15°C.

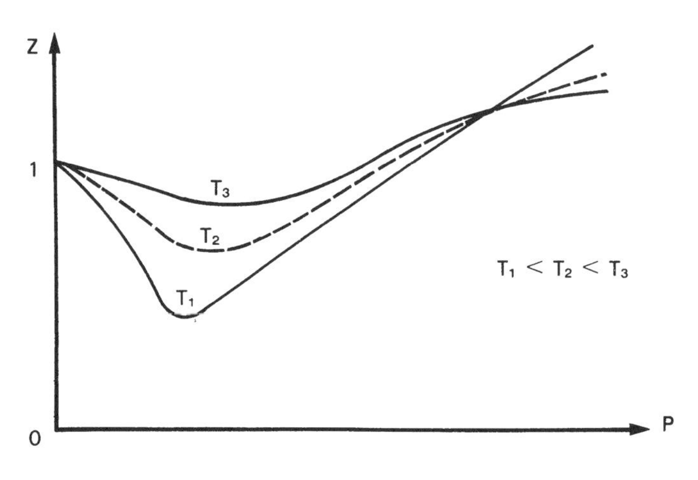

Fig. 3.6

3.2.2 Volume Factor of a Gas B_g

The quantities of gas in place, reserves, flow rates and cumulative production are expressed as standard volumes, corresponding to given pressures and temperatures, of which the most commonly used are the following:

P_{std} = 76 cmHg, T_{std} = 0°C (**standard** conditions, *GdF*),

P_{std} = 75 cmHg, T_{std} = 15°C *(SNEA (P))*,

P_{std} = 76 cmHg, T_{std} = 15°C (or 15.56°C ≈ 60°F, US standard).

For these conditions, it is assumed that $Z_{std} = 1$.

To go from the volume at P and T to the standard volume, the volume factor B_g is used:

$$B_g = \frac{\text{Volume occupied at P and T by a mass of gas m}}{\text{Standard volume of the mass m}}$$

$$B_g = \frac{P_{std}}{P} \cdot Z \cdot \frac{T}{T_{std}}$$

The symbol B_g stands for the gas bulk volume.

N.B.: Standard volume of mass m = volume that mass m would occupy at P_{std} and T_{std} if it was in the vapor state in these conditions.

3.2.3 Determination of Z

3.2.3.1 Experimental

P/V analysis of a mass of gas m at T (M known).

3.2.3.2 Calculations and Charts

These methods are based on the law of corresponding states.

A chart has been prepared giving Z as a function of:

$$\text{Pseudoreduced pressure} = \frac{\text{Absolute pressure}}{\text{Absolute pseudocritical pressure}}$$

and of:

$$\text{Pseudoreduced temperature} = \frac{\text{Absolute temperature}}{\text{Absolute pseudocritical temperature}}$$

based on studies of many gases (see Chart A2 at the end of the chapter).

The pseudo critical pressure and temperature of a given gas (different from the critical pressure and temperature) are obtained by calculation from the composition, or by chart from the specific gravity (see Chart A1 at the end of the chapter).

If the gases contain nitrogen, the foregoing methods apply, possibly with corrections.

If the gases contain small amounts of carbon dioxide (less than 4 to 5%), these methods also apply.

There is no calculation method with hydrogen sulfide.

3.2.4 Condensables Content of a Gas (GPM)

This is the C_3^+ (propane and higher), C_4^+ (butane and higher) or C_5^+ (pentanes and higher) content of a gas, i.e. the proportion of condensables for which the market is different from that of the actual gas.

The GPM is expressed:

(a) In gallons of liquid condensables per 1000 cubic feet of total standard gas (USA): liquid C_3^+ is measured at 76 cmHg and 15.6°C (60°F), the total standard gas includes the gaseous equivalent of the condensables.

(b) In grams of condensables per cubic meter of total standard gas (Europe, etc.).

The GPM of Saint Marcet is about 100 g/m³, that of Hassi R'Mel about 300 g/m³, and that of Frigg is 4 g/m³.

3.2.5 Viscosity of a Gas

At low pressures (close to atmospheric pressure), the viscosity of a gas rises with temperature (greater agitation of the molecules).

At other pressures, it rises as the pressure increases and the temperature decreases.

Order of magnitude: 0.01 to 0.03 cP (1 to 3.10^{-5} Pa.s).

The viscosity is determined by calculation and chart (see Chart A4 at the end of the chapter).

3.2.6 Well Effluent Composition of Gas Reservoirs

Examples:

Composition (mol %)		**Condensate Gas**	**Dry gas**
Hydrogen sulfide	H_2S	-	-
Nitrogen	N_2	4.27	2.12
Carbon dioxide	CO_2	2.58	0.75
Methane	C_1	77.02	95.47
Ethane	C_2	6.70	1.02
Propane	C_3	3.24	0.33
Butanes	C_4	2.65	0.16
Pentanes	C_5	1.33	0.05
Hexanes	C_6	0.79	0.03
Heptanes	C_7	0.30	0.02
	C_8	0.25	0.01
	C_9	0.20	0.01
	C_{10}	0.15	0.01
	C_{11}^+	0.52	0.02
		100.00	**100.00**

3.3 OILS

Oils have a specific gravity between 0.75 and 1. The Americans use API gravity:

$$°API = \frac{141.5}{sp.\ gr.\ 60°/60°F} - 131.5$$

3.3.1 Behavior in the One-Phase Liquid State and the Two-Phase State (Oil and Gas)

For an under-saturated oil ($P > P_b$), the isothermal compressibility factor is defined as:

$$C_o = -\frac{1}{V}\left(\frac{\Delta V}{\Delta P}\right)_T$$

This varies with the pressure, generally $10^{-4} < C_o < 4.10^{-4}$ (bar)$^{-1}$ or $7.10^{-6} < C_o < 30.10^{-6}$ (psi)$^{-1}$, depending on the oil and its gas content. For a given oil, however, its value changes very slightly as a rule between the initial pressure and the bubble point.

At $P < P_b$, the bubble point, the hydrocarbon becomes two-phase. An increasing fraction of the complex hydrocarbon mixture gradually goes into the gas phase (as the pressure falls). The total volume increases faster. This is illustrated by Fig. 3.7.

3.3.2 Formation Volume Factor and Gas/Oil Ratio

The quantities of oil in place, reserves, flow rates and cumulative production are expressed in volumes (or masses) of stock tank oil.

The gas in place (dissolved and free) is measured in standard volumes.

A ratio is set up between the volume (or flow rate) of gas produced and the volume (or flow rate) of the stock tank oil produced.

3.3.2.1 Variation between Reservoir Oil and Stock Tank Oil at a Given Reservoir Pressure

Let us consider a volume of one-phase oil in the reservoir, that flows to the borehole, rises in the tubing, passes into the gathering line and the proces-

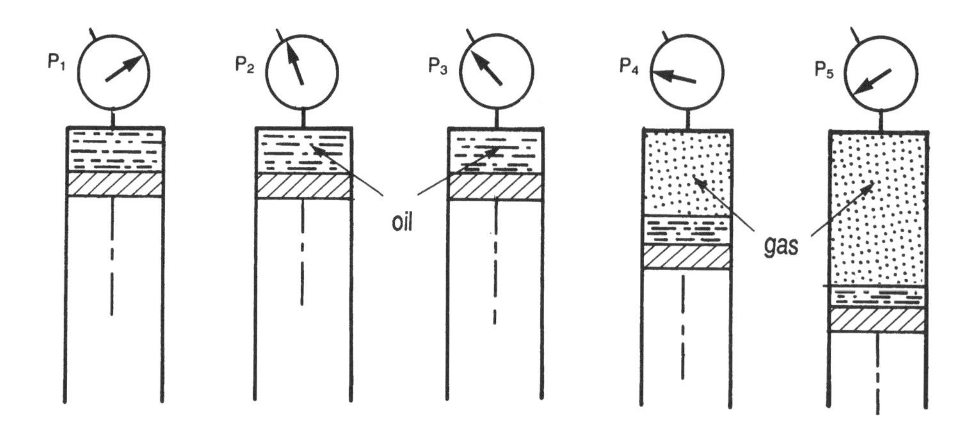

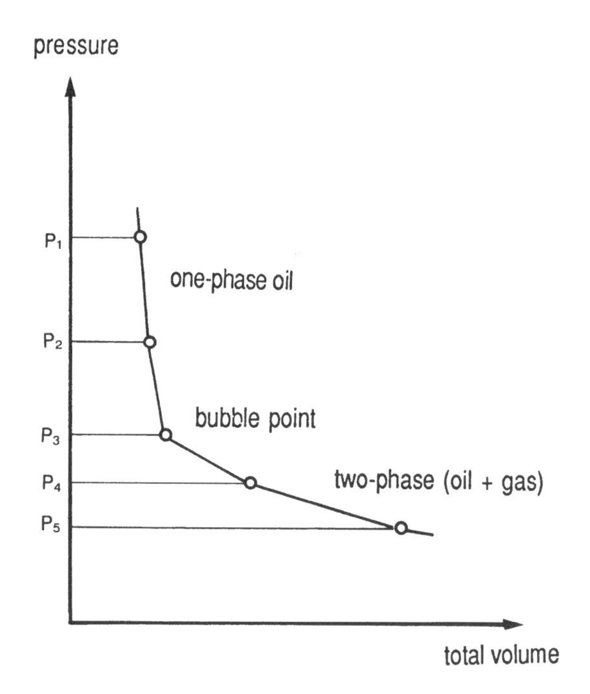

Fig. 3.7

Variation in hydrocarbons with decreasing pressure.

sing installations, to become a certain volume of stock tank oil. A number of phenomena occur: drop in temperature and pressure, and initially-dissolved gas comes out of solution. This results in the recovery of a smaller stock tank liquid volume than the volume leaving the reservoir.

The following terms are employed:

(a) **Formation Volume Factor (FVF)** B_o: the volume of reservoir liquid phase that has yielded a unit volume of oil in stock tank conditions (the symbol B_o corresponds to the oil bulk volume).
(b) Solution **Gas Oil Ratio (GOR)** R_s: the standard volume of gas recovered with a unit volume of stock tank oil (the symbol R_s corresponds to the solution ratio).

The FVF is expressed in barrels per barrel or in cubic meters per cubic meter, and the GOR in cubic feet per barrel or in cubic meters per cubic meter. Note that:

$$1 \text{ ft}^3/\text{bbl} = 0.178 \text{ m}^3/\text{m}^3$$

This can be represented schematically as follows:

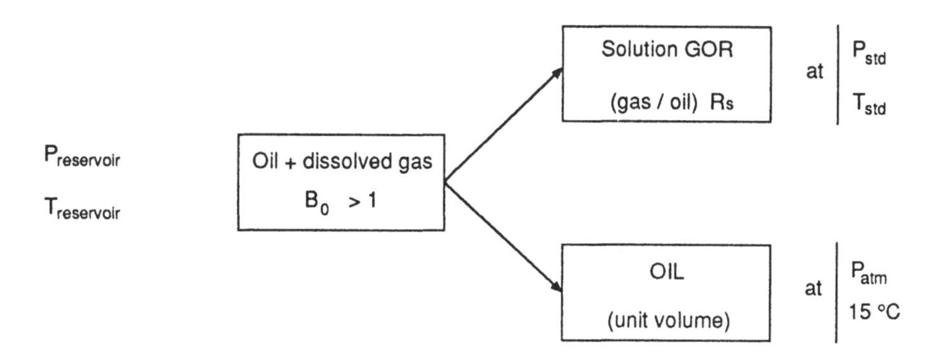

N.B.:

(a) The FVF is always higher than 1, even in the absence of gas. This is because the effect of temperature is greater than that of pressure.
(b) The FVF and GOR depend on the processing conditions. Good processing yields maximum stable stock tank oil and hence minimum FVF and GOR.

3.3.2.2 Variation in FVF B_o and Solution GOR R_s with Reservoir Pressure, Production GOR R

An important pressure can be observed in both figures, the bubble-point pressure below which the gas is released from solution.

(a) **Figure 3.8:** The FVF of the oil B_o decreases as the pressure falls for $P < P_B$ because of the decrease in volume due to the gas coming out of solution.

This figure also shows the total volume B_t occupied (for $P < P_B$) in a cell when the pressure decreases, with a unit volume of stock tank oil in the cell. This two-phase volume factor B_t is equal to the volume (oil + dissolved gas), i.e. B_o, to which the volume of gas released from solution is added:

$$B_t = B_o + (R_{si} - R_s)\, B_g$$

where R_{si} is the initial solution GOR at $P = P_i$.

(b) **Figure 3.9:** R the production GOR is the ratio, at a given time, of the total flow rate of standard gas produced to the flow rate of stock tank oil.

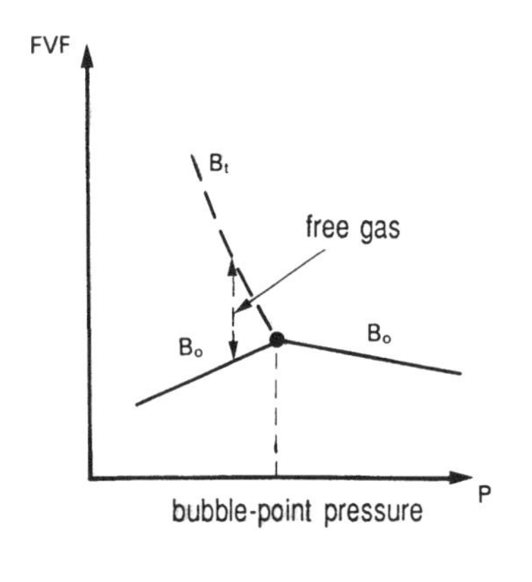

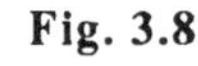

Fig. 3.8

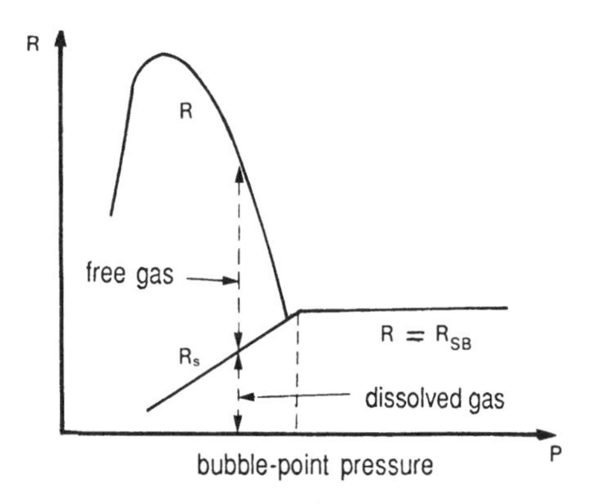

Fig. 3.9

For $P > P_B$, the gas produced can originate only in dissolved gas. The production GOR is constant and equal to the initial solubility R_{si}.

For $P < P_B$, part of the gas produced originates in dissolved gas. This part decreases with the pressure. The rest derives from free gas: when a sufficient saturation (critical gas saturation S_{gc}) is reached, this gas begins to flow.

The gas flow rate in reservoir conditions increases with saturation, hence with decreasing pressure. But the GOR corresponds to gas measured in standard conditions. We have:

$$\frac{Q \cdot P}{Z \cdot T} = \frac{Q_{std} \cdot P_{std}}{1 \cdot T_{std}}$$

(P, Q, T, Z in reservoir conditions reproduced by the laboratory measurement cell, and subscript std = standard).

Q standard hence depends on Q, which increases, and P which decreases. This explains that R reaches a peak when the influence of P becomes preponderant.

Order of magnitude: $R = R_{SB}$ can rise, for example, from 80 m^3/m^3 to more than 400 m^3/m^3.

This means that, if nothing is done to prevent pressure from falling below the bubble point, each cubic meter of oil produced costs more and more gas (hence "energy"). The recovery is low.

B_o and R_s are determined in the PVT laboratory.

Charts are also available to determine B_o and R_s (charts A7 and A8 at the end of the chapter).

The two examples shown in Fig. 3.10 help to evaluate the variations in B_o, R_s and C_o as a function of pressure (a., b. and c. in Fig. 3.10).

Figure 3.11 indicates a correlation between the solution GOR and the factor FVF (*Source:* R. Monicard).

Expression of production gas/oil ratio

When the pressure of an initially one-phase oil reservoir becomes lower than the bubble-point pressure, an oil phase and a gas phase are present in the pores of the reservoir. When the gas phase begins to flow, the gas produced consists partly of free reservoir gas and partly of gas dissolved in the oil.

The reference stock tank oil flow rate is written Q_o.

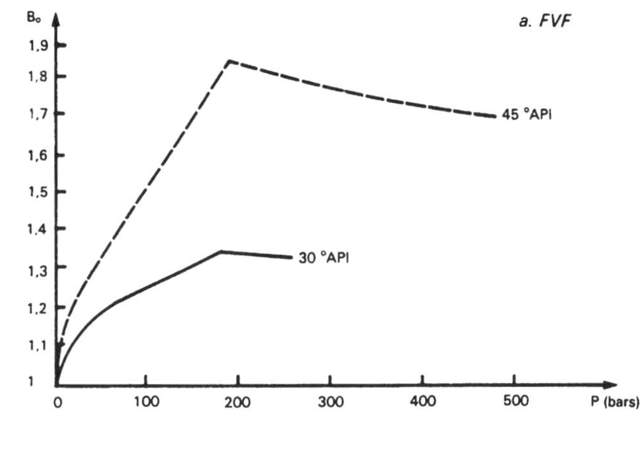

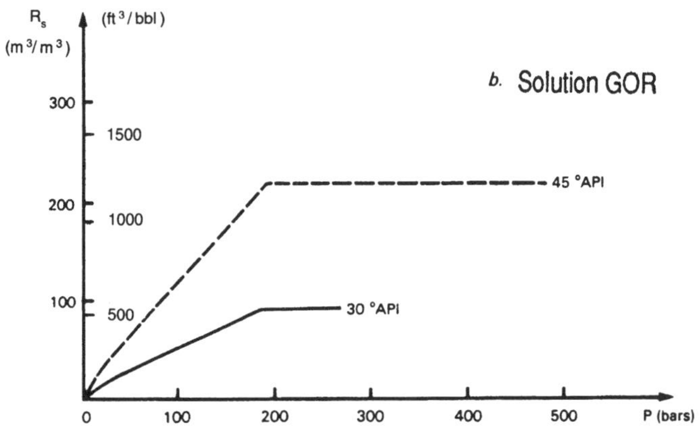

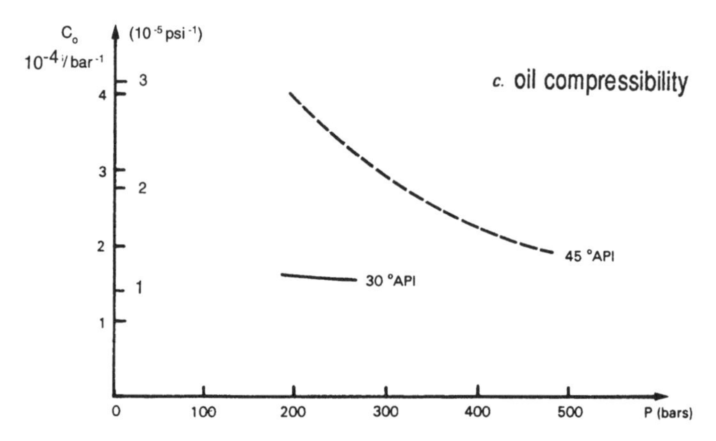

Fig. 3.10

Two examples of properties: light oil (45°API) and medium oil (30°API).

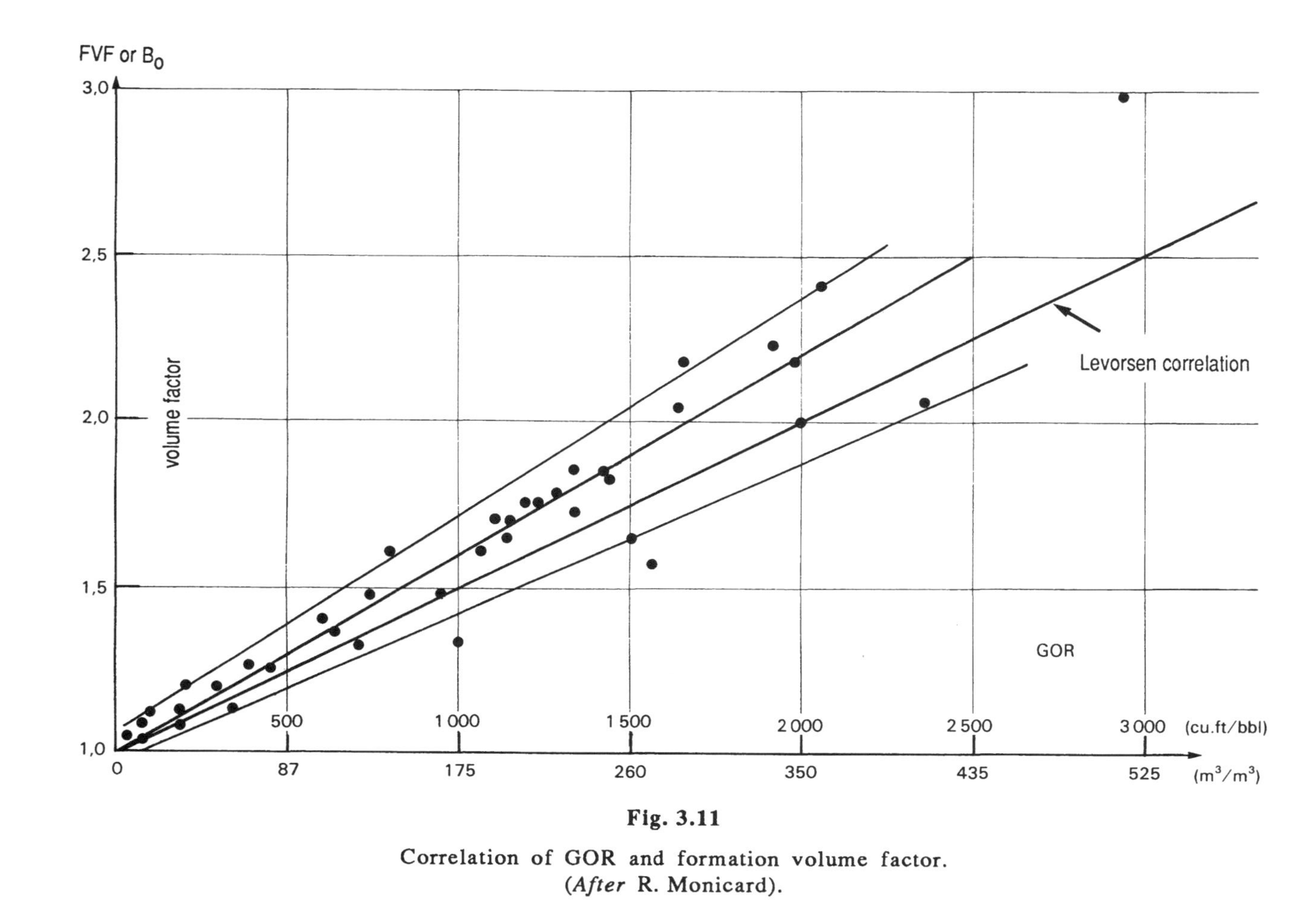

Fig. 3.11

Correlation of GOR and formation volume factor.
(*After* R. Monicard).

The oil flow rate in reservoir conditions is:

$$Q_o B_o = \frac{A\,k_o}{\mu_o} \cdot \frac{\Delta P}{\Delta l}$$

It gives the following:

(a) Stock tank oil flow rate:

$$Q_o = \frac{A\,k_o}{B_o \mu_o} \cdot \frac{\Delta P}{\Delta l}$$

(b) Standard dissolved gas flow rate:

$$Q_{gd} = Q_o\,R_s$$

The free gas flow rate in reservoir conditions (ignoring the gas/oil capillary pressure) is:

$$Q_{gf} B_g = \frac{A\,k_g}{\mu_g} \cdot \frac{\Delta P}{\Delta l}$$

It gives a standard free gas flow rate of:

$$Q_{gf} = \frac{A\,k_g}{\mu_g B_g} \cdot \frac{\Delta P}{\Delta l}$$

The production GOR is the ratio of the flow rate of total standard gas produced ($Q_{gd} + Q_{gf}$) to the flow rate of reference stock tank oil Q_o:

$$R = \frac{Q_{gd} + Q_{gf}}{Q_o} = R_s + \frac{k_g \cdot \mu_o \cdot B_o}{k_o \cdot \mu_g \cdot B_g}$$

In this expression, the viscosities and FVF depend on the pressure. By contrast, the ratio of the effective permeabilities k_g/k_o depends on the saturation with gas (or oil), which depends primarily on the cumulative production of oil (see Section 7.3.3 in Chapter 7).

Hence of the GOR:

$$\boxed{R = R_s + \frac{k_g}{k_o} \frac{\mu_o\,B_o}{\mu_g\,B_g}}$$

3.3.3 Viscosity

The viscosity varies with the pressure, temperature and quantity of dissolved gas. In the reservoir, the following prevails for the hydrocarbon liquid phase.

Order of magnitude: from 0.2 cP (very light oil) to 1 P (2.10^{-4} to 10^{-1} Pa.s), called **heavy** oil above 1 P, up to about 100 P (10 Pa.s).

Determination:

(a) Laboratory:

- Fall of a ball in a calibrated tube filled with oil at a given temperature.
- Capillary tube viscometer.

(b) Chart (see Chart A9 at the end of the chapter).

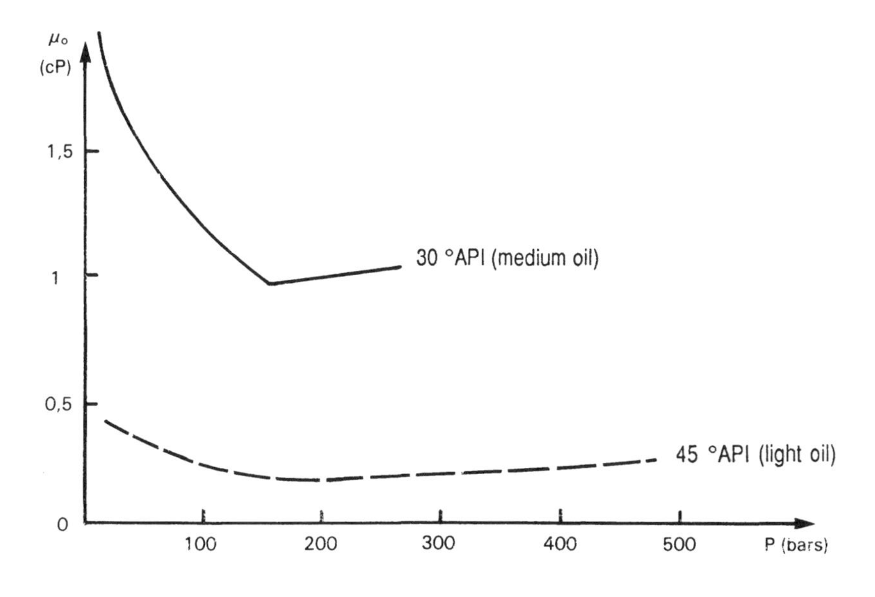

Fig. 3.12

Two examples.

3.3.4 Well Effluent Composition of Oil Reservoirs

Examples:

Composition (mol %)		North Sea (light oil)	Middle East (medium oil)
Hydrogen sulfide	H_2S	-	0 (to 25)
Nitrogen	N_2	0.41	0.10
Carbon dioxide	CO_2	1.24	5.73
Methane	C_1	46.54	23.32
Ethane	C_2	7.40	5.56
Propane	C_3	7.54	6.32
Butanes	C_4	4.57	5.36
Pentanes	C_5	3.01	2.33
Hexanes	C_6	2.27	2.75
Heptanes	C_7	3.37	3.17
	C_8	3.14	3.81
	C_9	2.32	3.32
	C_{10}	1.97	2.57
	$C_{11}+$	16.22	35.66
		100.00	**100.00**

3.3.5 Exercise

An exploration well has identified an oil-impregnated reservoir. The measurements taken have given the following main characteristics:

(a) Initial pressure: P_i = (2552 psi), 176 bar.
(b) Temperature: 87°C.
(c) Flow rate: (815 bpd), 130 m^3/d.
(d) GOR: 494 ft^3/bbl = 88 m^3/m^3.
(e) Gravity: γ_o = 36°API.
(f) Gas gravity: γ_g = 0.7 (estimated).
(g) Compressibility: C_o = (14.10^{-6} psi^{-1}), 2.10^{-4} bar^{-1} (estimated).

Using the charts, determine the bubble point, the formation volume factor B_o, and the oil viscosity μ_o (see Section 3.5).

Answer

(a) The bubble point is determined from Chart A.5.

This gives:

$$P_b = 2200 \text{ psi} \approx \mathbf{152\ bar}$$

Even if this value is approximate, the oil identified is one-phase ($P_i > P_b$).

(b) B_{ob} is obtained from Chart A.7:

$$B_{ob} = 1.27$$

In fact:

$$B_o = B_{ob}\ (1 + C_o\ (P_b - P))$$

for $P > P_b$.

For P_i:

$$B_{oi} = 1.27 \times 0.9952 = 1.264 \text{ m}^3/\text{m}^3$$

take $\mathbf{B_{oi} = 1.26}$.

(c) Viscosity μ_{ob} is given by Chart A.9:

$$\mu_{ob} \approx 0.59 \text{ cP} \qquad \mu_{oi} = \mu_{ob} + 0.01$$

$$\boldsymbol{\mu_{oi} \approx 0.60\ \text{cP} \approx 6.0 \cdot 10^{-4}\ \text{Pa.s}}$$

3.4 FORMATION WATER

The main characteristics are the following:

3.4.1 Composition

This is of interest to the geologist (water origin) and the reservoir specialist for the construction of a reservoir model. The presence of two or more compositions may indicate several aquifers, and, for water flooding, the **compatibility** with the injected water must be investigated (Chapter 8).

3.4.2 Compressibility

The isothermal compressibility factor of water is about:

$$C_w \approx 0.5 \cdot 10^{-4}\ (\text{bar})^{-1} \quad (\text{or } 3.5 \cdot 10^{-6}\ (\text{psi})^{-1})$$

This is the property that enables the water from an aquifer to drain a reservoir by its expansion, in view of the often wide extension of the aquifer in comparison with the size of the reservoir.

3.4.3 Viscosity

The viscosity is determined in the laboratory or from a chart (Chart A.10). Its value is approximately:

$$\mu_w = \mathbf{0.3\ to\ 0.7\ cP}\ (3 \cdot 10^{-4}\ \text{to}\ 7 \cdot 10^{-4}\ \text{Pa.s})$$

3.4.4 Water and Hydrocarbons

Formation water may also give rise to difficulties during production:

(a) With oil: emulsion problems.
(b) With gases: since the gases in the reservoir are in the presence of water, they are saturated with it. Production causes condensation. This incurs the risks of the formation of gas hydrates with methane, ethane, propane, butanes, CO_2 and H_2S in certain pressure and temperature conditions. These hydrates are liable to clog the lines at the surface.

3.5 CHARTS

The reservoir specialist does not always enjoy the availability of complete PVT studies to provide him with detailed knowledge of the fluids identified in a given reservoir.

This makes it necessary to use charts, above all in the preliminary reservoir survey phase. These charts are generally constructed from American **correla-**

tions which have **a fairly high degree of reliability**, and are used to determine certain properties of the fluids.

The charts below (A.1 to A.10) make it possible to determine the requisite values.

(a) **For a gas:**

A.1 and A.2	Compressibility factor Z (P).
A.3	Formation volume factor B_g (P).
A.4	Viscosity μ_g (P).

Beforehand it is necessary to know the:

- Gas gravity γ_g.
- Initial reservoir pressure P_i.
- Reservoir temperature T.

(b) **For a crude oil:**

A.5 and A.6	Bubble-point pressure P_b.
A.7 and A.8	Formation volume factor B_o (P).
A.8	Solution GOR R_s (P).
A.9	Viscosity μ_o (P).

Beforehand it is necessary to know the:

- Specific gravity of stock tank oil γ_o.
- Production GOR.
- Gravity of dissolved gas γ_g.
- Initial reservoir pressure P_i.
- Reservoir temperature T.

(c) **For water:**

A.10	Viscosity of the water μ_w.

Beforehand it is necessary to know the:

- Salt content (NaCl).
- Reservoir temperature T.

3.6 LIQUID/VAPOR EQUILIBRIA, EQUATION OF STATE

Production very often brings the hydrocarbons into the state of a mixture of two phases, one liquid and the second vapor, as we have pointed out. This may take place in the reservoir, the well, or the process plant. It may involve the appearance of a gas phase for an oil reservoir, or a liquid phase for a gas reservoir. If both fluids coexist, they may also break down into different phases.

The equilibrium between the two phases can be investigated experimentally. Calculation methods can also be used, based on an **equation of state**, employed to determine the quantities and properties of the phases at equilibrium in given pressure and temperature conditions.

For the reservoir, these calculations are necessary because of the indispensable need to be able to determine the characteristics of the phases prior to the study of different development configurations, based on the reservoir fluid **composition** alone.

For the surface, these calculation methods directly concern the "process" (surface production calculations), and help to determine the optimal operating conditions of the separators. The general **equation of state** can be written:

$$\frac{P}{RT} = \frac{1}{V - b} - a \,.\, f(V, T, b)$$

where a and b are constants related to the internal pressure and the volumes of the molecules respectively.

The most commonly-used forms are the **Redlich-Kwong** and **Peng-Robinson** equations. They are ideal for the study of petroleum fluids.

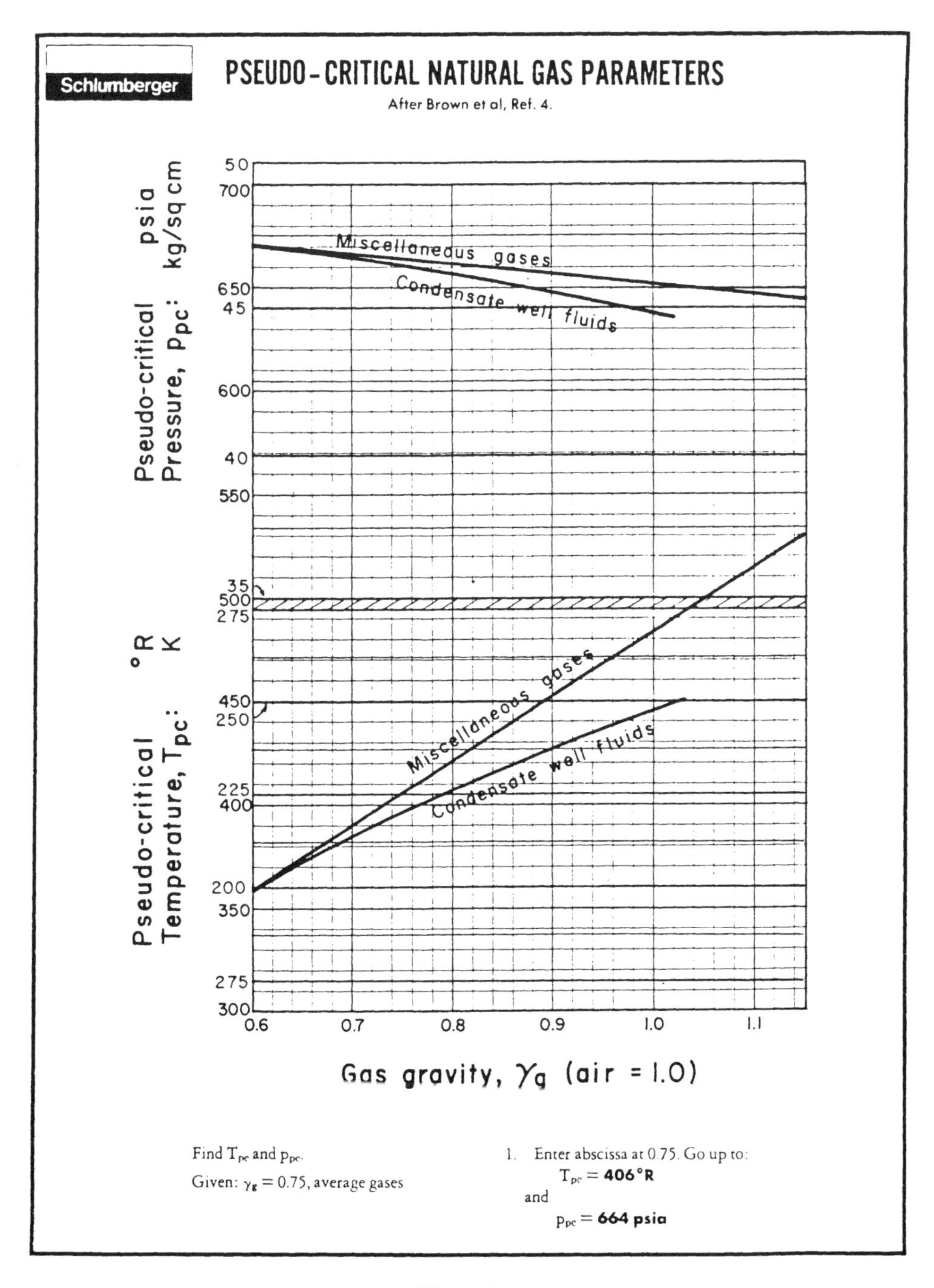
Schlumberger
PSEUDO-CRITICAL NATURAL GAS PARAMETERS
After Brown et al, Ref. 4.
Pseudo-critical Pressure, Ppc: psia kg/sq cm
Miscellaneous gases
Condensate well fluids
Pseudo-critical Temperature, Tpc: °R K
Miscellaneous gases
Condensate well fluids
50
700
650
45
600
40
550
35
500
275
450
250
225
400
200
350
275
300
0.6
0.7
0.8
0.9
1.0
1.1
Gas gravity, γg (air = 1.0)
Find Tpc and ppc.
Given: γg = 0.75, average gases
1. Enter abscissa at 0.75. Go up to:
Tpc = 406°R
and
ppc = 664 psia

Chart A.1

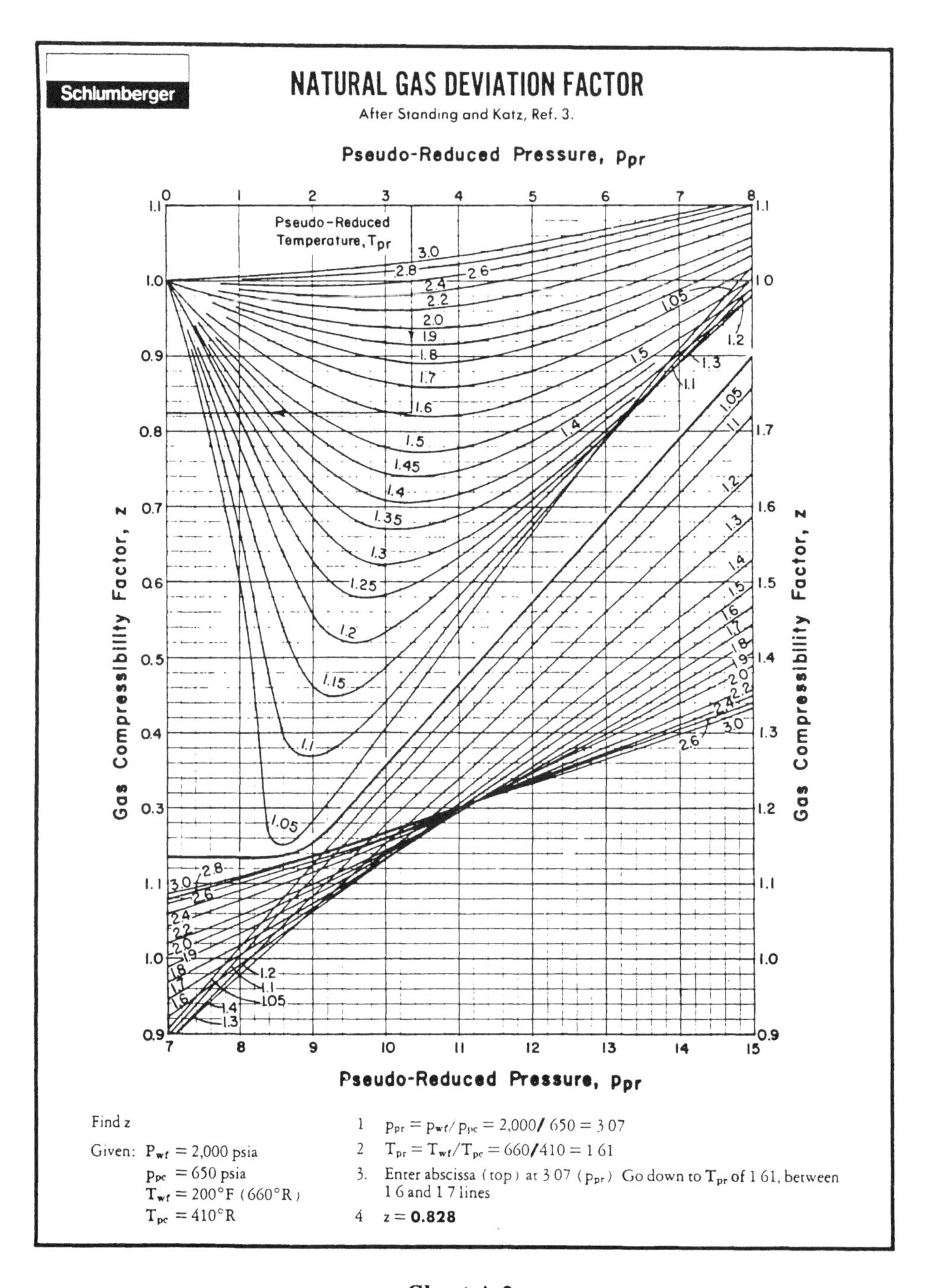

Schlumberger
NATURAL GAS DEVIATION FACTOR
After Standing and Katz, Ref. 3.
Pseudo-Reduced Pressure, p_{pr}
Pseudo-Reduced Temperature, T_{pr}
Gas Compressibility Factor, z
Pseudo-Reduced Pressure, p_{pr}
Find z
Given: P_{wf} = 2,000 psia
p_{pc} = 650 psia
T_{wf} = 200°F (660°R)
T_{pc} = 410°R
1 $p_{pr} = p_{wf}/p_{pc}$ = 2,000/650 = 3.07
2 $T_{pr} = T_{wf}/T_{pc}$ = 660/410 = 1.61
3. Enter abscissa (top) at 3.07 (p_{pr}). Go down to T_{pr} of 1.61, between 1.6 and 1.7 lines
4 z = **0.828**

Chart A.2

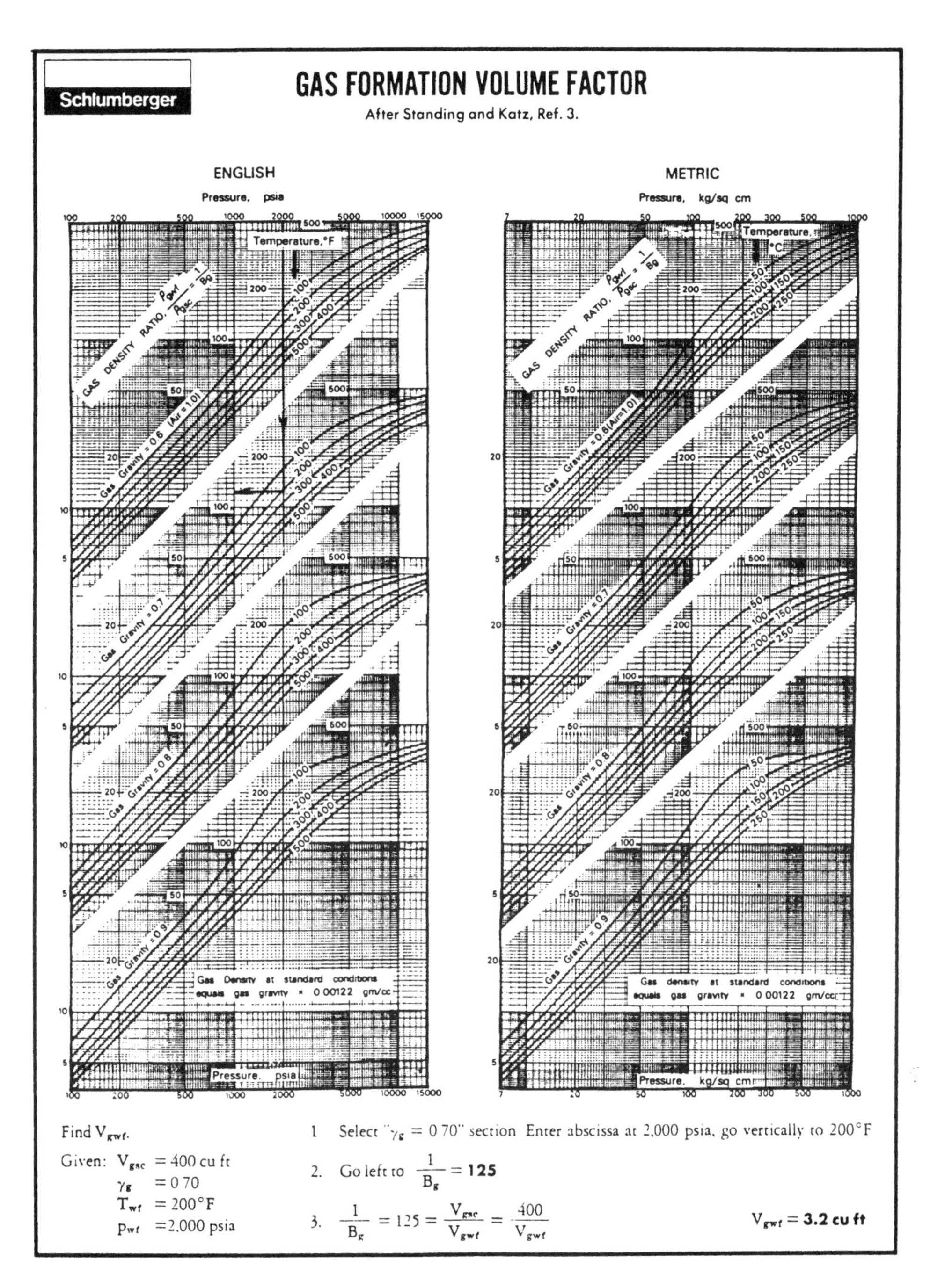

Schlumberger
GAS FORMATION VOLUME FACTOR
After Standing and Katz, Ref. 3.
ENGLISH
Pressure, psia
Temperature, °F
GAS DENSITY RATIO, $\frac{\rho_{gwf}}{\rho_{gsc}} = \frac{1}{B_g}$
Gas Gravity = 0.6 (Air = 1.0)
Gas Gravity = 0.7
Gas Gravity = 0.8
Gas Gravity = 0.9
Gas Density at standard conditions equals gas gravity × 0.00122 gm/cc
METRIC
Pressure, kg/sq cm
Temperature, °C
Gas density at standard conditions equals gas gravity × 0.00122 gm/cc
Find V_{gwf}.
Given: V_{gsc} = 400 cu ft
γ_g = 0.70
T_{wf} = 200°F
p_{wf} = 2,000 psia
1. Select "γ_g = 0.70" section. Enter abscissa at 2,000 psia, go vertically to 200°F
2. Go left to $\frac{1}{B_g}$ = **125**
3. $\frac{1}{B_g} = 125 = \frac{V_{gsc}}{V_{gwf}} = \frac{400}{V_{gwf}}$
V_{gwf} = **3.2 cu ft**

Chart A.3

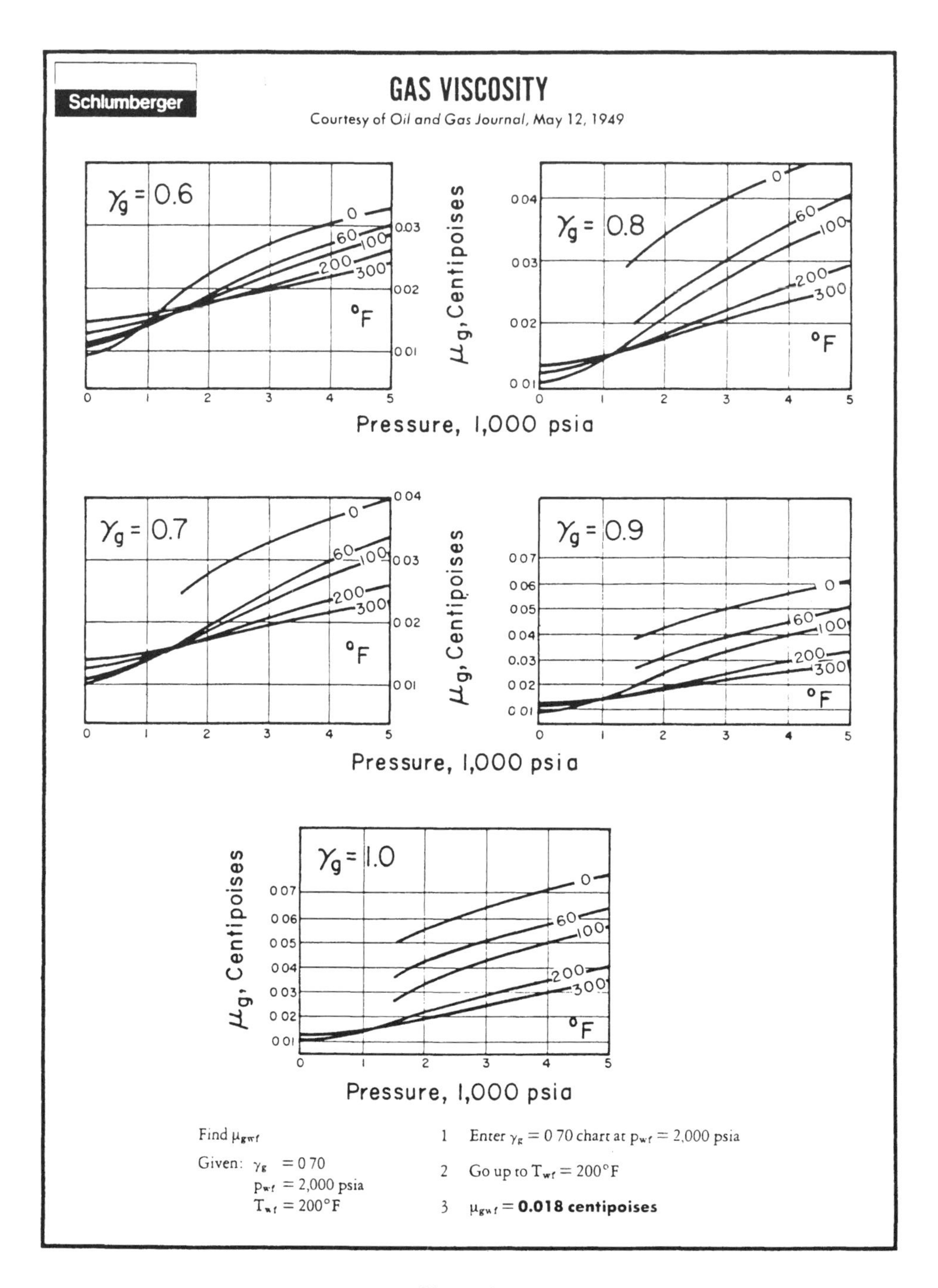
Schlumberger
GAS VISCOSITY
Courtesy of Oil and Gas Journal, May 12, 1949
γ_g = 0.6
γ_g = 0.8
γ_g = 0.7
γ_g = 0.9
γ_g = 1.0
μ_g, Centipoises
Pressure, 1,000 psia
°F
Find μ_{gwf}
Given: γ_g = 0.70
p_{wf} = 2,000 psia
T_{wf} = 200°F
1 Enter γ_g = 0.70 chart at p_{wf} = 2,000 psia
2 Go up to T_{wf} = 200°F
3 μ_{gwf} = **0.018 centipoises**

Chart A.4

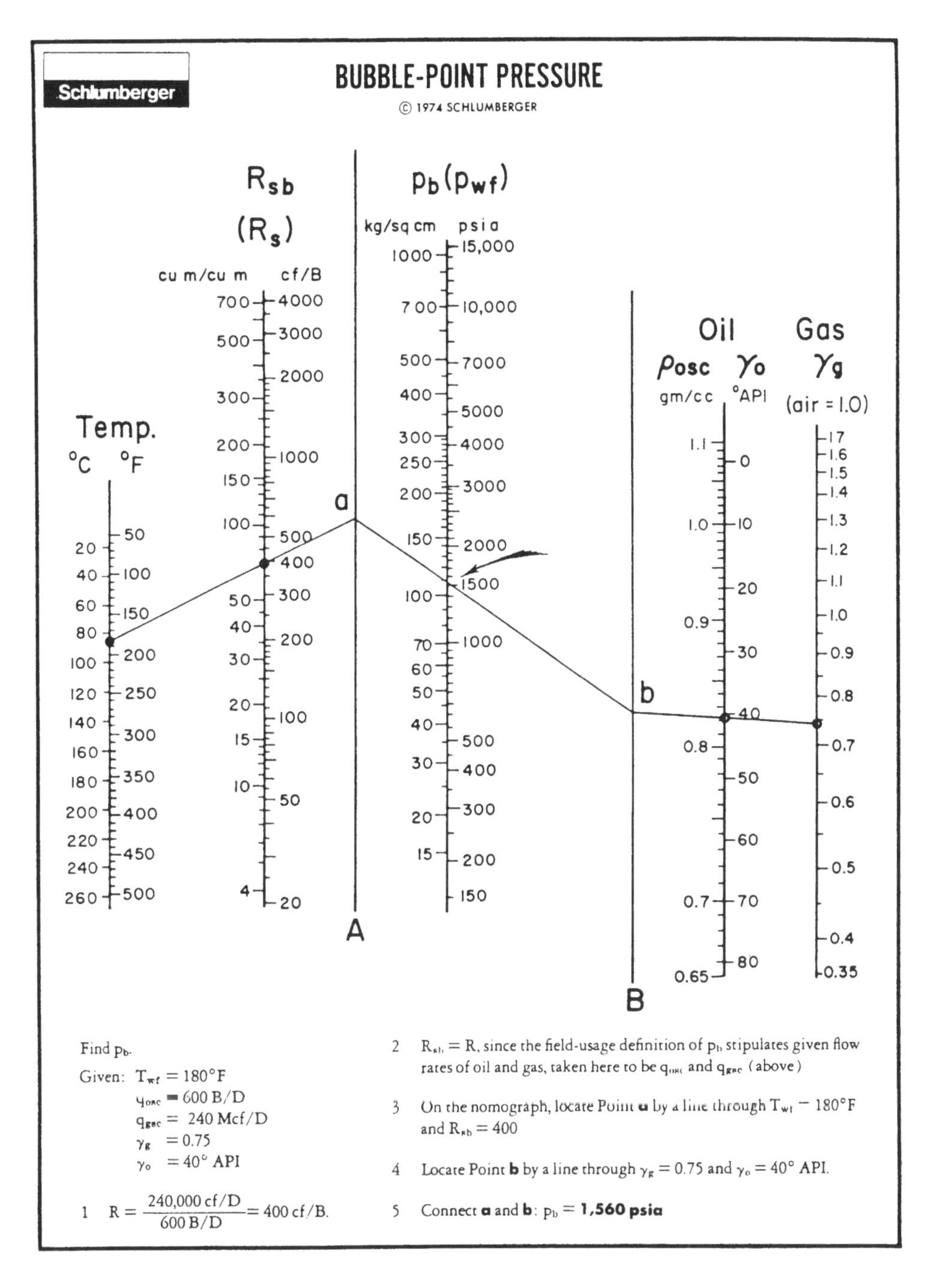
Schlumberger
BUBBLE-POINT PRESSURE
© 1974 SCHLUMBERGER
Temp.
°C °F
Rsb
(Rs)
cu m/cu m cf/B
pb(pwf)
kg/sq cm psia
Oil
ρosc γo
gm/cc °API
Gas
γg
(air = 1.0)
a
b
A
B
Find pb.
Given: Twf = 180°F
qosc = 600 B/D
qgsc = 240 Mcf/D
γg = 0.75
γo = 40° API
1 R = 240,000 cf/D / 600 B/D = 400 cf/B.
2 Rsb = R, since the field-usage definition of pb stipulates given flow rates of oil and gas, taken here to be qosc and qgsc (above)
3 On the nomograph, locate Point a by a line through Twf = 180°F and Rsb = 400
4 Locate Point b by a line through γg = 0.75 and γo = 40° API.
5 Connect a and b: pb = 1,560 psia

Chart A.5

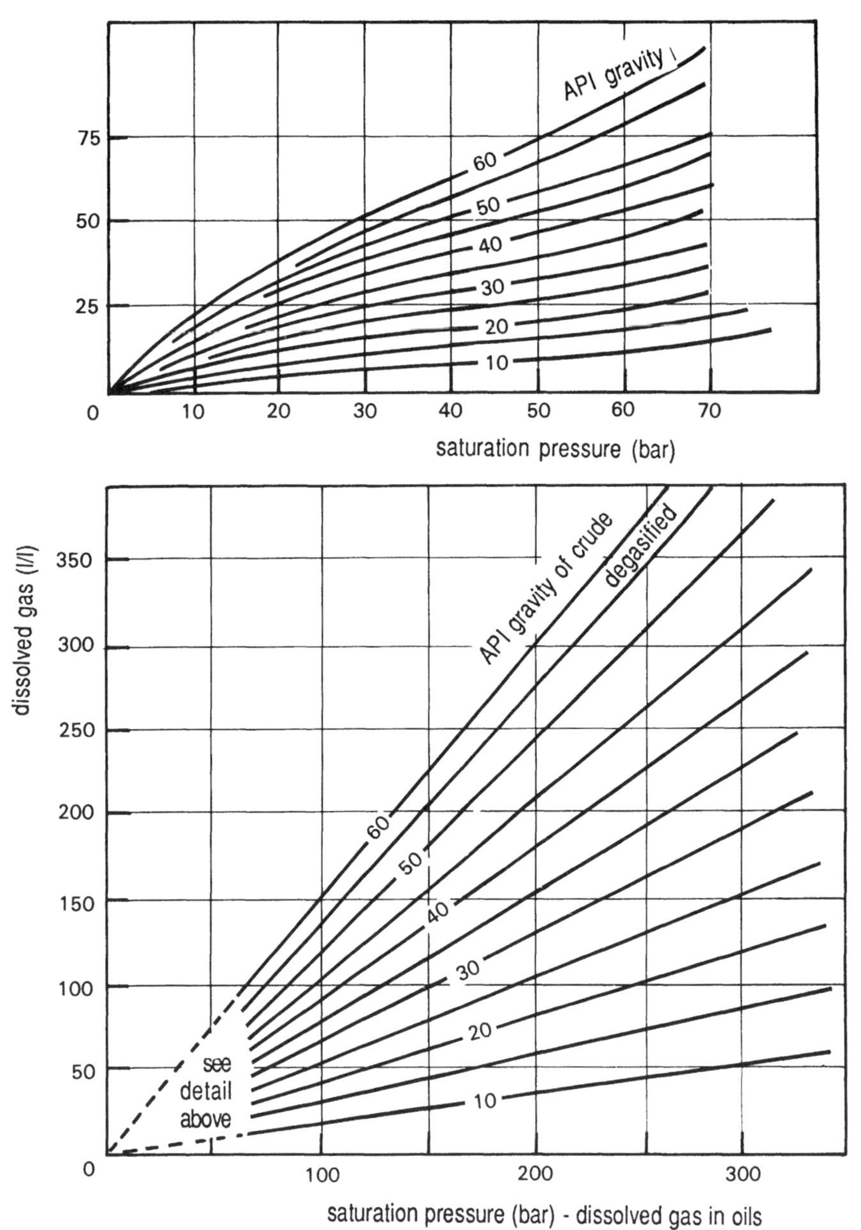

This diagram does not take account of temperature (*After* Katz).

Chart A.6

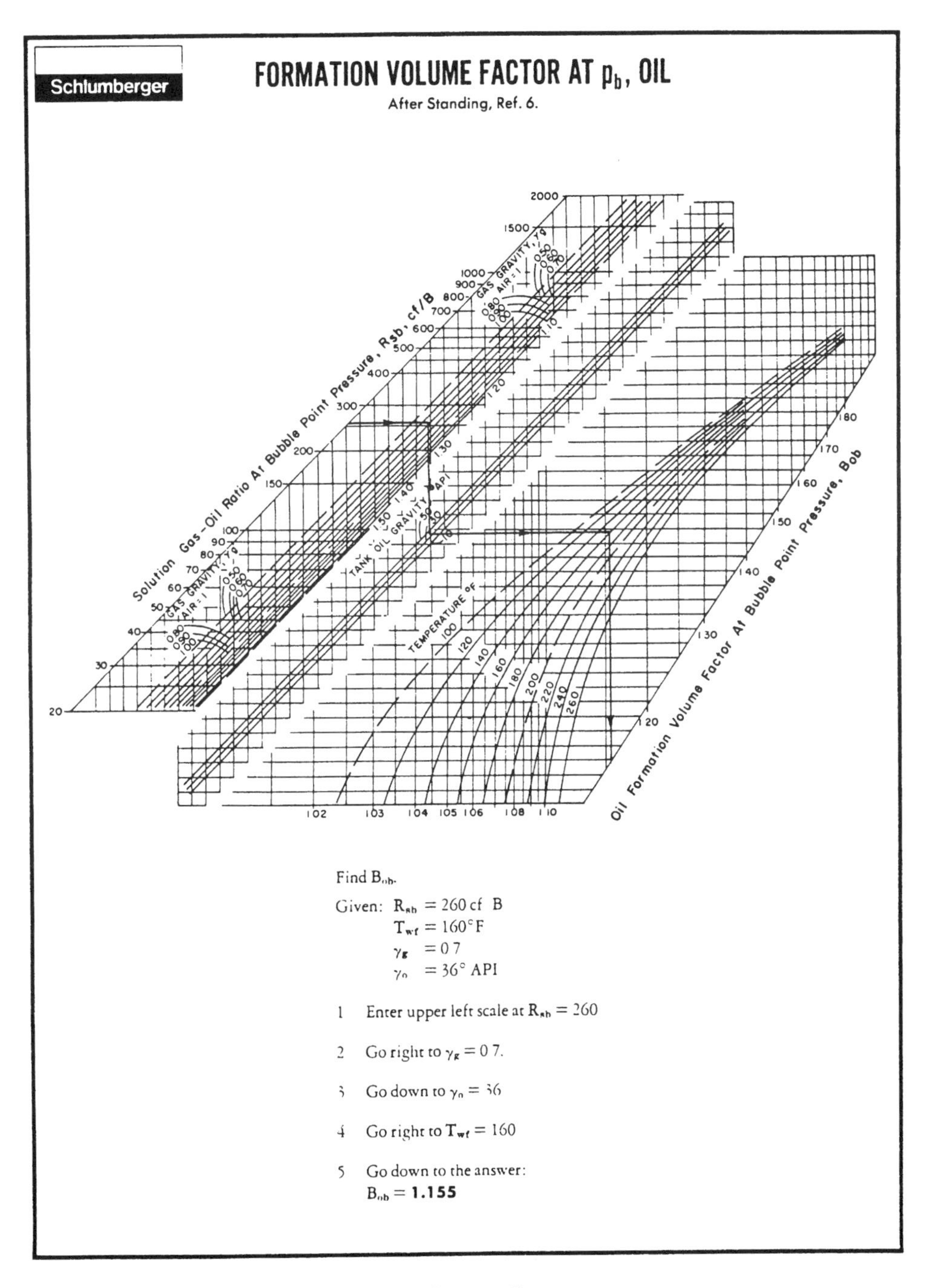
Schlumberger
FORMATION VOLUME FACTOR AT p_b, OIL
After Standing, Ref. 6.
Solution Gas-Oil Ratio At Bubble Point Pressure, Rsb, cf/B
2000
1500
1000
900
800
700
600
500
400
300
200
150
100
90
80
70
60
50
40
30
20
GAS GRAVITY, γg
AIR = 1
TANK OIL GRAVITY °API
TEMPERATURE °F
100
120
140
160
180
200
220
240
260
Oil Formation Volume Factor At Bubble Point Pressure, Bob
1 02
1 03
1 04
1 05
1 06
1 08
1 10
1 20
1 30
1 40
1 50
1 60
1 70
1 80
Find B_{ob}.
Given: R_{sb} = 260 cf B
T_{wf} = 160°F
γ_g = 0 7
γ_o = 36° API
1 Enter upper left scale at R_{sb} = 260
2 Go right to γ_g = 0 7.
3 Go down to γ_o = 36
4 Go right to T_{wf} = 160
5 Go down to the answer:
B_{ob} = 1.155

Chart A.7

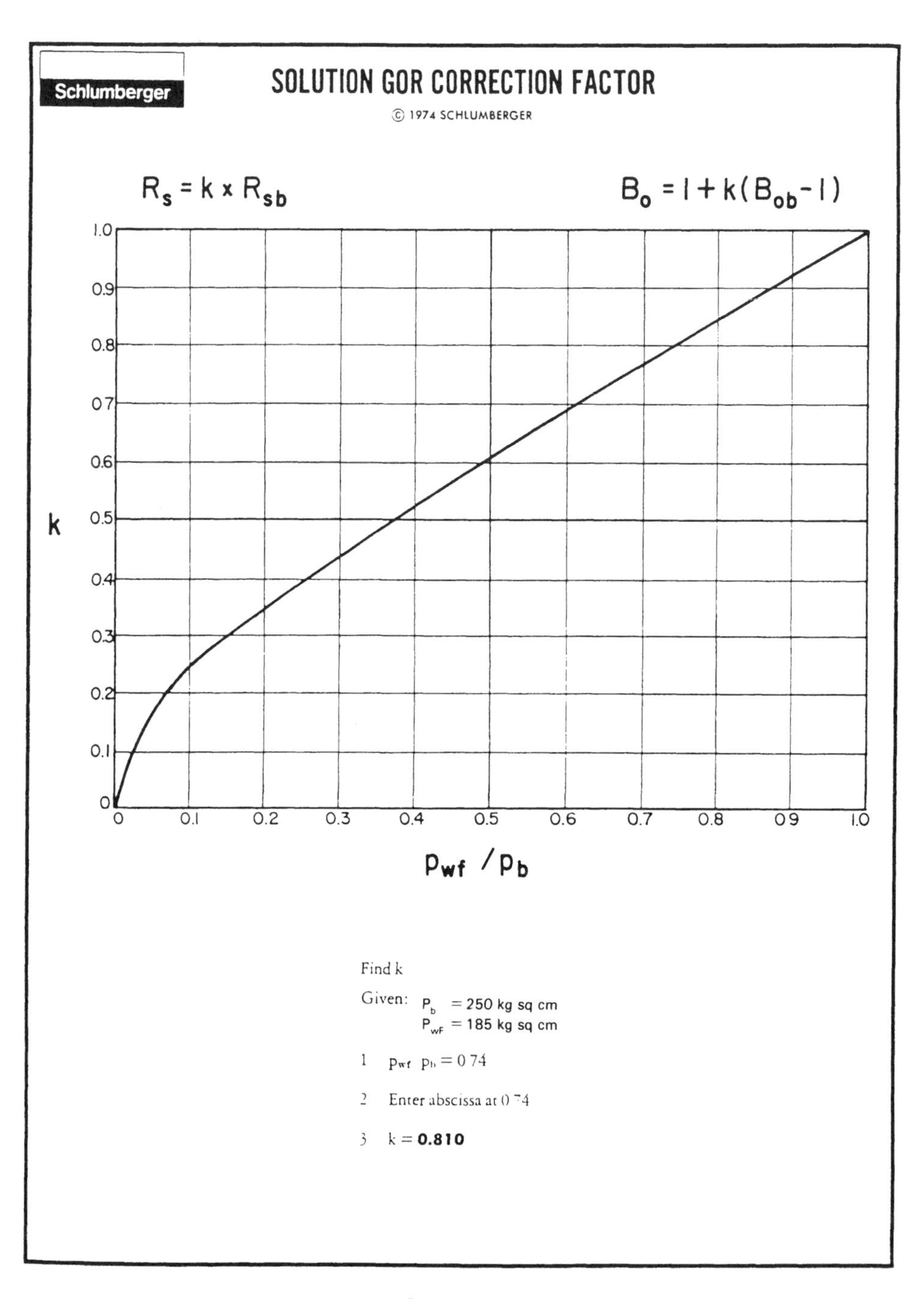
Schlumberger
SOLUTION GOR CORRECTION FACTOR
© 1974 SCHLUMBERGER
$R_s = k \times R_{sb}$
$B_o = 1 + k(B_{ob} - 1)$
k
0
0.1
0.2
0.3
0.4
0.5
0.6
0.7
0.8
0.9
1.0
p_{wf} / p_b
Find k
Given: P_b = 250 kg sq cm
P_{wF} = 185 kg sq cm
1 p_{wf} p_b = 0 74
2 Enter abscissa at 0 74
3 k = **0.810**

Chart A.8

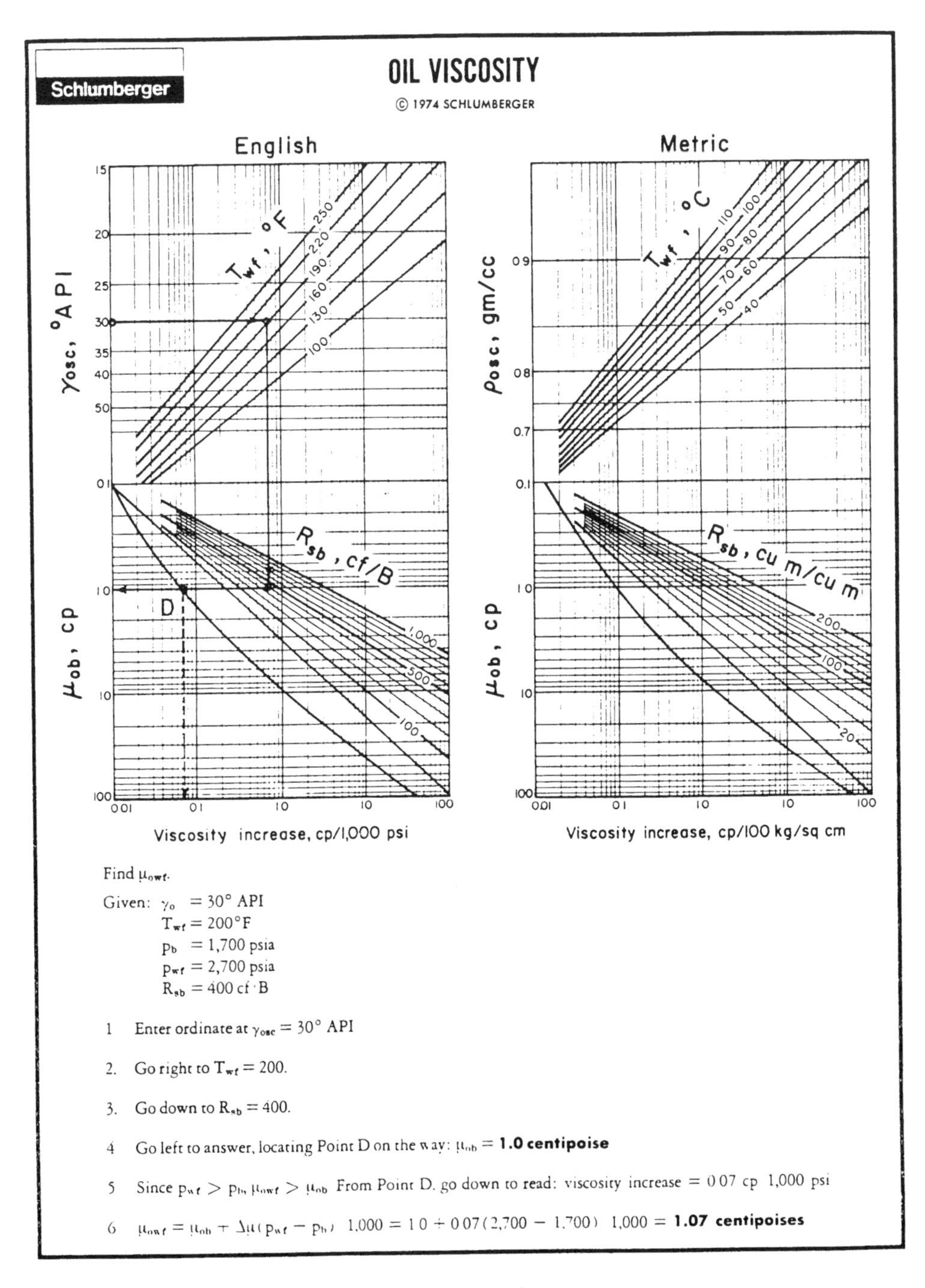
Schlumberger
OIL VISCOSITY
© 1974 SCHLUMBERGER
English
Metric
γosc, °API
Twf, °F
250
220
190
160
130
100
Rsb, cf/B
1,000
500
100
D
μob, cp
Viscosity increase, cp/1,000 psi
ρosc, gm/cc
Twf, °C
110
100
90
80
70
60
50
40
Rsb, cu m/cu m
200
100
20
Viscosity increase, cp/100 kg/sq cm
Find μowf.
Given: γo = 30° API
Twf = 200°F
pb = 1,700 psia
pwf = 2,700 psia
Rsb = 400 cf/B
1 Enter ordinate at γosc = 30° API
2. Go right to Twf = 200.
3. Go down to Rsb = 400.
4 Go left to answer, locating Point D on the way: μob = 1.0 centipoise
5 Since pwf > pb, μowf > μob From Point D, go down to read: viscosity increase = 0.07 cp/1,000 psi
6 μowf = μob + Δμ(pwf − pb)/1,000 = 1.0 + 0.07(2,700 − 1,700)/1,000 = 1.07 centipoises

Chart A.9

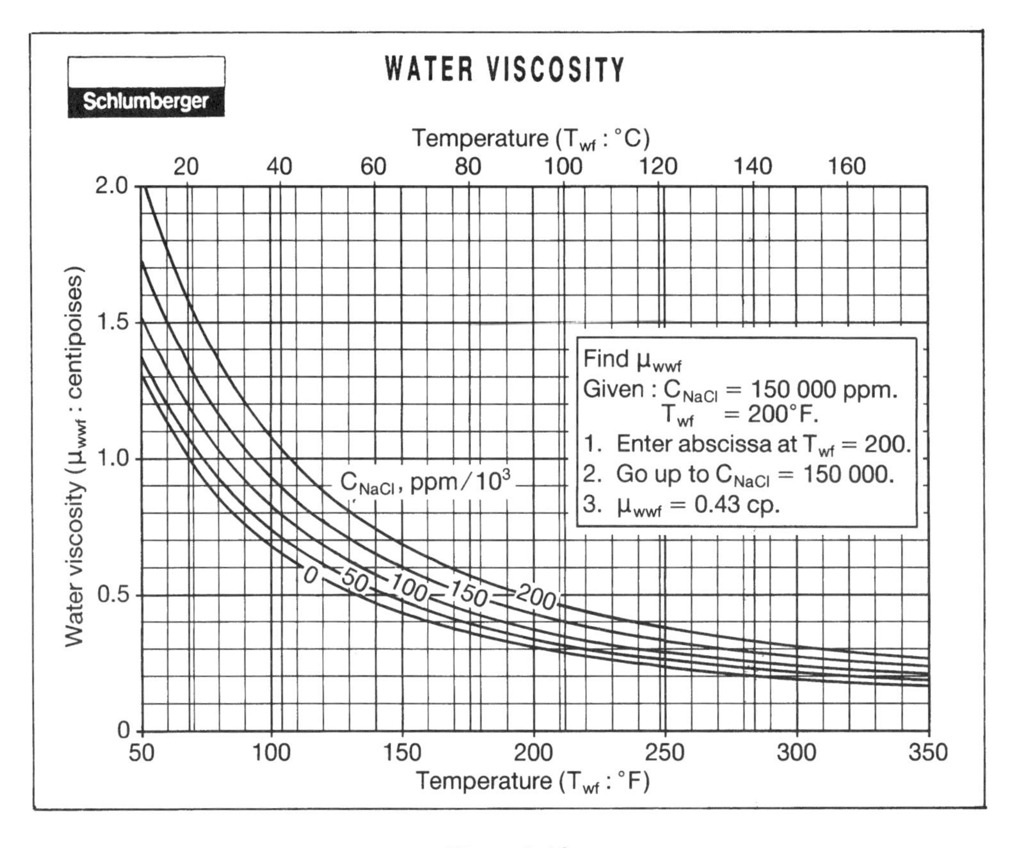

Schlumberger
WATER VISCOSITY
Temperature (T_{wf} : °C)
20 40 60 80 100 120 140 160
2.0 1.5 1.0 0.5 0
Water viscosity (μ_{wwf} : centipoises)
C_{NaCl}, ppm/10^3
0 50 100 150 200
Find μ_{wwf}
Given : C_{NaCl} = 150 000 ppm.
T_{wf} = 200°F.
1. Enter abscissa at T_{wf} = 200.
2. Go up to C_{NaCl} = 150 000.
3. μ_{wwf} = 0.43 cp.
50 100 150 200 250 300 350
Temperature (T_{wf} : °F)

Chart A.10

Chapter 4

VOLUMETRIC EVALUATION OF OIL AND GAS IN PLACE

4.1 GENERAL INTRODUCTION

Knowing the volume of hydrocarbons in place in a reservoir is of fundamental importance. Any development project on a field obviously depends on the oil and/or gas in place in the reservoir rocks. After having been called "reserves in place" for many years, these volumes of hydrocarbons are now more commonly called:

(a) **Oil and gas in place**: OOIP (Original Oil In Place), OGIP (Original Gas In Place).
(b) Or accumulations.

The term of reserves in place in fact allows for the possibility of confusion with the term "reserves" employed in English-speaking countries, and implying "**recoverable** reserves", namely the volumes of oil and/or gas that will be **produced** during the lifetime of the field. This concept is more difficult to clarify, because it depends on the technical and economic conditions of production, which are also linked to the energy policy of the producing country.

Before speaking of the different methods used to estimate these volumes, it is important to clarify the concept of oil and gas in place, which are calculated in standard conditions.

4.2 THE DIFFERENT CATEGORIES OF OIL AND GAS IN PLACE

Volumes in place are classed according to different criteria that vary with time as the reservoir is better understood, and are obtained essentially from the wells drilled, together with supplementary geophysical and geological surveys.

When a reservoir is discovered, **a rapid preliminary calculation** is made to estimate the approximate volume of hydrocarbons in place. Since very few data are available from a single borehole, they can provide only a preliminary and very rough assessment. This estimation is reviewed when all the results subsequent to the discovery have been analyzed: thorough interpretation of logs, petrophysical measurements, PVT analyses, geophysical and, if necessary, geological reinterpretation.

The first and/or second estimations lead to the decision to drill one or more **extension wells**, which are primarily intended to clarify the general image of the reservoir and then, if possible, to participate in production.

Following these new boreholes, the new estimation of the oil and gas in place is much less rough, and the assessment of volumes, as well as of petrophysical properties and fluids (especially the position of the interfaces) is significantly improved.

The **development of the reservoir** then begins, with each new well providing its harvest of new information that is added to the image of the reservoir and further clarifies the estimation already made of the volumes in place. In actual fact, therefore, these assessments are made more or less continuously during the life of the field, in the form of corrections, and the estimation of the accumulations is finally clear only at the end of the field development phase.

The different categories of quantities in place are represented roughly by concepts of "proven", "probable" and "possible" fluids for a **level** or a **reservoir**. These terms are used in general by European and American oil companies. On the scale of a **basin**, however, the concepts may be different: the concept of accumulations anticipated from reservoirs unknown today.

Figure 4.1 provides an example of these three categories:

(a) **Proven oil in place**: considered as certain (zones penetrated by wells particularly).

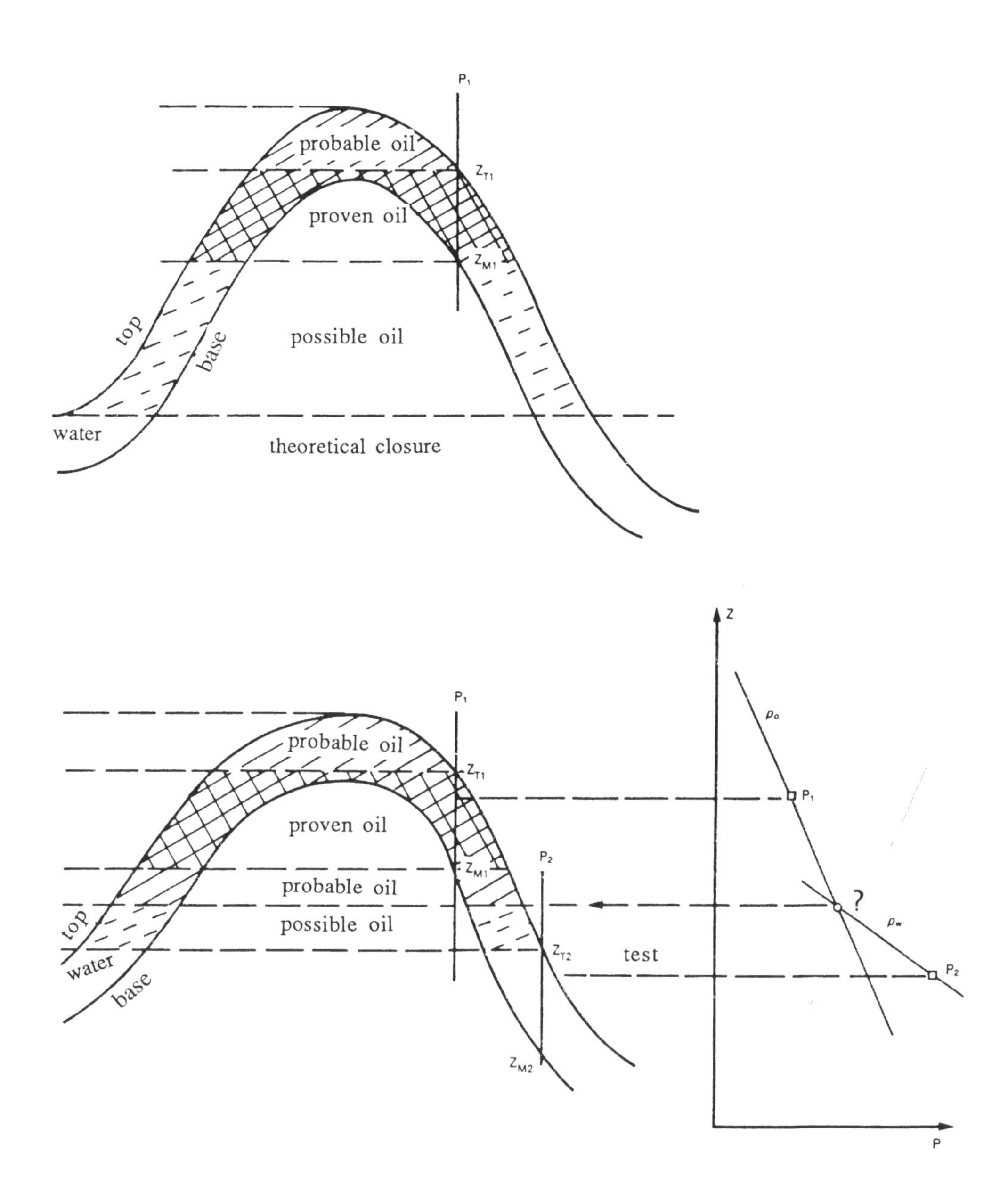

Fig. 4.1
Different categories of oil in place.

(b) **Probable oil in place**: the structural data and the interpretations of the logs and pressures indicate the zones may be impregnated, but without absolute certainty.
(c) **Possible oil in place**: the lack of data on the fluid interfaces and the extension of the facies in certain zones allows considerable uncertainty to subsist, but the presence of hydrocarbon-saturated rocks is not discarded.

Let us examine Fig. 4.1. The three categories are defined for one and then two wells. As a rule, at the start of the life of a field:

$$\text{Proven oil} < \text{Actual oil}$$

$$\text{Proven} + \text{Probable} + \text{Possible oil} > \text{Actual oil}$$

Additional boreholes are drilled to refine the image of the reservoir, locate the interface(s), and the values of proven oil in place progressively approach the values of the actual oil.

4.3 VOLUMETRIC CALCULATION OF OIL AND GAS IN PLACE

Two very different groups of methods are available for assessing the volumes of hydrocarbons in reservoirs:

(a) **Volumetric methods discussed in this section.**
(b) **Dynamic** methods discussed in Chapter 7 with material balances, and in Chapter 9 with models.

However, the second group of methods is applicable only if the reservoir **has already produced** for some time (at least one to two years), and they serve to confirm the values obtained by volumetric methods.

4.3.1 Principle of Volumetric Methods

The assessment of the accumulations is difficult because of the complexity of the porous medium: uncertainty as to the exact shape of the reservoir and, in general, little sampling for petrophysical data (porosity/saturation), normally only a few boreholes in areas measuring dozens of square kilometers.

Hence the difficulty resides in determining the parameters characterizing the volume of hydrocarbons in place rather than in calculating this volume. Calculation is reduced to the following simple operations.

(a) Volume (**reservoir conditions**) =

$$\text{volume of impregnated rock } V_R \times \frac{\text{Useful thickness}^{1}}{\text{Total thickness}} \text{ or } \frac{\text{net pay}}{\text{gross pay}}$$

$$\times \text{ porosity } \times \text{ saturation with hydrocarbons}$$

(b) Volume (**surface conditions**) =

$$\frac{\text{Volume in reservoir conditions}}{\text{Formation volume factor}}$$

Hence, for example, for oil:

$$\boxed{N = V_R \cdot \frac{h_u}{h_t} \cdot \phi \cdot \left(1 - S_{wi}\right) \cdot \frac{1}{B_o}}$$

In practice, the volumes in place N (oil) are often indicated in 10^6 m^3 and G (gas) in 10^9 m^3, these volumes being expressed in **standard conditions** (N is also expressed in 10^6 t). These volumes of oil and gas in place are adjusted to "surface" conditions for easy comparison with cumulative hydrocarbon production.

The tools and methods employed are listed in the table page 121.

4.3.2 Calculation of Volume of Impregnated Rock V_R

This calculation can be made considering the reservoir as a whole, or as composed of several sectors or compartments (faults, different levels, facies variations). Hence the first job is to "subdivide" the reservoir vertically and horizontally.

---

1. The "useful" thickness (h_u) corresponds to the beds with reservoir characteristics. It is equal to the total thickness (h_t), minus shale or other beds with very low porosities and permeabilities, and which are not considered as contributing to production. The terms "net pay" and "gross pay" are commonly used for useful thickness and total thickness respectively.

Horizontal Subdivision

The first subdivision is made automatically when the structure straddles two or more permits or leases. It is in fact indispensable to clarify the distribution of the volumes involved as part of a joint development used to allocate the capital and operating expenditures as well as production revenues.

Another, more problematic, horizontal subdivision is the one related to structural and faciologic data. Different sectors are individualized on the basis of faults, structural saddles, facies variations, and data proving the existence of independent sectors (different oil/water interfaces, distinct initial pressures). By identifying the distribution of the volumes in place in these different compartments, the production rates can be defined by sector and, based on this, the number and location of development wells.

Vertical Subdivision

This subdivision depends essentially on the geological model adopted. If several large units have been identified from logs and/or sedimentological analyses, for example, it is essential to determine the volumes in place in each of them, and from there to try to determine a well balanced and logical production scheme, which is reflected in a certain development policy, in other words a policy of bringing in the wells. As a rule, this subdivision is made by units that can be identified on the entire sector. If the distribution of sedimentary bodies is anarchic, no subdivision is feasible, and the reservoir must be considered as a whole.

It is worthwhile to subdivide the reservoir to a reasonable degree in order to minimize the errors incurred by weighting the reservoir characteristics needed to make an estimation. Yet the number of subdivisions should not be increased willy-nilly, because this would lead to vast calculations. In fact, the top and the base of each unit must be mapped and then plotted by planimeter. Beyond a certain stage, if the subdivision is continued, the computer time is significantly increased without any appreciable gain in accuracy.

The example illustrated by Fig. 4.2 shows that this subdivision, which is based on the analysis of logs and laboratory measurements (cuttings and cores), should help to determine whether the reservoirs are independent or not, which is not an easy task at the very start of production. With the subsequent production records, the analysis of the changes in pressure and production in each level can provide very important data on this subject.

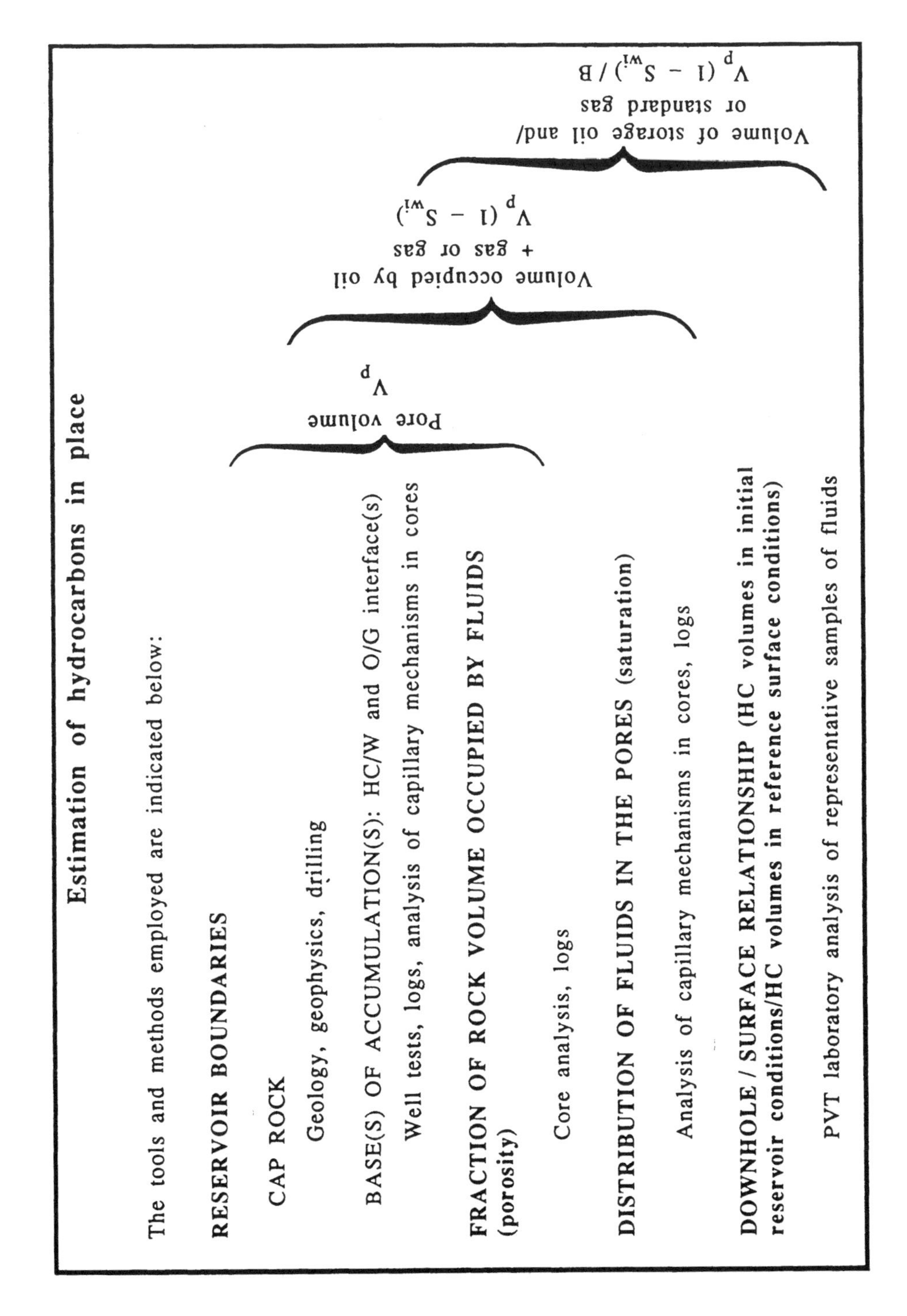
Estimation of hydrocarbons in place
The tools and methods employed are indicated below:
RESERVOIR BOUNDARIES
CAP ROCK
Geology, geophysics, drilling
BASE(S) OF ACCUMULATION(S): HC/W and O/G interface(s)
Well tests, logs, analysis of capillary mechanisms in cores
Pore volume
V_p
FRACTION OF ROCK VOLUME OCCUPIED BY FLUIDS (porosity)
Core analysis, logs
Volume occupied by oil + gas or gas
$V_p (1 - S_{wi})$
DISTRIBUTION OF FLUIDS IN THE PORES (saturation)
Analysis of capillary mechanisms in cores, logs
Volume of storage oil and/ or standard gas
$V_p (1 - S_{wi}) / B$
DOWNHOLE / SURFACE RELATIONSHIP (HC volumes in initial reservoir conditions/HC volumes in reference surface conditions)
PVT laboratory analysis of representative samples of fluids

Now that we have clarified the concept of subdivision, we shall restrict ourselves below to a clearly-defined level to which the term reservoir will be applied.

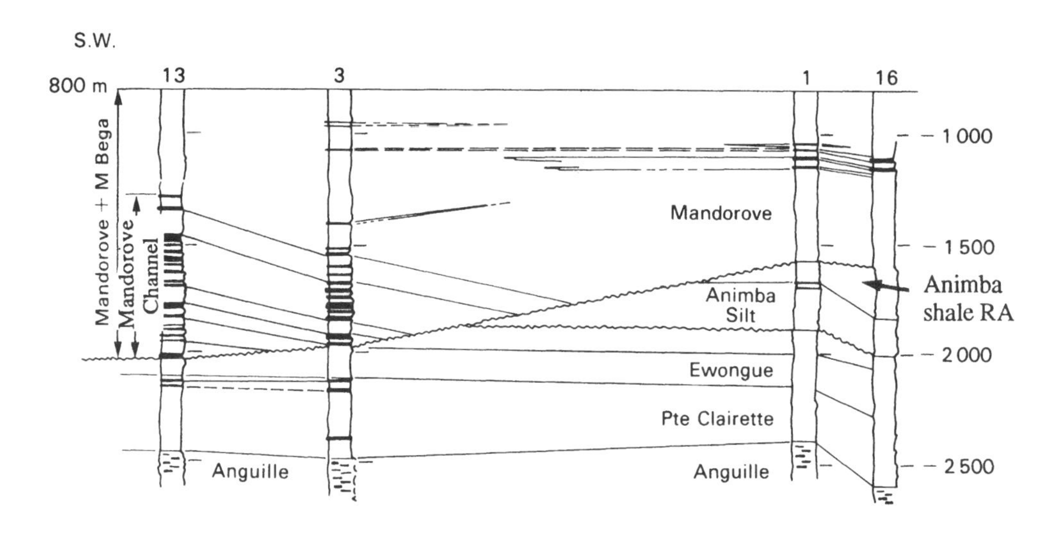

Fig. 4.2

Gabon (Anguille) AGM correlations 13, 3, 1, 16.

Fluid Interfaces

For each level thus identified, the exact position of any O/W, G/W and G/O interfaces must be clarified. These interfaces are identified by means of logs, core analyses, and production tests. Note that large transition zones (low permeabilities) are considered as special units. In fact, the S_w is variable and different from S_{wi}.

Once the levels (or units) are clearly defined, two methods are available for calculating the volume of impregnated rock V_R.

4.3.2.1 Calculation of the Volume of Rocks from Isobaths Cubic Content or Area/Depth Method

It now remains to calculate the volume of each unit. The geological and geophysical surveys furnish isobath maps, generally at the top and the base of the reservoir. A planimeter is used on these two maps to calculate the volume of rock (Fig. 4.3).

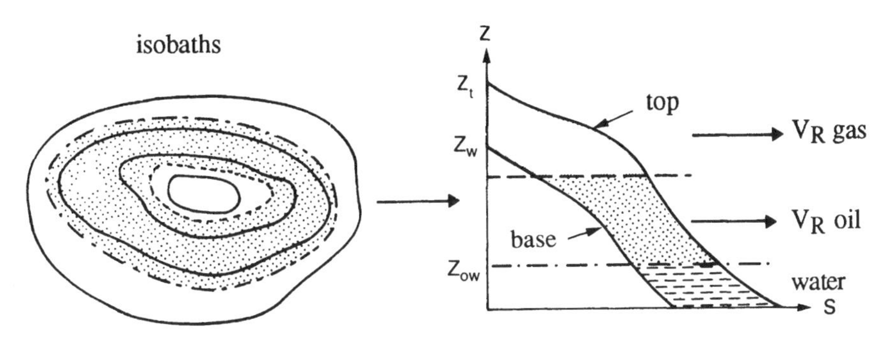

Fig. 4.3

If the planimeter areas of the top and base isobaths are plotted on a **depth/area diagram**, together with the O/W interface, for example, planimeter tracing the subtending area between the two curves gives the volume of impregnated rock:

$$V_R = \int_{Z_{OW}}^{Z_t} S\ (\text{top})\ .\ d_z - \int_{Z_{OW}}^{Z_b} S\ (\text{base})\ .\ d_z$$

If a gas cap exists, the calculation is made for the volumes of oil and for the volumes of gas in the gas cap, the boundary being the G/O interface.

4.3.2.2 Rapid Calculation Method

If the structure is poorly known at the time of the discovery well, it is sometimes sufficient to make a rapid estimate to obtain an order of magnitude. If so, the area/depth method is not used, and the structure is treated as a spherical dome or a trapezoidal shape, for example (Figs 4.4a, 4.4b and 4.4c).

4.3.3 Calculation of the Volume of Oil from Isopach Maps

This calculation is made later, when a minimum number of wells has been drilled.

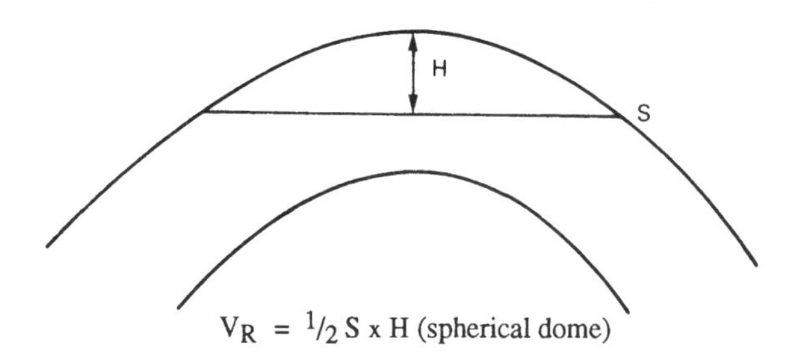

Fig. 4.4a

Bed impregnated with oil or gas over a thickness H less than the bed thickness h.

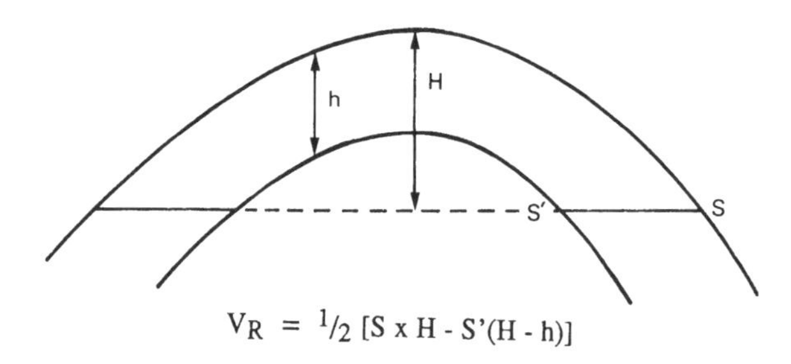

Fig. 4.4b

Bed impregnated with oil or gas over a thickness H greater than the bed thickness h.

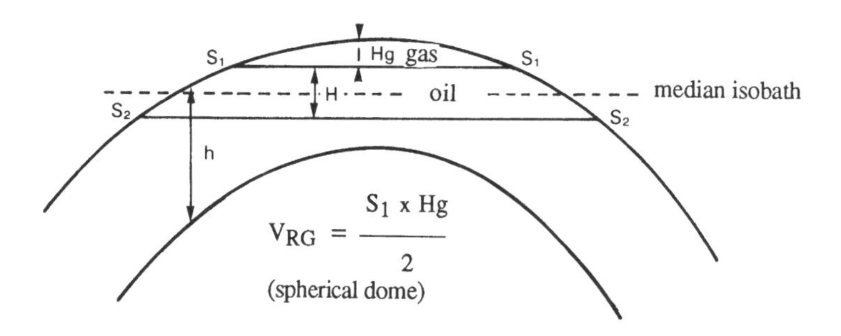

$V_{RO} = {}^1/_2\,(S_1 + S_2) \times H$ (trapezium)

Fig. 4.4c

Bed containing oil and gas over a total thickness H + Hg less than the bed thickness h.

The set of two isobath maps at the top and the base of the reservoir gives us an isopach map of the reservoir. Knowing the porosities for each well, we can chart a map of porosities. By combining the two maps, an "iso-hϕ" map is obtained.

Planimeter tracing the areas S located between each curve allows the calculation of the pore volume (Fig. 4.5):

$$v_p = S \cdot h_u \cdot \phi \qquad V_p = \sum_{i=1}^{n} v_{pi}$$

This method is more accurate if the lateral variations in thickness and porosity are substantial.

To calculate N, it is also necessary to determine the average values of B_o and S_{wi} (or S_w in the transition zones). This requires the plotting of **"iso-oil"** (or iso-gas) maps. These maps are plotted as soon as the number of wells (for example 4, 5 or 6) allows.

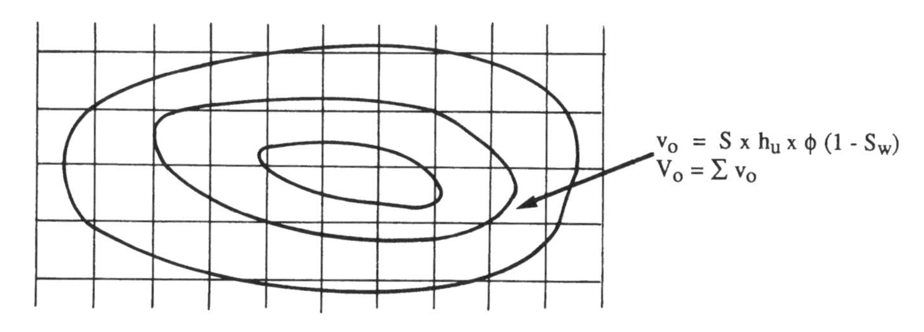

Fig. 4.5

4.4 CHOICE OF AVERAGE CHARACTERISTICS, UNCERTAINTIES

4.4.1 Choice of Average Characteristics

We showed in Section 4.3.1 of this chapter that the calculation of the oil and gas in place entailed the choice of average characteristics for the different terms h_u/h_t, ϕ, S_w and B_o.

If the reservoir has just been discovered, the characteristics obtained from the first well are extrapolated to the entire reservoir, giving assumed values of h_u/h_t (and B_o). For the other parameters ϕ and S_{wi}, a vertically-weighted mean is employed:

$$\overline{\phi} = \frac{\Sigma\ \phi_i\ hu_i}{\Sigma\ hu_i}$$

and

$$\overline{S_{wi}} = \frac{\Sigma\ S_{wi}\ \phi_i\ hu_i}{\Sigma\ hu_i\ .\ \phi_i}$$

Figure 4.6 shows an example of this calculation, on a well in which three reservoir levels are distinguished, of which two are impregnated with oil, and the third with oil and water.

The calculations give the oil thickness, or oil equivalent.

Note that two assumptions are made in calculating $\alpha = H_u / H_t$ (as well as for ϕ and S_o, which has not been developed): either involving a future "low well" structurally located at the same depth as this well, or a future "high well" located so that the third reservoir is completely impregnated with oil, as shown by Fig. 4.7.

Subsequently, with the data obtained from other wells, an attempt is made to weight the values found for the different wells as before. This weighting is sometimes arbitrary and depends on the geographic location of the wells.

Note also that the characteristics of the fluids in place may vary significantly from one point of the field to another. A **"variable bubble-point"** oil is referred to when in reality **the oil composition varies** in space, vertically and horizontally, as do all the parameters P_b, B_o, R_s, ρ_o and μ_o. Two characteristic fields can be noted in this respect:

(a) Hassi Messaoud (Algerian Sahara): $155 < P_b < 210$ bar: very vast structure, where the variations observed are lateral.
(b) Grondin Marine (offshore Gabon): $160 < P_b < 230$ bar.

These values vary vertically between – 2300 and – 2175 m/SL respectively, representing a variation of 0.5 bar/m, which is high. Note an average "US" variation of 1 psi/m, or 0.07 bar/m, vertically (which is about the order of magnitude of the variation in static pressure of a light oil... but in the reverse direction).

GR	reservoir	H_u	Φ	S_w	$H_u\,\Phi$	$H_u\,\Phi\,(1 - S_w)$
2 305	shale, top					
		4,5 m	18 %	32 %	0,81	0,551
	shale					
		2 m	16 %	30 %	0,32	0,224
	oil, compact					
		4 m	20 %	24 %	0,80	0,608
2 320	oil / water					
	aquifer	6 m	20 %	100 %	—	—
2 326	base					
		Σ = 16,5 m	—	—	1,93	1,383

Average values:

$$\alpha = \frac{\Sigma\, H_u}{H_T} = \frac{16.5}{21} = 0.79 \text{ (high well)} \quad \text{or} \quad \frac{10.5}{15} = 0.70 \text{ (low well)}$$

$$\overline{\Phi} = \frac{\Sigma\, hu\, \Phi}{\Sigma\, hu} = \frac{1.93}{10.5} = 0.184$$

$$\overline{S_o} = \frac{\Sigma\, Hu\, \Phi\, (1 - S_w)}{\Sigma\, Hu\, \Phi} = \frac{1.383}{1.93} = 0.72$$

$$\Sigma\, Hu \,.\, \overline{\Phi} \,.\, \overline{S_o} = 10.5 \,.\, 0.184 \,.\, 0.72 = 1.39 \text{ m oil equivalent}$$

Fig. 4.6 Average parameters on a well.

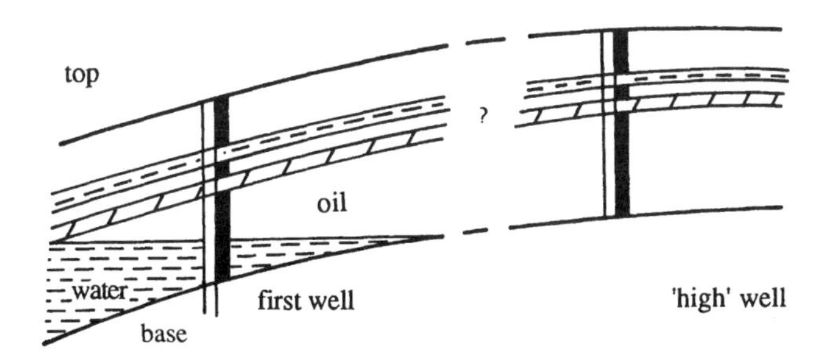

Fig. 4.7

4.4.2 Uncertainties and Probabilistic Methods

As we shall show with a numerical example, the lack of information about the reservoirs can lead to major numerical errors.

A considerable effort has been made in this area in the past decade, with the use of **probabilistic** methods applied to all the parameters characterizing the reservoir.

Among the methods used are the **Monte Carlo** and **"krigeing"** method. The result gives an average value and a probabilized "range" of volumes in place.

Uncertainties of two origins can be distinguished:

(a) **Systematic**, because related to the techniques: for example, uncertainty in the seismic picking, the acoustic velocity and consequently in the value of the isobaths, etc.
(b) **Occasional**, because related to the reservoir itself: for example, concerning the water level(s), correlations (function of the type of sedimentation), etc.

These uncertainties can be quantified in two ways:

(a) In a field of "soft" knowledge, the case of the porous reservoir, the probabilities are derived from a "light" statistical approach, and especially from the recommendations drawn from the experience of specialists (reservoir geologists). They can be called **subjective uncertainties**.
(b) At the end of development of the fields, the body of data is much larger, and more valid statistics are compiled: these concern **objective uncertainties**.

However, decisions are mainly taken at the start of development, and the uncertainties are hence essentially subjective (Fig. 4.8).

Must we conclude that a good evaluation of the quantities in place is very difficult to make in the appraisal phase? Not necessarily, because a skilful analyst will use his data accurately. Yet it is necessary to be fully aware of the extent of the uncertainties in this field, and hence work with the concept of a "range" within reasonable limits.

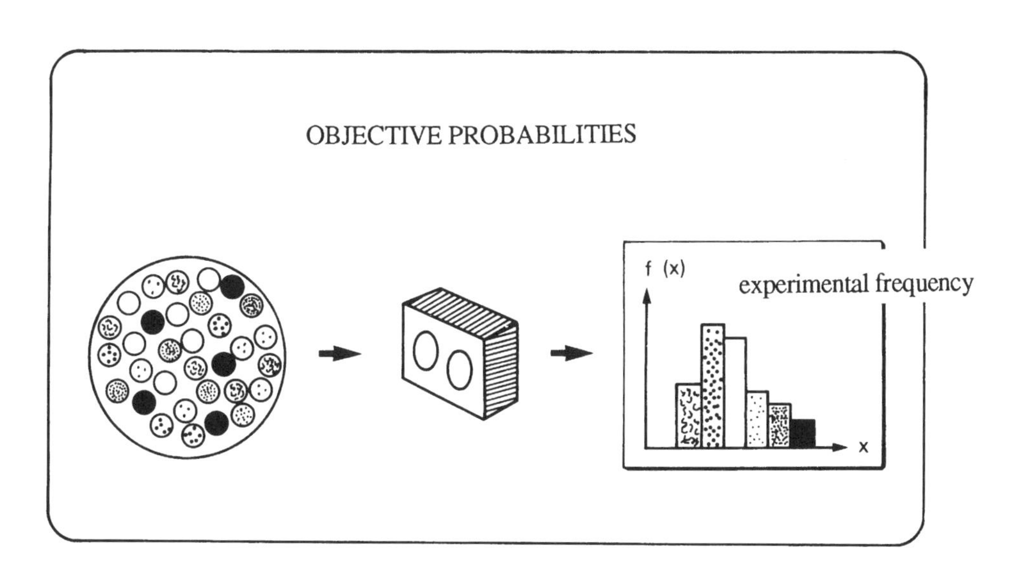

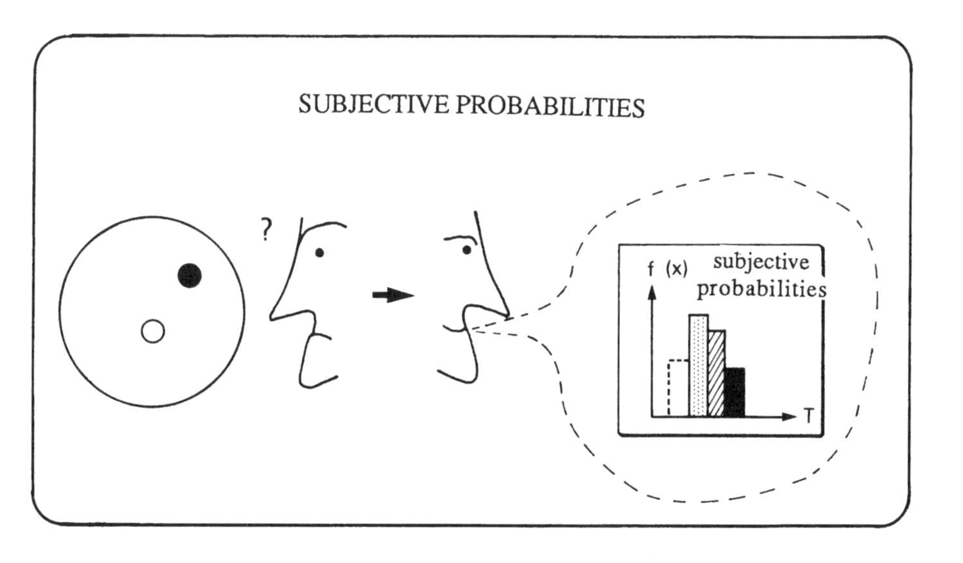

Fig. 4.8

(With the kind authorization of *SNEA(P)*).

4.4.3 Example of Cubic Content or Area/Depth Calculation

An oil reservoir has just been identified. Only one well has been drilled, the discovery well. We shall use the area/depth method.

Planimeter tracing the isobaths at the top of the reservoir gives the following results:

Isobath		Area
2820 m/SL	≈	21.2 km^2
2840 m/SL	≈	44.4 km^2
2860 m/SL	≈	73.3 km^2
2880 m/SL	≈	99.4 km^2

The following useful data about the reservoir are known (**average** values):

(a) Thickness of reservoir: 25 m (vertical).
(b) Oil/water interface: – 2876 m/SL.
(c) Useful thickness/total thickness: 0.85, (net pay/gross pay).
(d) Porosity ϕ: 18%.
(e) Saturation S_{wi}: 25%.
(f) FVF of oil B_o: 1.34.

We shall now calculate the volume of oil in place in standard conditions and, estimating that the relative error possible on each of the parameters h_u/h_t, ϕ, $(1 - S_{wi})$, B_o is about 5% (optimistic case), also calculate the possible error in the volume obtained.

Answer

On a depth/area graph, the curve at the top of the reservoir and the curve at the base of the reservoir are plotted by a 25 m translation along the depth (ordinates). The horizontal line representing the oil/water interface intersects these two curves. The surface to be traced by planimeter, which represents the volume of rock, corresponds substantially in this case to a trapezoidal shape that can be calculated easily (difference between two triangles) (Fig. 4.9).

We obtain:

$$V_r = (0.5 \,.\, 95.10^6 \,.\, 72) - (0.5 \,.\, 62.10^6 \,.\, 47)$$

$$V_r = 1.96 . 10^9 \, m^3$$

$$N = V_R . \frac{h_u}{h_t} . \phi . \left(1 - S_{wi}\right) . \frac{1}{B_o}$$

$$N = 1.96 . 10^9 . 0.85 . 0.18 . \frac{0.75}{1.34} = 167.8 . 10^6 \, m^3$$

$$N \approx 168 . 10^6 \, m^3$$

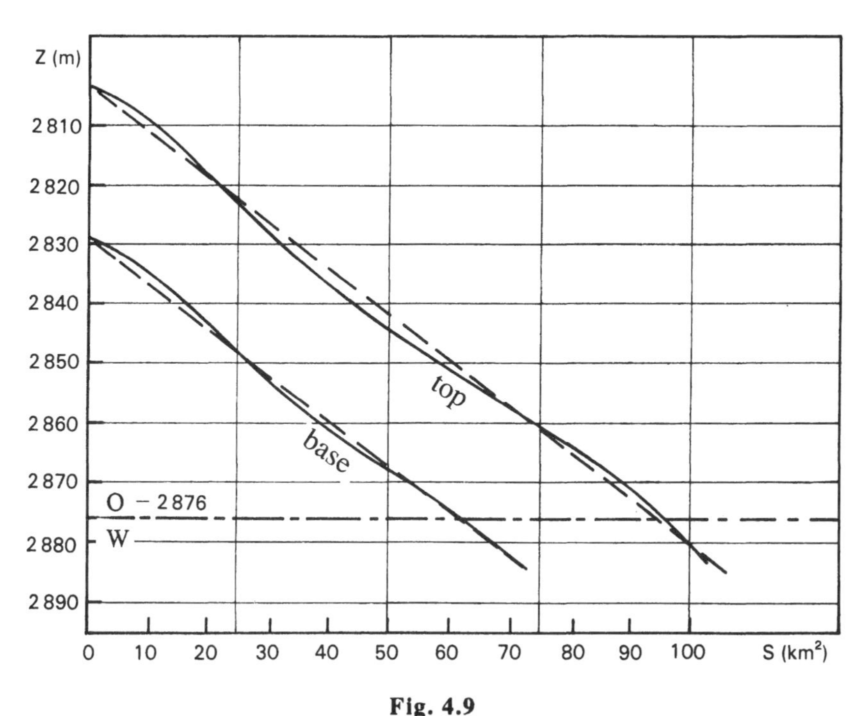

Fig. 4.9

Depth/area graph.

Since the equation is linear, the relative errors are added together, hence 4 x 5% = 20%. The volume of oil in place is thus:

$$135 . 10^6 \, m^3 < N < 201 . 10^6 \, m^3$$

We have applied a deterministic error calculation here. In general, if a **probabilistic** calculation was applied, the errors found would be smaller and more realistic. Assuming that the errors are random and independent for each parameter, this probabilistic calculation consists in calculating the variance σX_i^2 of each variable x_i (here ϕ, S_w, etc.) and the resulting variance on N, hence the standard deviation:

$$\sigma N = \left(\Sigma \left(\frac{\partial N}{\partial X_i} \right)^2 . \sigma X_i^2 \right)^{1/2}$$

The simple numerical calculation above was made to familiarize the reader with the uncertainties concerning a reservoir, particularly in the exploratory period, and without taking account of the **possible error in the calculation of the volume of rocks** (inaccuracy of isobath maps), **which is often the larger**.

In actual fact, when the field is better known (drilling of other wells), the errors are smaller.

Chapter 5

ONE-PHASE FLUID MECHANICS AND WELL TEST INTERPRETATION

5.1 GENERAL INTRODUCTION

The following two alternatives can be expected to occur in reservoirs:

(a) Flow of a fluid either alone in the layer or in the presence of another immobile fluid (one-phase flow).
(b) Simultaneous flow of two or three fluids (multiphase flow, see Chapter 6).

The **laws of one-phase fluid mechanics** are relatively simple, and serve to compile a fair body of data on the reservoir and the the connection between the pay zone and the borehole. They relate the flow rates and pressures in space as a function of time, and also of a number of properties of the fluids and of the rock.

Well tests essentially consist of **pressure** measurements in the well, at the reservoir level, at the start of and during production.

They are intended to identify the following:

(a) The **production capacity** of the well.
(b) **"Static" pressure** of the reservoir (or of the area drained by the well).
(c) The **"hk"** product (producing thickness, multiplied by the **permeability** of the formation beyond the zone near the hole).
(d) The degree of change in the properties of the zone close to the well (damage due to drilling and completion, or improvement due to stimulation, called **skin effect**).

(e) Well drainage radius R.
(f) The existence of heterogeneities in the rocks and in the structure (such as faults).
(g) The types of and changes in the fluids produced.
(h) If applicable, oil and gas in place and drive mechanisms.

At the onset of the life of a reservoir, during the tests, **sampling** is also carried out to conduct **PVT studies** and an analysis to determine how the effluents will be processed.

The tests also serve to define or improve well **completion** (equipment for production).

5.2 OIL FLOW AROUND WELLS

5.2.1 Diffusivity Equation

Let us consider the simplest case, that of a homogeneous isotropic reservoir, in other words with a given porosity and constant permeability unrelated to the direction concerned. In this section, we shall state that the volume of the fluid is equal to the pore volume (only one fluid present).

The temperature of a reservoir is practically constant due to the large volume of rock with high heat capacity.

We shall assume that the rock is incompressible, and that the compressibility of the liquid, which is low, can be considered constant in the pressure interval corresponding to the area drained by the well. This also applies for the viscosity:

$$c = c_o \quad [1] \qquad \mu = \mu_o$$

The following are also used.

1. The application of the laws obtained with these assumptions to actual reservoirs entails the use of the following parameters : porosity ϕ $(1 - S_W)$, effective permeability to liquid k_1, and apparent compressibility of the liquid taking account of that of the rock and that of the interstitial water.

Darcy's Law

This is written in differential form:

$$\vec{V} = -\frac{k}{\mu}\,\overrightarrow{\text{grad}}\,(P + \rho g z)$$

where

V = the filtration rate, i.e. the ratio of the flow rate Q_F passing through an element to the area A of this element,

Q_F = flow rate in reservoir conditions ($Q_F = Q_o B_o$).

Law of Conservation of Mass

For a small sample $\partial x\ \partial y\ \partial z$, the mass variation of the fluid per unit time is equal to the difference in input and output mass flow rates during the time interval concerned, or, by simplifying:

$$\phi \frac{\partial \rho}{\partial t} = -\left[\frac{\partial}{\partial x}(\rho V_x) + \frac{\partial}{\partial y}(\rho V_y) + \frac{\partial}{\partial z}(\rho V_z)\right]$$

where V_x, V_y and V_z are the components of the filtration rate.

Equation of State

$$\rho = \rho_o\, e^{c_o (P - P_o)} \qquad \text{or} \qquad \rho = \rho_o \left[1 + c_o (P - P_o)\right]$$

The **diffusivity equation** is obtained by taking account of these different equations and hypotheses:

$$\frac{\partial^2 p}{\partial x^2} + \frac{\partial^2 p}{\partial y^2} + \frac{\partial^2 p}{\partial z^2} = \frac{1}{K}\frac{\partial P}{\partial t}$$

with

$$K = \frac{k}{\phi\, \mu_o\, c_o}$$

K is called the **hydraulic diffusivity**.

In circular radial flow (in a plane), which actually occurs around the wells (Fig. 5.1), we obtain:

$$\frac{\partial^2 p}{\partial r^2} + \frac{1}{r}\frac{\partial P}{\partial r} = \frac{1}{K}\frac{\partial P}{\partial t}$$

Many solutions are available for this equation. Each one is characterized by its boundary conditions (time and space).

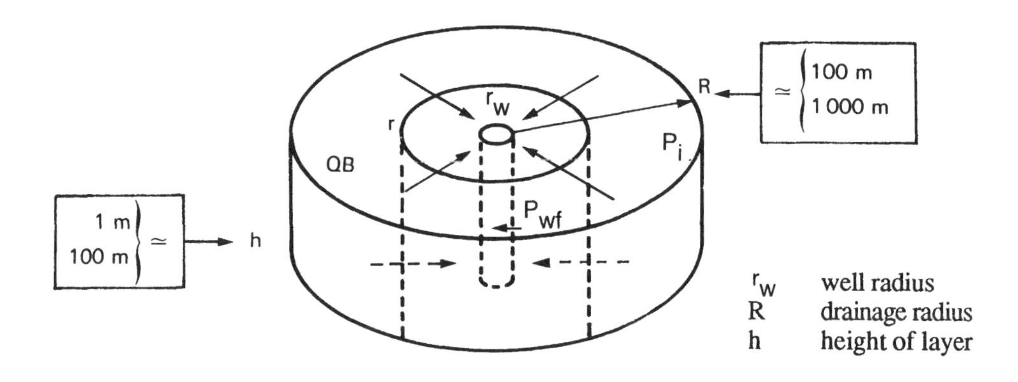

Fig. 5.1
Schematic circular radial flow.

5.2.2 Standard Solutions to the Diffusivity Equation

Putting a well on stream or changing its flow rate create a disturbance in the porous medium which is propagated over an increasingly wide area.

Until the disturbance reaches the boundary of the reservoir or the boundary of the well drainage zone, the pressure at any point depends not only on the position of this point, but also on the time. **The fluid movement is transient.**

The influence of the boundary does not yet make itself felt and matters proceed as if the reservoir was of infinite extent. The solution to the diffusivity equation is obtained with the boundary conditions corresponding to an **"infinite extent reservoir"**. When the disturbance reaches the reservoir boundary (or the boundary of the well drainage zone), its influence is felt at every point. Following a transition period, a new set of flow conditions is established. The solutions to the diffusivity equation are then given for two cases only: constant pressure outer boundary reservoir and bounded cylindrical reservoir. The influence of the boundary occurs after a time $t_R \approx 0.25\ R^2/K$ (Fig. 5.2a).

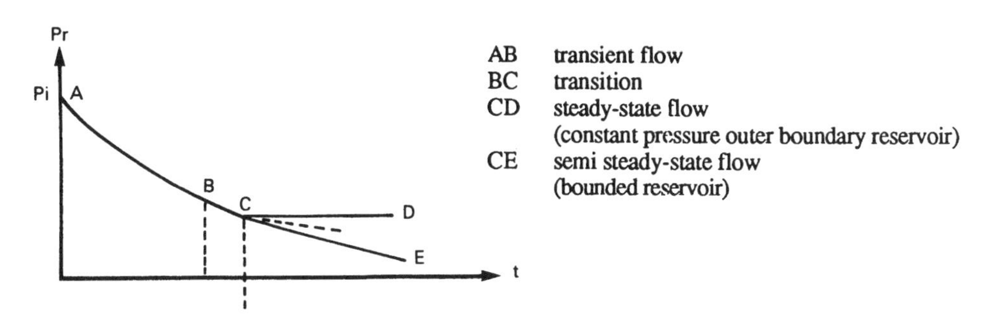

Fig. 5.2a

Constant Pressure Outer Boundary Reservoir

This situation is rarely encountered. Following the transient and transition periods, the pressure no longer depends on the point concerned, but remains constant at every point. The flow is said to be steady state.

(a) **First example** of a constant pressure outer boundary reservoir: the aquifer is vast and displays very good characteristics, so that every volume of hydrocarbons leaving the reservoir is replaced by an **equal** volume of water. This is rarely encountered (Fig. 5.2b).

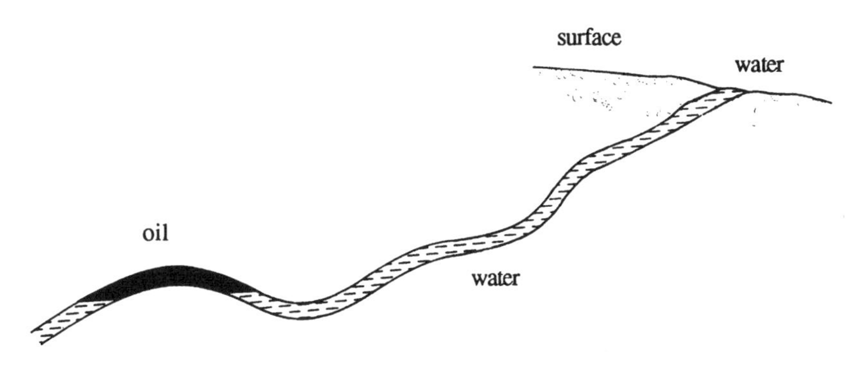

Fig. 5.2b

Cross-section.

(b) **Second example**: production with water flooding and pressure maintenance (volumes injected equal to volumes produced). This case is more frequently encountered.

Bounded Reservoir

Following the transient period, the reservoir produces by decompression of its entire mass. The pressure drop rate is the same at every point, and the pressure difference between two given points remains constant. For this reason, the **flow** is said to be semi **steady state**.

Examples of bounded reservoirs: one example is sandstone lenses in shale. Another example is that of a reservoir drained by many wells. Each of the wells of the central zone drains a zone that can be considered as a bounded reservoir area. The edge wells drain the reservoir areas partially fed by water drive (Fig. 5.2c).

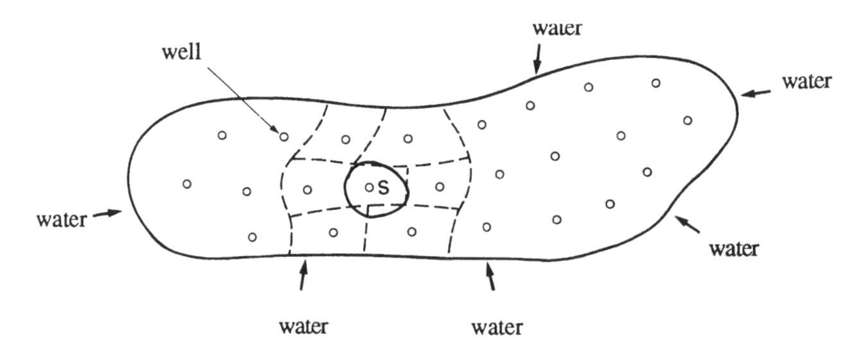

Fig. 5.2c

Plane view.

5.2.3 Pressure Drawdown Equations (Flow Rate)

A well is assumed to be at the center of a cylindrical reservoir.

5.2.3.1 Constant Flow Rate

Transient flow. It can be shown that:

$$P_i - P_{(r,t)} = \frac{\mu_o Q_o B_o}{4 \pi h k} \left[- E_i \left(- \frac{r^2}{4 Kt} \right) \right]$$

where

P_i = **initial pressure of reservoir area**[1],
$P_{(r,\,t)}$ = pressure at a distance r from the well at time t,
Q_o = oil flow rate (assumed constant),
E_i = "exponential integral" function,
h = thickness of producing layer.

A logarithmic approximation can be used very rapidly for the well ($r = r_w$) of which the radius r_w is small:

$$P_i - P_{wf} = \frac{\mu_o Q_o B_o}{4 \pi h k} \left[0.81 + \ln \frac{Kt}{r_w^2} \right]$$

where $P_{wf\,(t)}$ **is the well flowing pressure ($r = r_w$) at time t.**

This is valid as long as the reservoir behaves like an "infinite extent" reservoir.

After a period of time, approximately $t_R \approx 0.25\ R^2 / K$, the disturbance reaches the boundaries of the reservoir or of the well drainage zone (at distance R from the well). The two typical solutions in this case are the following.

(a) **Semi steady-state flow** (bounded cylindrical reservoir).

It can be shown that:

$$\overline{P} - P_{wf} = \frac{\mu_o Q_o B_o}{4 \pi h k} \left(\ln \frac{R}{r_w} - \frac{3}{4} \right)$$

where $\overline{P}$ **is the average "static" pressure in the drainage area.**

This is called "static" because it is obtained with the well closed (Section 5.2.4).

This pressure difference $\overline{P} - P_{wf}$ is constant. It is called the drawdown or **"production ΔP"**.

---

1. In a large number of reservoirs, this initial pressure is the same order of magnitude as the hydrostatic pressure P (bar) ≈ Z / 10 (m).

The relationship between production and fluid decompression is given by the compressibility factor:

$$c_o = -\frac{1}{V} \cdot \frac{dV}{dP}$$

Hence:

$$\frac{dP}{dt} = \frac{-1}{c_o V} \frac{dV}{dt}$$

with

$$\frac{dV}{dt} = -Q_o \, Bo \qquad \text{and} \qquad V = \pi R^2 h \phi$$

Hence:

$$\frac{dP}{dt} = \frac{Q_o B_o}{\pi R^2 h \phi c_o} = \text{constant}$$

Since the flow rate Q_o, the pressure drop rate dP/dt as well as c_o are known, the volume of the reservoir V (or of the zone drained by the well) can be determined, as well as R, the radius of the well drainage area.

(b) **Steady-state flow** (constant pressure outer boundary reservoir).

In this case, we have:

$$P_i - P_{wf} = \frac{\mu_o Q_o B_o}{2 \pi h k} \ln \frac{R}{r_w}$$ [1]

and the pressure, which is different at each point of the drainage area, is **constant over time.**

1. This equation derives directly from Darcy's Law by integration between the limits r_w and R:

$$Q_o B_o = \frac{k}{\mu_o} S \frac{dP}{dr} \qquad \text{with} \qquad S = 2 \pi r h$$

Hence:

$$dP = \frac{\mu_o Q_o B_o}{2 \pi h k} \cdot \frac{dr}{r} \qquad \int_{P_F}^{P_i} dP = \frac{\mu_o Q_o B_o}{2 \pi h k} \int_{r_w}^{R} \frac{dr}{r}$$

and

$$P_i - P_{wf} = \frac{\mu_o Q_o B_o}{2 \pi h k} \ln \frac{R}{r_w}$$

General representation of pressures as a function of time (Fig. 5.3)

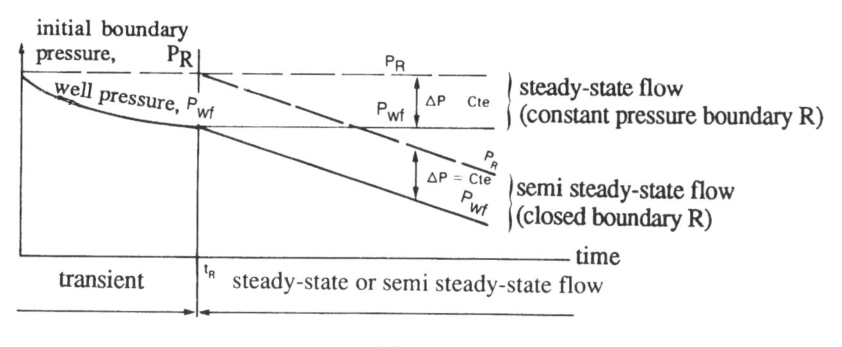

Fig. 5.3

Up to t_R, the reservoir behaves like an infinite extent reservoir. This is the time elapsed until a boundary is reached:

t_R = time for steady-state or semi steady-state flow conditions to be established,

$$t_R \approx 0.25 \frac{R^2}{K}.$$

5.2.3.2 Variable Flow Rate

During a well test, several successive flow rates are often imposed on the well to achieve a more reliable confirmation of the results. To be able to interpret these tests, equations are needed involving these different flow rates. These equations are obtained by using the **superposition principle**: the pressure drop at time t is the sum, at this time, of the pressure drops due to all the variations in flow rate (pulses), each measured from its origin. Figure 5.4 shows an example with two flow rates.

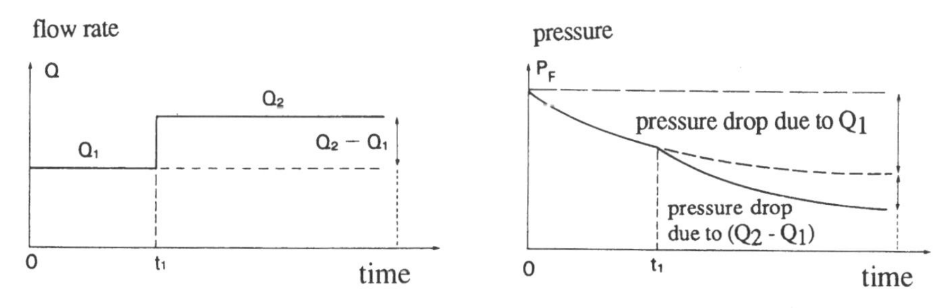

Fig. 5.4

Let the flow rate variations be $(0 \rightarrow Q_1)$ at time 0, and $(Q_1 \rightarrow Q_2)$ at time t_1. If the conditions remain transient, we have:

$$P_i - P_{wf(t)} = \frac{\mu_o Q_1 B_o}{4 \pi h k}\left(0.81 + \ln \frac{K t}{r_w^2}\right) + \frac{\mu_o (Q_2 - Q_1) B_o}{4 \pi h k}\left(0.81 + \ln \frac{K (t - t_1)}{r_w^2}\right)$$

5.2.4 Equations of Pressure Build-Up after Well Shut-In

With the well shut-in, the pressure approaches an equilibrium pressure at every point of the reservoir. The superposition theorem is used. Following a period of flow rate Q_0, the shut-in corresponds to a second zero flow rate (Fig. 5.5).

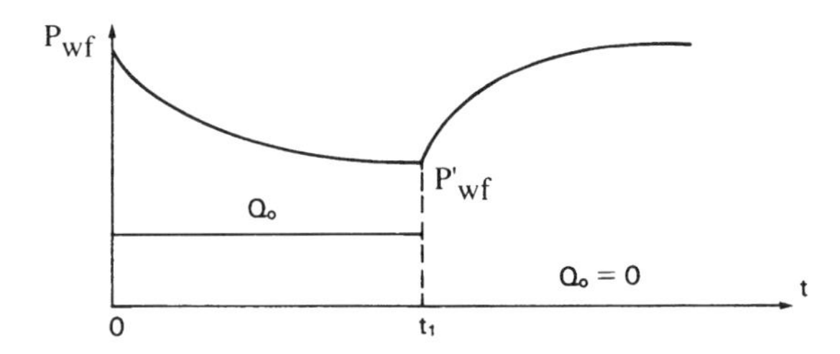

Fig. 5.5

"Infinite Extent" Reservoir

We obtain:

$$P_i - P_{ws} = \frac{\mu_o Q_o B_o}{4 \pi h k} \ln \frac{t}{t - t_1}$$

where t_1 is the time when the well is shut-in.

$P_{ws\,(t)}$ is the well shut-in pressure at time t.

If P_{ws} is plotted as a function of $\log t / t - t_1$, the points are in a line.

All Reservoirs

Provided that $t - t_1 \leq t_1$ and that, for noninfinite reservoirs, the steady-state or semi steady-state conditions are reached during the flow period:

$$P_{ws(t)} - P'_{wf} = \frac{\mu_o Q_o B_o}{4 \pi h k}\left[0.81 + \ln \frac{K (t - t_1)}{r_w^2}\right]$$

where **P'_{wf} is the bottom hole pressure at shut-in** (see practical units in Section 5.6).

If the following curves are plotted:

(a) P_{ws} as a function of $\ln (t - t_1)$, (1)

(b) P_{ws} as a function of $\ln \frac{t}{t - t_1}$, (Horner's method) (2)

curves (1) and (2) give **hk**, since the slope:

$$m = \frac{\mu_o Q_o B_o}{4 \pi h k}$$

of curve (2) indicates that the reservoir behaves like an "infinite extent" reservoir (points remaining in a line). In this case, extrapolation to an infinite shut-in time, i.e. $t/(t - t_1) = 1$, gives P_i (Fig. 5.6). Thus for $t/(t - t_1) \rightarrow 1$, $P_{ws\,(t)} \rightarrow P_i$.

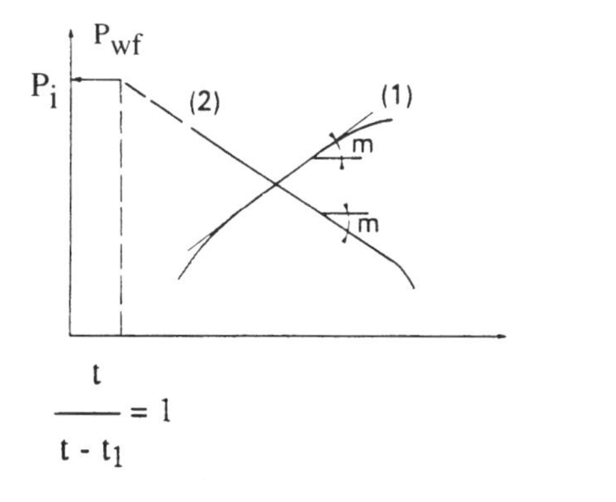

Fig. 5.6

We can also note that $t_1 = T$, and that $t - t_1 = \theta$ shut-in time, hence the curves of P_{ws} as a function of $\ln \theta$ and $\ln = T + \theta / \theta$.

The interpretation of the pressure build-up (well shut-in) thus serves to obtain the **reservoir pressure P_i** (extrapolated) and a **mean permeability** of the drainage area.

Bounded Reservoir (semi steady-state flow)

It can be shown that the mean pressure $\overline{P}$ in the drainage area considered is obtained, on the pressure build-up line corresponding to the transient period, by extrapolating this line to the value $\theta = 0.1\,R^2/K$ (Fig. 5.7). This is the Dietz method.

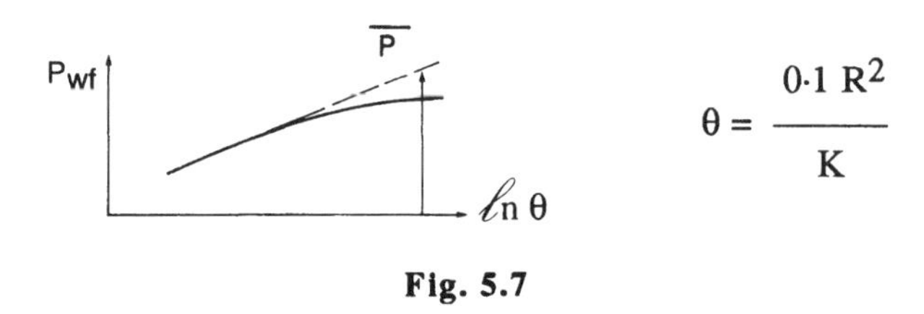

Fig. 5.7

If the well spacing pattern is uniform, in practice R = d/2 (d = well spacing).

Important Remark

A given flow rate is obtained by placing a given choke at the well head. It is stabilized only after a time interval. In fact, it can be demonstrated that the reservoir forgets the "earlier" flow rate variations. This is why it is preferable to interpret the pressure build-up curves rather than the drawdown pressure drop curves. For this interpretation, the stabilized flow rate and an equivalent time of flow are used:

$$T = \frac{\text{Cumulative production}}{\text{Stabilized flow rate}}$$

5.2.5 Skin Effect or Damage

The skin effect derives mainly from the drilling mud which, with the dual effect of the cake and the filtrate, causes partial plugging of the invaded zone, and hence a **decrease in permeability** in this zone.

It can be demonstrated that semi steady-state conditions are established fairly rapidly in the zone close to the borehole (at constant flow rate), and then increasingly farther away. When these conditions are reached, the change in well pressure reflects the influence of the more distant zone.

This is used in particular to investigate the degree of change in permeability in the neighborhood of the well due to drilling, completion or stimulation.

In these conditions, during flow, the pressure loss variation in this zone is constant. It can be written:

$$\delta P = 2S \cdot \frac{\mu_o Q_o B_o}{4 \pi h k}$$

with
- S = skin effect coefficient,
- $S > 0$ = if the layer near the well is damaged (additional pressure loss),
- $S < 0$ = if the layer near the well is improved (reduced pressure loss) (Fig. 5.8).

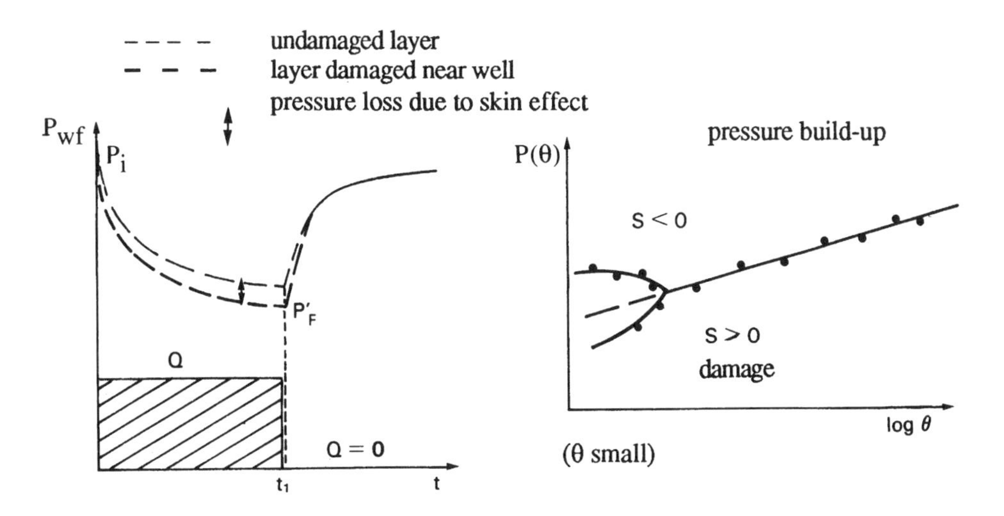

Fig. 5.8

The pressure drawdown equation is accordingly written:

$$P_i - P_{wf} = \frac{\mu_o Q_o B_o}{4 \pi h k} \left(0.81 + \ln \frac{K t}{r_w^2} + 2S \right)$$

The slope of the straight part of the curve (P_{wf}, $\ln_t$) thus serves to calculate the permeability of the untouched zone beyond the immediate vicinity of the borehole.

When a **well is shut in**, this is done at the surface, and the reservoir does not stop flowing instantly (**afterflood** recompressing the fluids in the well). The pressure loss due to the skin effect thus gradually tapers off, and the curve joins the one obtained in the absence of a skin effect. It can accordingly be concluded that, with or without skin effect, the plot of the straight part gives the permeability of the untouched zone. The equation becomes:

$$P_{ws} - P'_{wf} = \frac{\mu_o \, Q_o \, B_o}{4 \pi h k} \left[0.8 + \ln \frac{K \left(t - t_1\right)}{r_w^2} + 2 S \right]$$

This equation serves to calculate S by selecting a point on the straight part of the pressure build-up $P_{wf\,(t)}$.

The coefficient thus calculated is a total skin effect S_t, which includes that due to **perforations** S_p and, if applicable, that due to the **partial penetration effect** S_e. This latter effect exists when the entire layer has not been perforated, which applies in particular when an undesirable fluid (gas or water) exists in the layer, and a cone is liable to be formed (Section 4 of Chapter 6) (Figs 5.9 and 5.10). These two effects due to perforations S_p and partial penetration S_e are obtained from charts (Figs 5.11 and 5.12) or formulas.

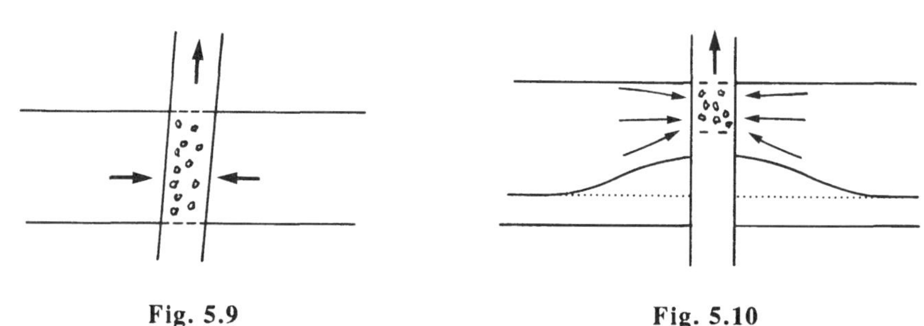

Fig. 5.9
Perforation effect.

Fig. 5.10
Partial penetration effect.

Knowing $S_t = S_c + S_p + S_e$, we can determine S_c (plugging).

If coefficient S_c is positive (additional pressure loss), an attempt can be made to improve the **productivity of the well by stimulation**, and this is the **purpose** of calculating the coefficient S_c. As a rule, the layer is **acidized** by the injection of very dilute hydrochloric acid or a mixture of hydrochloric and hydrofluoric acids (mud acid) to restore its initial characteristics and, if possible, to improve on them. This improvement in **production** can be spectacular in some situations.

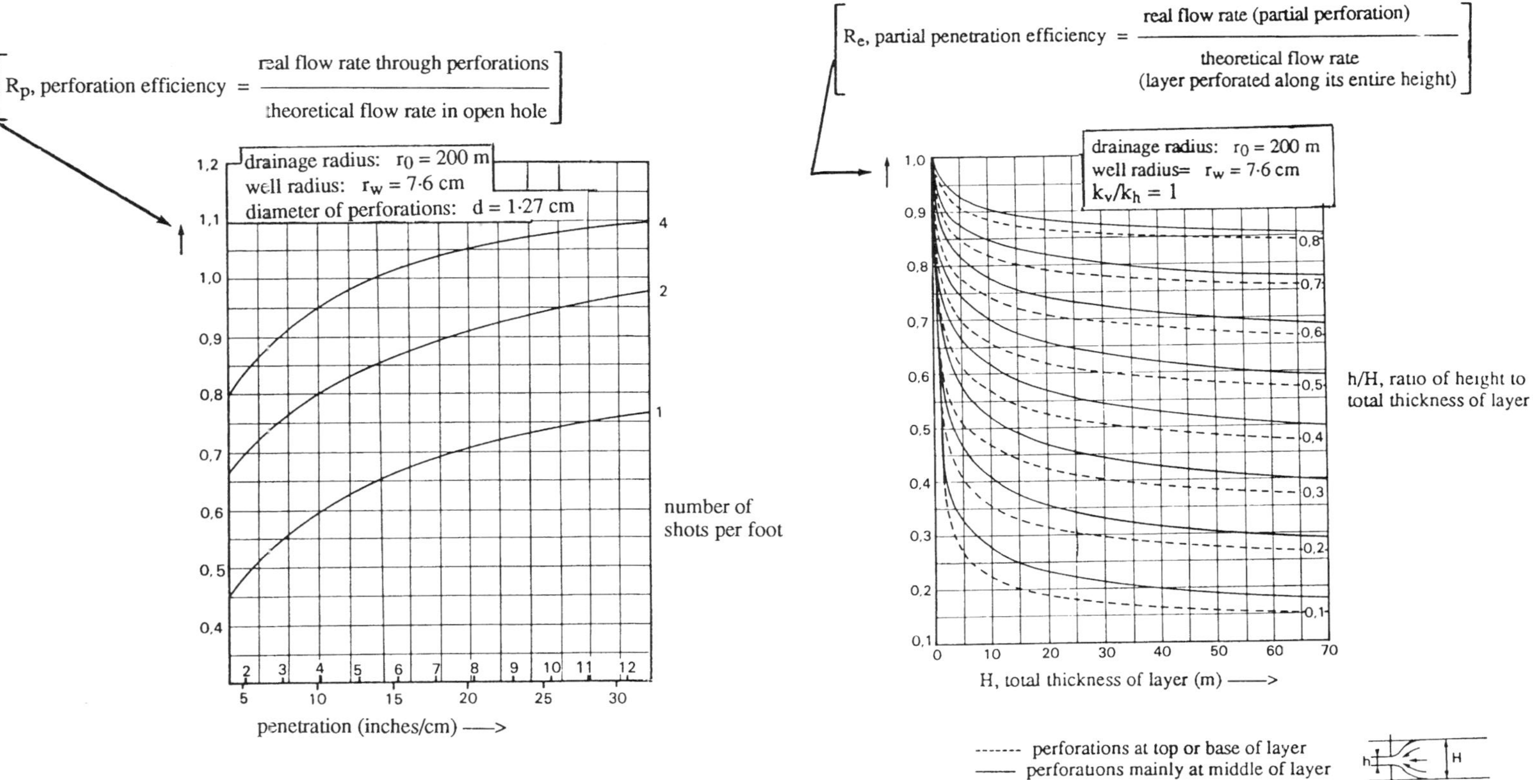

Fig. 5.11

Perforation effect on flow efficiency.
(Source: Harris (JPT April 1966)).

Fig. 5.12

Partial penetration effect on flow efficiency.
(Source: A.D. Odeh (AIME 1967)).

5.2.6 Productivity Index

The productivity or production index is defined as:

$$PI = \frac{Q_o}{P_i - P_{wf}} \qquad \text{or} \qquad \frac{Q_o}{\overline{P} - P_{wf}}$$

for constant pressure outer boundary and bounded cylindrical reservoirs respectively.

For a reservoir behaving as if of "infinite extent" (transient flow), the first expression is used, where P_{wf} is the bottom hole pressure of the flow period (also called P'_{wf}).

This index defines the **production capacity** of the well. It must be calculated in order to determine the completion of a well for a given flow rate (pumping or gas-lift) or the pressure loss that must be set at the wellhead (naturally flowing well).

5.2.7 Various Problems of Fluid Flow

Radial Heterogeneity

In transient flow, we have shown that the slope m of the straight line serves to calculate the mean permeability beyond the immediate vicinity of the hole:

$$k = \frac{\mu_o Q_o B_o}{4 \pi h \cdot m}$$

The reservoirs are more or less heterogeneous, and if the variations in permeability indicate a different average tendency between a fairly close zone and another zone more distant from the well, this is materialized by a change in the slope of the line. The following is obtained in pressure build-up, for example (Fig. 5.13).

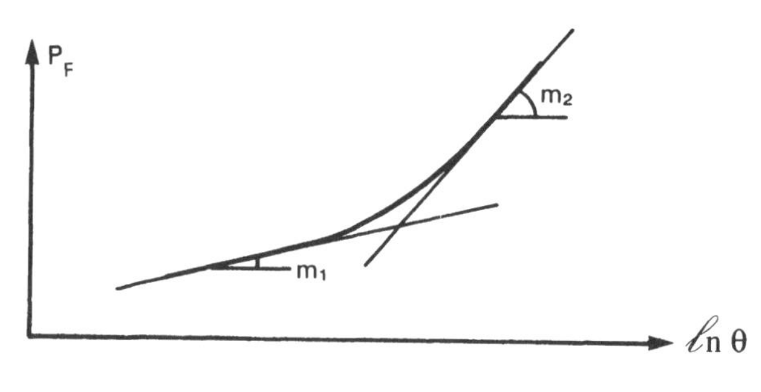

Fig. 5.13

Case of an Impermeable Fault

This can be examined by considering that the fault is a zero flow plane, namely by imagining a well, which is symmetrical to the actual well with respect to the fault, producing at the same flow rate.

On the first well, it is necessary to add the pressure drops due to the well itself and to its image. It can be demonstrated that the curves in semilog co-ordinates display two linear parts with one **slope double** the other. This also helps to determine the distance from the fault to the well.

N.B.: The same change in slope could result from certain radial heterogeneities or from the presence of superimposed multiple layers.

5.2.8 Two-Phase Flows

In a saturated oil reservoir, the pressure falls below the bubble-point pressure from the start of production. This also occurs after a period of time in undersaturated reservoirs. The flows are two-phase flows in this case (oil and gas).

However, all the equations developed for one-phase flow can be used, provided the compressibility and mobility (permeability/viscosity) of the oil are replaced by composite compressibility and mobility. This is also valid for a well producing oil with water.

Experience proves that the simple application of the above formulas often yields acceptable results. It is sufficient for the flow rate of undesirable fluid to be **low** enough in comparison with that of the main fluid. Yet it must be pointed out that the **productivity index of the well drops significantly** due to the influence of the effective permeabilities ($k_o = k \cdot k_{ro}$ and k_{ro} decreases) (Fig. 5.14).

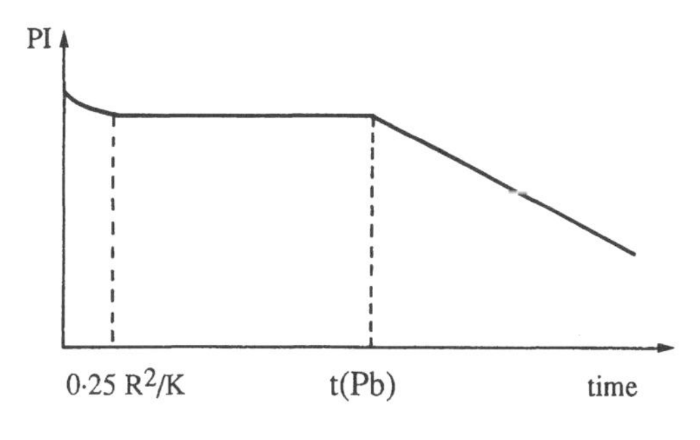

Fig. 5.14

5.2.9 Special Case of Nonflowing Wells

The above formulas cannot be used if it is possible to measure only the rise of the fluid level in the well (measurements by echometers on **pumping closed wells**).

If the effluent is one-phase, $\overline{P}$ can be obtained by graphic extrapolation by plotting $\Delta P/\Delta t$ as a function of pressure P_{wf}:

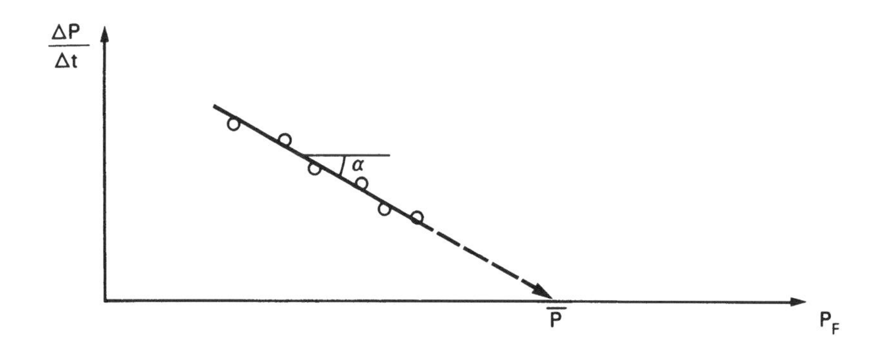

In the well:

$$\frac{\Delta P}{\Delta t} = \frac{\Delta h}{\Delta t} \rho g \qquad \text{and} \qquad Q = S \frac{\Delta h}{\Delta t}$$

with
- Δh = rise in level during Δt,
- ρ = density of effluent,
- S = well cross-section.

Hence:

$$\frac{\Delta P}{\Delta t} = \frac{Q}{S} \rho g$$

But PI $(\overline{P} - P_{wf}) = Q$ (the flow is considered as consisting of a series of steady-state flows), hence:

$$\frac{\Delta P}{\Delta t} = PI \cdot \frac{\rho g}{S} \cdot \left(\overline{P} - P_{wf}\right) = \alpha \left(\overline{P} - P_{wf}\right)$$

The permeability is calculated by the formula corresponding to semi steady-state flow:

$$k = \frac{\mu Q B}{2 \pi h (\bar{P} - P_{wf})} \left(\ln \frac{R}{r_w} - \frac{3}{4} \right)$$

5.3 GAS FLOW AROUND WELLS

In the discussion on permeabilities, we showed that the elementary law of pressure loss in porous media is written:

$$\rho \, dP = \frac{\mu Q_m}{S k} \left(1 + \frac{\bar{u} Q_m}{\mu S} \right) dr$$

with

Q_m = mass flow rate ($Q \cdot \rho$),
ρ = density of fluid,
$\bar{u}$ = "shape parameter" characterizing the porous medium,
S = fluid flow cross-section.

If $\bar{u} Q_m / \mu S$ is large, the second term in parentheses cannot be disregarded, and **Darcy's Law is no longer applicable**. This applies in particular to gases (very low viscosity μg) and liquids if their flow rate is high (high permeability, fractures).

For gases, we obtain:

$$\rho dP = (A'Q_m + B'Q_m^2) \, dr$$

with

$$\rho = \frac{M P}{Z R T} .$$

As a first approximation, for pressure variations that are not too large, it can be estimated that $Z = Z_m$ (mean) = constant. Hence $\rho dP = bP \cdot dP$, with b = constant.

Steady-State Flow

a) Quadratic equation:

By integrating the equation of the elementary pressure loss law, a quadratic equation is obtained:

$$P_i^2 - P_{wf}^2 = A\,Q_{std} + B\,Q_{std}^2$$

(Q_{std} = standard gas flow rate = $\frac{Q_m}{\rho_{std}}$).

The plot of

$$\frac{P_i^2 - P_{wf}^2}{Q_{std}}$$

as a function of Q_{std} gives a straight line, the **deliverability curve**, which serves to determine A and B. From A, we obtain hk, where k is the mean permeability taking account of the skin effect.

b) Pseudopressure:

In actual fact, the foregoing integration does not always yield a law in P^2.

At high pressures, the product μZ is proportional to P. We have:

$$\frac{\rho\, dP}{\mu} = \frac{M}{RT} \cdot \frac{P}{\mu Z} \cdot dP \rightarrow \frac{M}{RT} \cdot \beta \cdot dP$$

with β constant, hence a **law in P**.

To simplify the calculations, it is often assumed (not always with justification) that it is valid to use:

(a) The pressure P, if $P > 3000$ psi.
(b) The pressure raised to the square P^2, if $P < 2000$ psi.
(c) The pseudo-pressure ψ in other situations,

with

$$\psi = 2 \int_{P_o}^{P} \frac{P}{\mu Z}\, dP$$

c) Empirical equation:

Another equation, which is empirical, is also used for semi-steady state flow:

$$Q_{std} = C \left(P_i^2 - P_{wf}^2\right)^n$$

where $0.5 < n < 1$.

This equation can be written:

$$\log Q = \log C + n \cdot \log\left(P_i^2 - P_{wf}^2\right)$$

Log Q is plotted as a function of log $(P_i^2 - P_{wf}^2)$. C and n are determined from the deliverability curve obtained.

The deliverability curve also helps to determine the **absolute open-flow potential of the well** (theoretical value of Q_{std} for P_{wf} = 1 atmosphere) (Fig. 5.15).

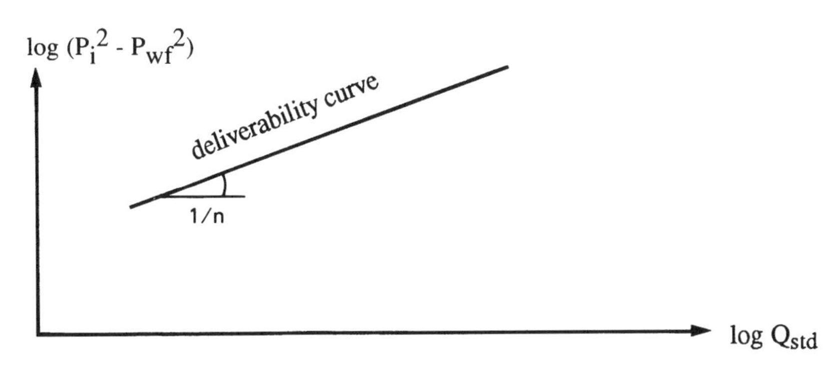

Fig. 5.15

Deliverability curve.

Transient Flow

In pressure build-up, the equation relating P_i and P_{wf} to the flow rate Q_{std} B_g is similar to that of a liquid as a first approximation. Hence hk is obtained in the same way as for a liquid. The skin effect can also be obtained, but this is more complex, because it is associated with a term in Q_{std}.

This gives:

$$h\,k = \frac{\mu_g Q_g B_g}{4\pi \cdot m}$$

(see practical units in Section 5.6.2.)

5.4 DIFFERENT WELL TESTS

5.4.1 Initial Tests

These tests concern both extension wells (including discovery wells) and development wells. They are intended to determine all the para-meters indicated at the beginning of this chapter (Section 5.1, General Introduction): **productivity, "hk", skin effect, static pressure, drainage radius, sampling** (PVT analysis), etc.

The performance of these tests generally includes the following:

(a) A **clearing period** after completion (about one day if the permeability is medium or high; or shorter, just adequate to ensure that the formation is not plugged if the permeability is low).
(b) A **shut-in period**: measurement of the initial static pressure (about two days if the permeability is medium or high; a few hours if the permeability is low).
(c) A **flow period** (two to three days per flow rate if the permeability is medium or high; a few weeks if the permeability is low).
(d) A **shut-in period** (two to three days if the permeability is medium or high; and a period approximately equal to the flow period if the permeability is low).
(e) and (f) Sometimes, a second flow and shut-in period is applied (Fig. 5.16).

Remarks concerning the foregoing points:

(a) **Clearing**: if clearing takes place, the quantities produced must be estimated.
(b) **First shut-in**: the downhole pressure gauge is run in before shut-in to achieve thermal stabilization.
(c) **Flow period**: one or generally more constant flow rates are applied to the well until semi steady-state flow or sometimes steady-state flow is reached to determine the productivity index (oil) or the absolute open-flow potential (gas).

Measurements

- *Flow rates*

Measurements are taken until stabilization, and then to check stabilization. Since the effluent is two-phase, a separator is needed to measure the one-

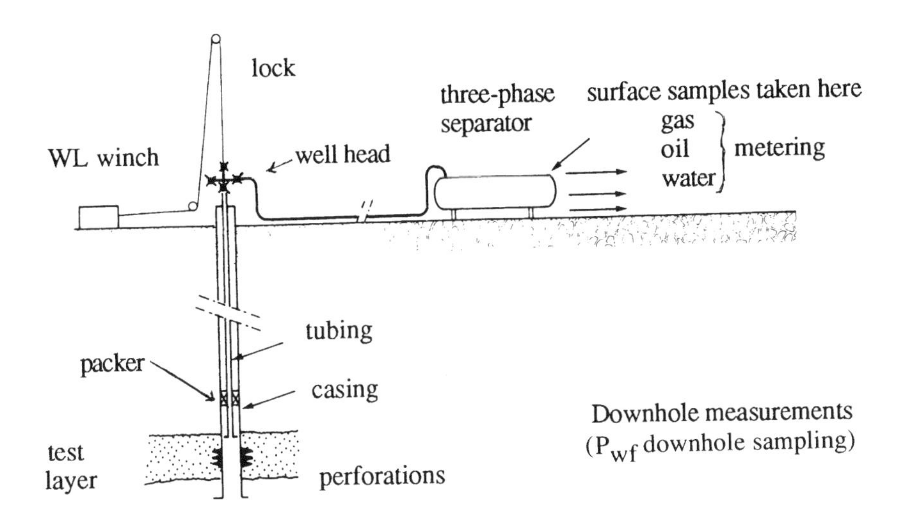

Performance of test, Oil well

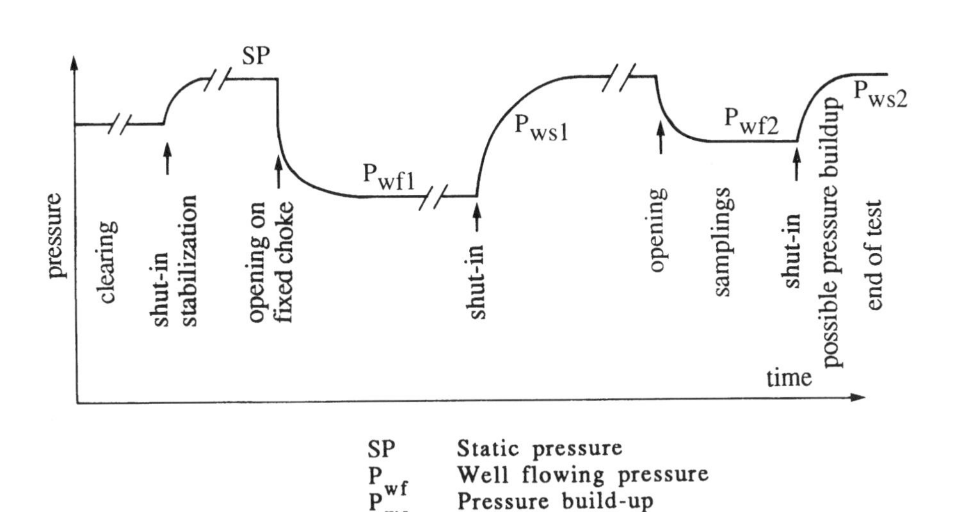

SP	Static pressure
P_{wf}	Well flowing pressure
P_{ws}	Pressure build-up
θ	Temperature

Fig. 5.16

Schematic representation of a well test.

phase oil and gas flow rates. It is also necessary for $P_{(sep)}$ and $T_{(sep)}$ to be constant.

• *Pressures*

The downhole pressure gauges are run in usually in tandem. They are generally pressure sensors featuring a strain gauge or a quartz crystal. Their accuracy is better than 10^{-3} . scale (for example, a few psi for 5000 psi). The temperatures are also recorded downhole and at the surface.

5.4.2 Tests Specific to Gas Wells

These tests are called **back-pressure tests** (water cushion). They involve **several flow rates (at least three)** for each of which steady-state flow conditions must be reached to obtain the deliverability curve.

Isochronal Tests

Back-pressure tests may be very lengthy. The **isochronal test** method is therefore often employed. This type of test is indispensable for relatively **impermeable formations**, because too much time would elapse before steady-state conditions were reached in each flow period.

A series of short flow periods is separated by pressure build-ups, with the flow and shut-in periods of equal length (Fig. 5.17). Only the last flow period is long enough to reach steady-state conditions, hence a considerable overall gain in time.

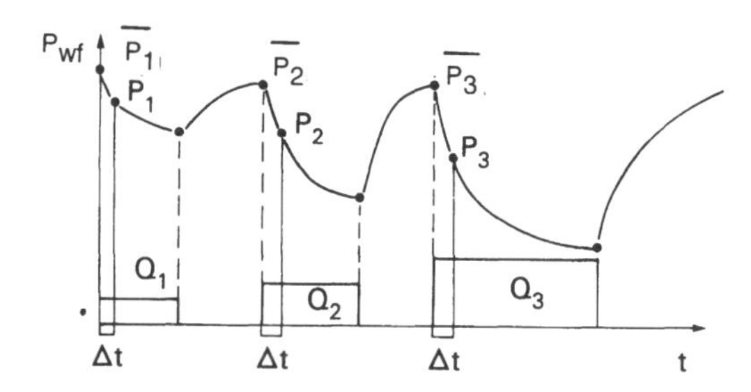

Fig. 5.17 Recording of isochronal test.

In the two representations shown in Section 5.3 (quadratic and empirical equations), the points corresponding to the same flow time Δt are on **lines parallel to the deliverability curve**. This curve is plotted using only the point corresponding to steady-state flow. Figure 5.18 shows the construction of the deliverability curve obtained by the empirical equation.

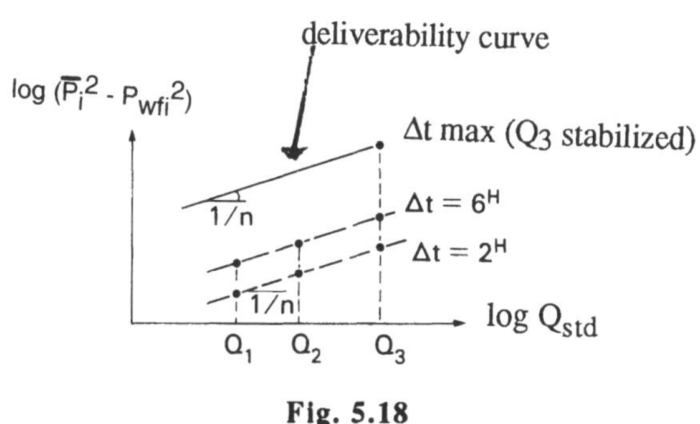

Fig. 5.18
Interpretation.

5.4.3 Periodical Tests

These tests are conducted like the initial tests, but are often simplified. They may be performed at a frequency of 3, 6 or 12 months, for example, depending on the rate of change in the well parameters.

They are intended to determine the following, at the time concerned:

(a) Static pressure.
(b) Well potential.
(c) kh far from the well, skin effect and effluent monitoring.

They involve a recording over a flow period, a shut-in period, or both.

5.4.4 Interference Tests

These tests usually involve the measurement on a well, called the receiver, of the effects of a pressure disturbance induced in the reservoir by varying the flow rate of a neighboring well called the transmitter (or test well). For example, new wells just brought on stream can be used (Fig. 5.19).

They are intended to achieve the following:

(a) Information on reservoir transmissibility ("directional" hk between two wells).
(b) More data on any water/hydrocarbon interfaces and an assessment of the activity of an aquifer.
(c) The analysis of the communication (or noncommunication) between reservoirs or parts of a reservoir.

All these details are of fundamental importance in understanding the reservoir, and especially for reservoir simulation on a model.

N.B.: In recent years, a method called **pulse** testing has also been used, which consists in regularly opening and shutting in a well for short periods lasting a few hours. This method is only valid for fairly high "hk" and short distances between the two wells.

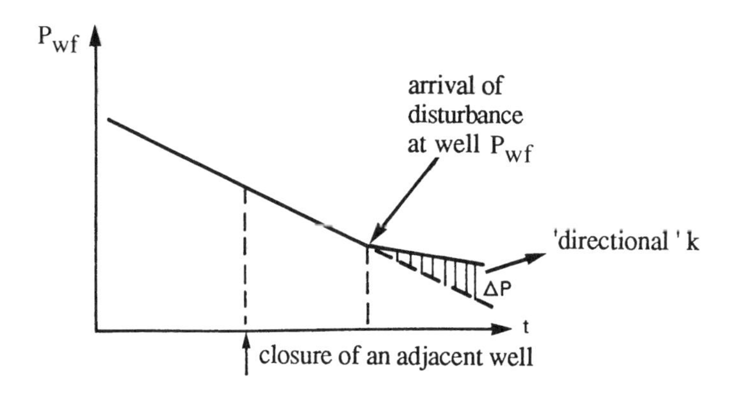

Fig. 5.19

Interference.

5.4.5 Water Flood Wells

Tests on injection wells are intended for the following:

(a) Determination of the injectivity index $II = Q_I / (P_I - \overline{P})$.
(b) The identification of absorption zones, with the use of downhole flow metering and temperature measurement.
(c) The determination of $(hk)_I$ and S.

These tests include the following:

(a) A clearing (if necessary) and a shut-in: during this period, the downhole pressure and temperature are recorded.
(b) Injection at constant flow rate, or with several successive flow rates: the downhole pressure and tubing head pressure are recorded.
(c) A fall off (injection stopped): the downhole pressure and temperature, and, if necessary, the tubing head pressure are recorded.

The "hk" and skin effect S can be calculated by analyzing the pressure drop after injection has been stopped, as in the case of pressure build-up for a producing well. However, the average pressure of the injection area cannot be calculated directly by extrapolating P_{wf} when $\frac{T + \theta}{\theta}$ approaches 1, due to the presence of two different fluids in the reservoir, in the event that injection takes place in the oil zone. If injection takes place in the aquifer, the extrapolation is valid.

5.5 TESTS DURING DRILLING (Drill Stem Tests - DST)

The diagram of a normal productive capacity test on an oil well with a water cushion is shown in Fig. 5.20.

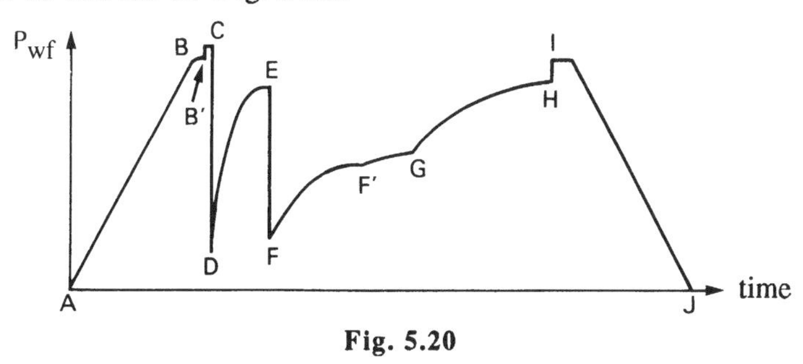

Fig. 5.20

AB Running in of tester into well filled with mud.

BB' Pause at bottom hole (variable length of time).

C Anchoring of packer, causing pressure surge (compression of mud located below packer).

CD Opening of a valve (C) allowing the layer to flow into the drillpipe. The bottom of the well is placed abruptly at the pressure of the water cushion (or gas/oil cushion) in the pipe (D). This is the **"preflow"**.

DE Pressure build-up after very small amount of production. The reservoir pressure is hence rapidly reached at E = **initial pressure**.

EF **Opening** of tester. The well bottom is placed in communication with the inside of the pipe.

FG The layer flows into the well. Since the oil level rises, the downhole pressure rises accordingly.

F' Arrival of water cushion at surface.

GH The tester is **closed**. The pressure rises. It would be stabilized at the reservoir pressure after a sufficiently long period of time (the waiting time depends on the productivity of the layer, the flow time and the condition of the hole).

HI The pressure of the mud column is again applied to the layer by equalization of the pressures above and below the packer which is then "unanchored".

IJ Pullout of tester.

The following data can be obtained from a test of this type.

Reservoir Pressure

(a) By the first pressure build-up (measurement of initial pressure).
(b) By the pressure build-up after flow. Since the quantity withdrawn is very small, the reservoir is virtually unaffected and behaves like an infinite extent reservoir. The curve of P as a function of log $T + \theta / \theta$ is plotted, and the straight part extrapolated to $T + \theta / \theta = 1$ to give P_i. To determine the straight part with sufficient accuracy, **the shut-in time must be approximately the same as the flow time.**

Product hk

From the slope of the straight part of the foregoing curve, the flow rate Q_F is taken as the cumulative flow divided by the flow time. This determination contains errors because the fluids produced are often mixtures of mud, gas, oil and water, and, above all, because the properties of the fluids are unknown or very slightly known at the time of the test.

Other data obtained:

(a) **Potential** (productivity index):

$$PI = \frac{Q}{P_i - P_{wf}}$$

based on the flow period, its determination is subject to the same criticism as above.

(b) **Skin effect before completion: standard calculation.**

5.6 MAIN EQUATIONS USED WITH PRACTICAL DRILLSITE UNITS

5.6.1 Equations for Oil (Metric p.u.)

<table>
<tr><th colspan="4">OIL</th></tr>
<tr><td rowspan="2">TRANSIENT FLOW</td><td>Flow</td><td>$P_i - P_{wf} = 21.5 \frac{\mu_o Q_o B_o}{h k}\left(\log 0.0192 \frac{K t}{r_w^2} + 0.87 S\right)$</td><td>(1)</td></tr>
<tr><td>Build-up</td><td>$P_i - P_{wf(\theta)} = 21.5 \frac{\mu_o Q_o B_o}{h k} \log \frac{T + \theta}{\theta}$
or (162.6 in US p.u.)*
$P_{ws(\theta)} - P'_{ws} = 21.5 \frac{\mu_o Q_o B_o}{h k}\left(\log 0.0192 \frac{K \theta}{r_w^2} + 0.87 S\right)$
$h k = 21.5 \frac{\mu_o Q_o B_o}{m_{10}}$ or (162.6 in US p.u.)*
$P_{ws(\theta)} \rightarrow P_i$
when $\frac{T + \theta}{\theta} \rightarrow 1$
Skin:
$S = 1.15 \frac{P_{wf(\theta)} - P'_{wf}}{m_{10}} - \log \frac{8 \cdot 10^{-4} K \cdot \theta}{r_w^2}$
with θ in hours</td><td>(2)

(3)
(4)

(5)</td></tr>
</table>

SEMI STEADY-STATE FLOW	Flow	$\overline{P}_{(t)} - P_{wf\,(t)} = 43\,\dfrac{\mu_o\, Q_o\, B_o}{h\,k}\left(\log 0.47\,\dfrac{R}{r_w} + 0.435\,S\right)$ or (325.2 in US p.u.)	(6)
	Build-up	Equation (3) provided $\theta << T$. kh calculation: equation (4). $\overline{P}$ calculation: reading on the pressure build-up line for $\theta = 11.7\,R^2 / K$. Skin[1]: equation (5).	
STEADY-STATE FLOW	Flow	$P_i - P_{wf} = 43\,\dfrac{\mu_o\, Q_o\, B_o}{h\,k}\left(\log \dfrac{R}{r_w} + 0.435\,S\right)$ or (325.2 in US p.u.)	(7)
	Build-up	Equation (3). kh calculation: equation (4). Skin[1]: equation (5).	

5.6.2 Equations for Gas (Metric p.u.)

Flow rate (steady-state flow)

$$P_i^2 - P_{wf}^2 = A\,Q + B\,Q^2 \qquad \text{with} \qquad A = 86\,\frac{\mu_g \,.\, B_g \,.\, \overline{P}}{h\,k}\,\log\frac{R}{r_w} \qquad (8)$$

where:

$$Q = c\,(P_i^2 - P_{wf}^2)^n \qquad \text{and} \qquad \overline{P} = \frac{P_i + P_{wf}}{2}$$

Pressure build-up (transient)

$$h\,k = 21.5\,\frac{\mu_g\, Q_g\, B_g}{m_{10}} \qquad (9)$$

(or 162.6 in US p.u.)

1. Skin effect: equations (3) and (5) are valid for any pressure build-up irrespective of the previous flow conditions.

The following **practical units** (p.u.) are employed:

Symbol	Definition	US Industry unit	Metric practical unit
A	Coefficient of gas deliverability curve	$psi^2/ft^3/d$	pa^2/m^3 per day
c	Compressibility	psi^{-1}	pa^{-1}
h	Reservoir thickness	ft	m
k	Permeability	mD	mD
K	Hydraulic diffusivity ($K = k/(\phi \mu c)$)	ft^2/day	m^2/day
m_{10}	Slope of line in semi-log (base 10)	psi/cycle	pa/cycle
P	Pressure	psi	pa
P'_{wf}	Downhole pressure at shut-in	psi	pa
Q	Flow rate (standard)	bpd	m^3/day
R	Radius of drained area	ft	m
r_w	Well radius	ft	m
S	Skin effect	–	–
t, θ	Time	hours	days
μ	Viscosity	cP	Pa.s
log	Base 10 logarithm	–	–

5.7 PRINCIPLE OF TYPE CURVES

5.7.1 Presentation

In recent years, many authors and investigators have developed and improved "type curves", and their use is now widespread. This interpretation technique has become routine today. However, research is continuing to achieve further improvements. These methods are complementary to conventional methods.

The type curves are an application of the following standard procedure. The equations are written for all possible situations: homogeneous reservoir, heterogeneous (fractured) reservoir, with boundary conditions (in well, outside well), as a function of the reduced variables: dimensionless time and pressure. This produces series of curves on graphs (Fig. 5.21).

The recorded curve is plotted in log/log coordinates and superimposed on the graph, to yield the group **hk**, C = storage coefficient[1], S = **skin** effect. Yet above all, the type-curves allow a **diagnosis** concerning the type of reservoir. Thus fractured wells are characterized by a different slope, 0.5 instead of 1 in log/log at the start of pressure build-up, because $\Delta P = \sqrt{(\Delta t)}$ for a fractured environment.

Derivative curves were recently developed, depending on $dP/d(\Delta t)$. They offer the advantage of a better differentiation of the $C_D\, e^{2S}$ groups, hence a closer approximation of the skin effect (see curves in Fig. 5.21). Moreover, in some cases the derivative allows boundaries to be determined more accurately (faults, channels, etc.). However, this presentation requires virtually error-free recordings. In fact, the measurements are generally affected by background noise. The data derivation algorithm must therefore eliminate most of the noise by smoothing. As a rule, however, the derivative is selected as the simple mean of the slopes of the chords adjacent to the point concerned on the curve.

5.7.2 Use

Pressure build-up curves are superimposed on type curves (general case). The values of the recording are transferred to log/log tracing paper. It is easy to go from the values plotted in ΔP and Δt to a type curve by simple translation, because the multiplication factors used to yield the dimensionless variables correspond to translations of the axes to the logarithmic scale.

Accordingly:

$$P_D = \frac{kh}{\alpha \cdot \mu\, QB} \cdot \Delta P \quad \text{dimensionless pressure}$$

1. The storage coefficient (or well storage capacity) $C = \Delta V_W / \Delta p$ characterizes the ability of the downhole flow to rise initially from the value 0 to the value QB. V_W = volume of effluent in the well.

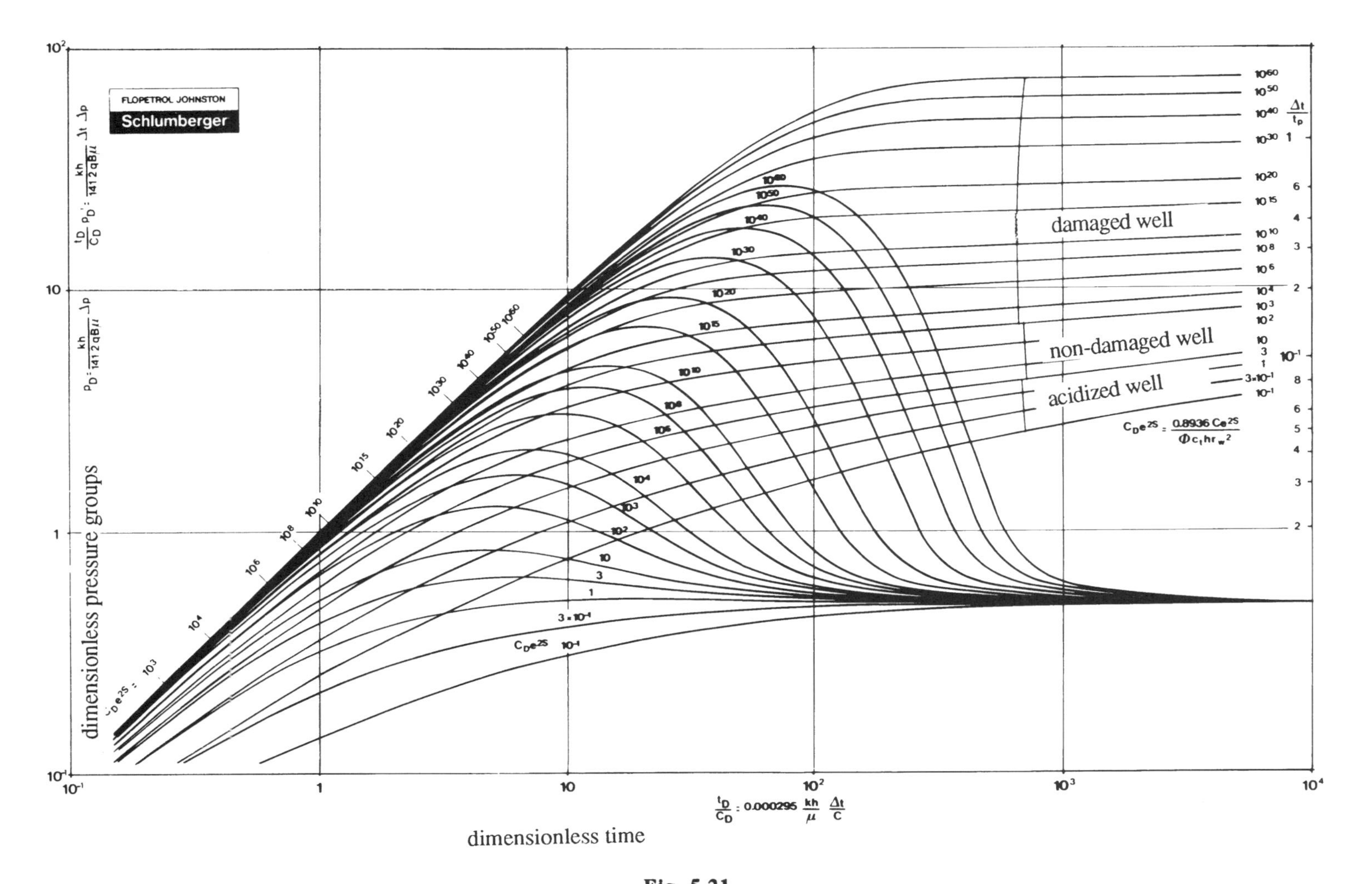

FLOPETROL JOHNSTON
Schlumberger
dimensionless pressure groups
dimensionless time
damaged well
non-damaged well
acidized well
$C_De^{2S} = \frac{0.8936\ Ce^{2S}}{\Phi c_t h r_w^2}$
$\frac{t_D}{C_D} = 0.000295 \frac{kh}{\mu} \frac{\Delta t}{C}$
$P_D = \frac{kh}{141.2 qB\mu} \Delta p$
$\frac{t_D}{C_D} P_D' = \frac{kh}{141.2 qB\mu} \Delta t\ \Delta p$
$\frac{\Delta t}{t_p}$
C_De^{2S}

Fig. 5.21

Hence:

$$\log P_D = \log \frac{k\,h}{\alpha\,.\,\mu\;Q\,B} + \log \Delta P$$

The interpretation method thus involves the following.

(a) Plot the measured pressures on tracing paper of the same log/log scale as the family of type curves selected.
(b) Find the superimposition of these measurement points on a type curve, only allowing for translations (the respective axes remain parallel to each other).
(c) Note the designation of this type curve ($C_D\,e^{2S}$).
(d) Arbitrarily select a match point, of which the coordinates are read in each system of axes (on the tracing paper and on the set of type curves). The multiplication factor corresponding to each axis is immediately identified (US practitioners speak of "pressure match" and "time match").
(e) Based on the data obtained in (3) and (4), calculate the desired parameters (kh, C, S, etc).

In practical US units (units of sets of type curves):

$$kh = 141.2\,q\,B\,\mu\,\frac{(P_D)_M}{(\Delta\,p\,)_M} \qquad C = 0.000295\,\frac{kh}{\mu}\,\frac{(\Delta t)_M}{\left(t_D/C_D\right)_M}$$

$$C_D = \frac{0.89\,C}{r_w^2\,.\,h\,\phi\,C_t} \qquad S = \frac{1}{2}\,\ln\,\frac{C_D\,e^{2S}}{C_D}$$

(see example in Section 5.8).

Units: mD, ft, bbl/day, cP and C_t in psi^{-1}.

$C_t = C_oS_o + C_wS_w + C_p$ total compressibility.

Other sets of type curves:

The most widely used concern "infinite extent" reservoirs. For a standard test, three categories essentially exist, depending on the reservoir/well configuration:

(a) **Homogeneous reservoir** and cylindrical well (the situation described above).
(b) Homogeneous reservoir and well with a **vertical hydraulic fracture** (the case of a horizontal fracture has also been dealt with, although rarely encountered).

(c) **Fractured reservoir** (double porosity).

Type curves also exist for **interference** tests in homogeneous and fractured environments. In every situation, the overall procedure is the same as the one described above.

5.7.3 Semiautomatic Test Analysis Programs: Well Models

Practically all the oil companies have set up more or less automatic interpretation programs, particularly for build-ups, giving kh, S, $\overline{P}$, PI, (R), etc., with comparison of the different methods. For periodical tests, the production periods are subdivided into sections in which the flow rate is fairly stable, and the superposition theorem is applied. This yields a summary analysis for each well. These programs are entering general use today (Fig. 5.22).

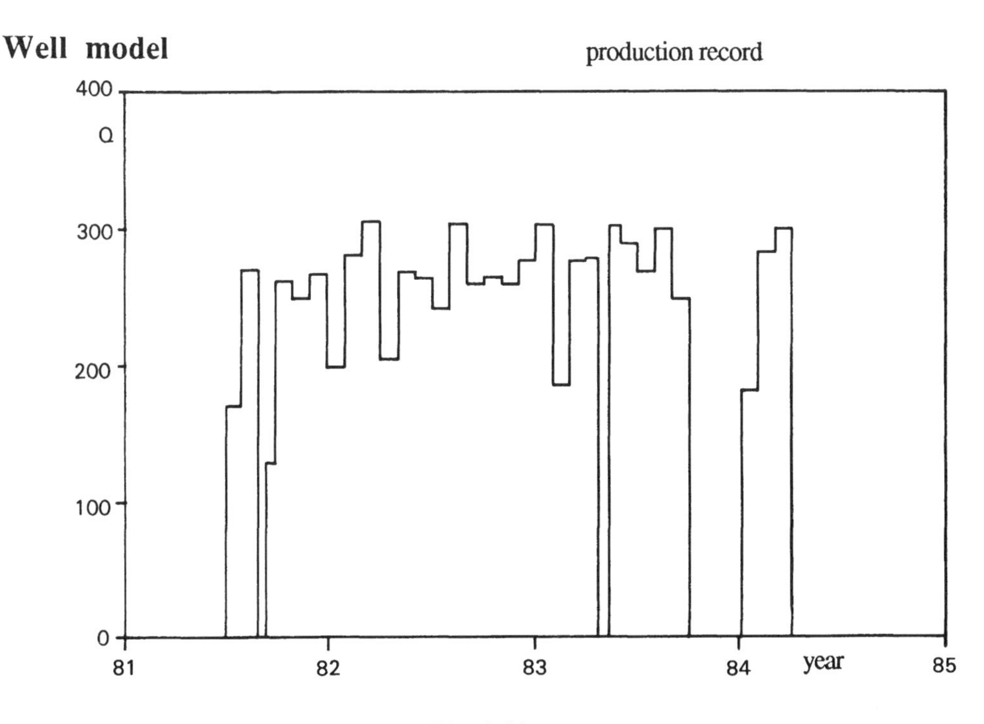

Fig. 5.22

Application of the **superposition** theorem.

In conclusion, the ideal approach to well test interpretation usually involves three steps:

(a) Diagnosis and rough calculation using type curves.
(b) Application of conventional methods.
(c) Use of a well model.

5.8 EXAMPLE OF TEST INTERPRETATION

Oil Reservoir

The following data are available:

$QB = 100 \text{ m}^3/\text{day}$	$\mu = 1$ cP
$h = 36$ m	$\phi = 0.15$
$C_t = 2.9 \cdot 10^{-4} \text{ bar}^{-1}$	$r_w = 0.09$ m (well)

Using practical US units:

$QB = 600$ bbl/day	$\mu = 1$ cP
$h = 120$ ft	$\phi = 0.15$
$C_t = 2 \cdot 10^{-5} \text{ psi}^{-1}$	$r_w = 0.3$ ft

The test is a build-up preceded by a **very long production period** at a single constant flow rate (type curves applicable). We directly give the values of:

$$\Delta p = (\text{pressure during build-up}) - (\text{pressure just before shut-in})$$

$$\Delta t = \text{time measured from well shut-in } (\Delta t \equiv \theta)$$

The pressures are measured downhole, but the shut-in is performed at the well head.

Δt (hours)	**Δp (psi)**
0.20	69
0.31	96
0.60	142
0.84	165
1.17	187
2.27	222
3.94	245
6.85	264
10.66	278
14.85	288
20.69	298
28.82	308
40.14	318
50.06	324

The reservoir **permeability**, **skin** effect of the well and **static pressure** must be determined.

Interpretation by Type Curve

In view of the knowledge about the reservoir indicating that it is not fractured in the zone near the well, the interpretation is normally made by comparison with the "homogeneous reservoir" set of curves.

The points are plotted on **tracing paper superimposed** on the type curve set. A minimum of two lines is drawn surrounding a cycle on the abscissa and similarly on the ordinate. In Fig. 5.23, for example, the lines of 1^H and 10^H for Δt, and 100 psi and 1000 psi for ΔP.

A match point M is selected, for example the point whose coordinates on the tracing paper are:

$$(\Delta P)_M = 100 \text{ psi} \qquad (\Delta t)_M = 1 \text{ h}$$

For the same point, the following are read on the type curve set:

$$(P_D)_M = 1.8 \qquad (t_D/C_D)_M = 9$$

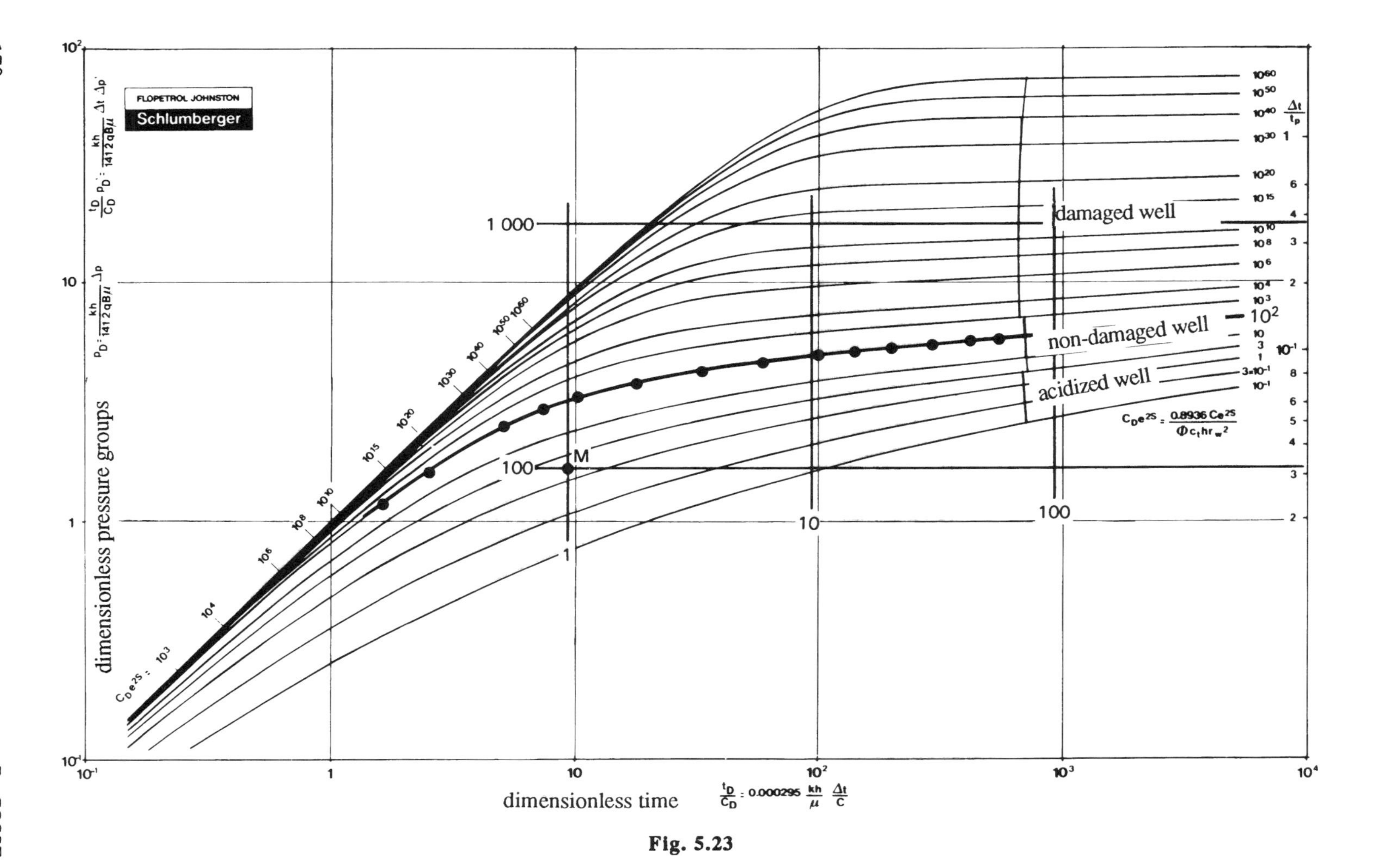

FLOPETROL JOHNSTON
Schlumberger
dimensionless pressure groups
dimensionless time
damaged well
non-damaged well
acidized well
M

Fig. 5.23

and the designation of the superimposed type curve is noted:

$$C_D\, e^{2S} = 10^2$$

The following is obtained using the formulas given in the text:

$$kh = 141.2 \times 600\ \frac{1.8}{100} = 1525 \text{ mD} \times \text{ft} \qquad \text{hence } \mathbf{k = 12.7 \text{ mD}}$$

$$C = 0.000295 \times 1525 \times \frac{1}{9} = 0.05 \text{ bbl/psi}$$

$$C_D = \frac{0.89 \times 0.05}{0.09 \times 120 \times 0.15 \times 2 \times 10^{-5}} = 1373$$

$$S = \frac{1}{2} \ln \frac{100}{1373} \sim \mathbf{-1.3}$$

The well has been "improved".

Conventional Interpretation

The straight line is identified and drawn (Fig. 5.24).

The slope m and the value of ΔP_{1h} on this line are read:

$$m = 68 \text{ psi/cycle} = 4.7 \text{ bar/cycle}$$

$$\Delta P_{1h} = 208 \text{ psi} = 14.3 \text{ bar}$$

The following is obtained using the practical formulas in Section 5.6:

$$kh = \frac{21.5 \times 100 \times 1}{4.7} = 457 \text{ mD} \times \text{m} \qquad \text{hence } k = 12.7 \text{ mD}$$

$$S = 1.15 \left[\frac{14.3}{4.7} - \log \frac{8 \,.\, 10^{-4} \times 12.7 \times 1}{1 \times 0.15 \times 2.9 \times 10^{-4} \times 81 \,.\, 10^{-4}} \right]$$

$$S \sim -1.6$$

Agreement between the results of the two methods is satisfactory. It is assumed that:

$$\mathbf{k = 13 \text{ mD}} \qquad \text{and} \qquad \mathbf{S = -1.5}$$

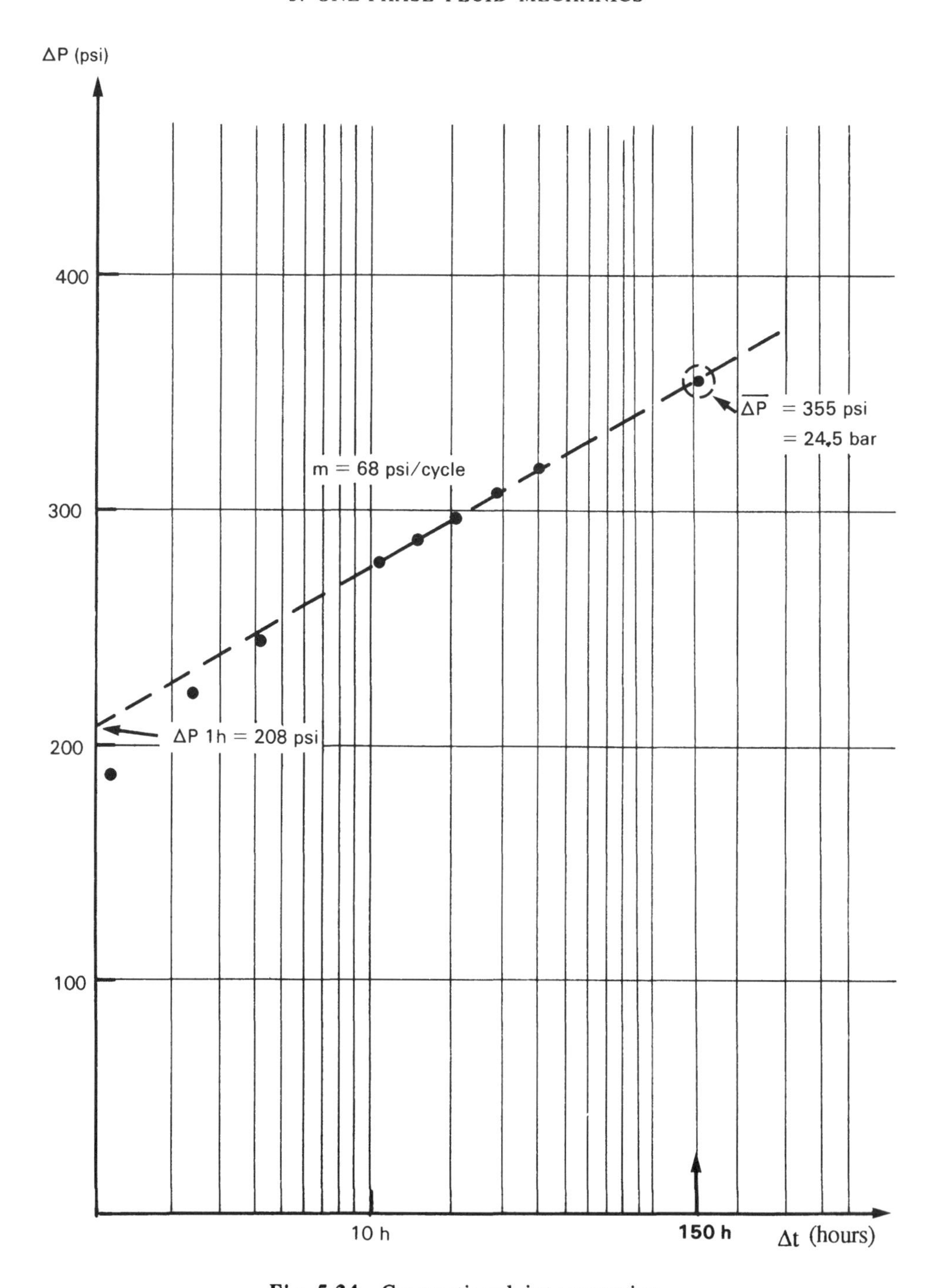

Fig. 5.24 Conventional interpretation.

Calculation of the Mean "Static" Pressure $\overline{P}$

Since the production period is very long, it is assumed in this example that the flow is semi steady-state.

$\overline{P}$ is obtained by direct reading by extrapolating the pressure build-up over time $\Delta t_s = 0.1\ R^2/K$ (see Section 5.2.4).

Hence, in metric practical units:

$$\Delta t_s = 11.7\ R^2/K$$

with Δt_s in days.

In the present case, thus:

$$R \approx 400\ m$$

and

$$K = k/\phi \mu c = 13/0.15\ .\ 1\ .\ 2.9.10^{-4}$$

$$K = 29.9\ .\ 10^4$$

Hence:

$$\Delta t_s = 11.7\ .\ \frac{16\ .\ 10^4}{29.9\ .\ 10^4} \sim 6.26\ \text{days} \approx 150\ h$$

Pressure extrapolation:

$$m = 4.7\ \text{bar} \qquad \text{and} \qquad \Delta P_{1h} = 14.3\ \text{bar}$$

$$\Delta\overline{P} = 14.3 + \log(150) \times 4.7 = 24.5\ \text{bar}$$

$\Delta\overline{P} = 24.5$ bar (with $\Delta P_s = \overline{P} - P'_{wf}$ end of production period)

In a **short duration test** (drill stem test or initial test), $\overline{P}$ is calculated in a different way. Horner's method must be applied in this case, for example, to calculate the group:

$$\frac{T + \Delta t}{\Delta t}$$

To illustrate this case, let us take the data above and assume a production time of $T = 48$ h (standard case $T \sim \Delta t$) at a constant flow rate.

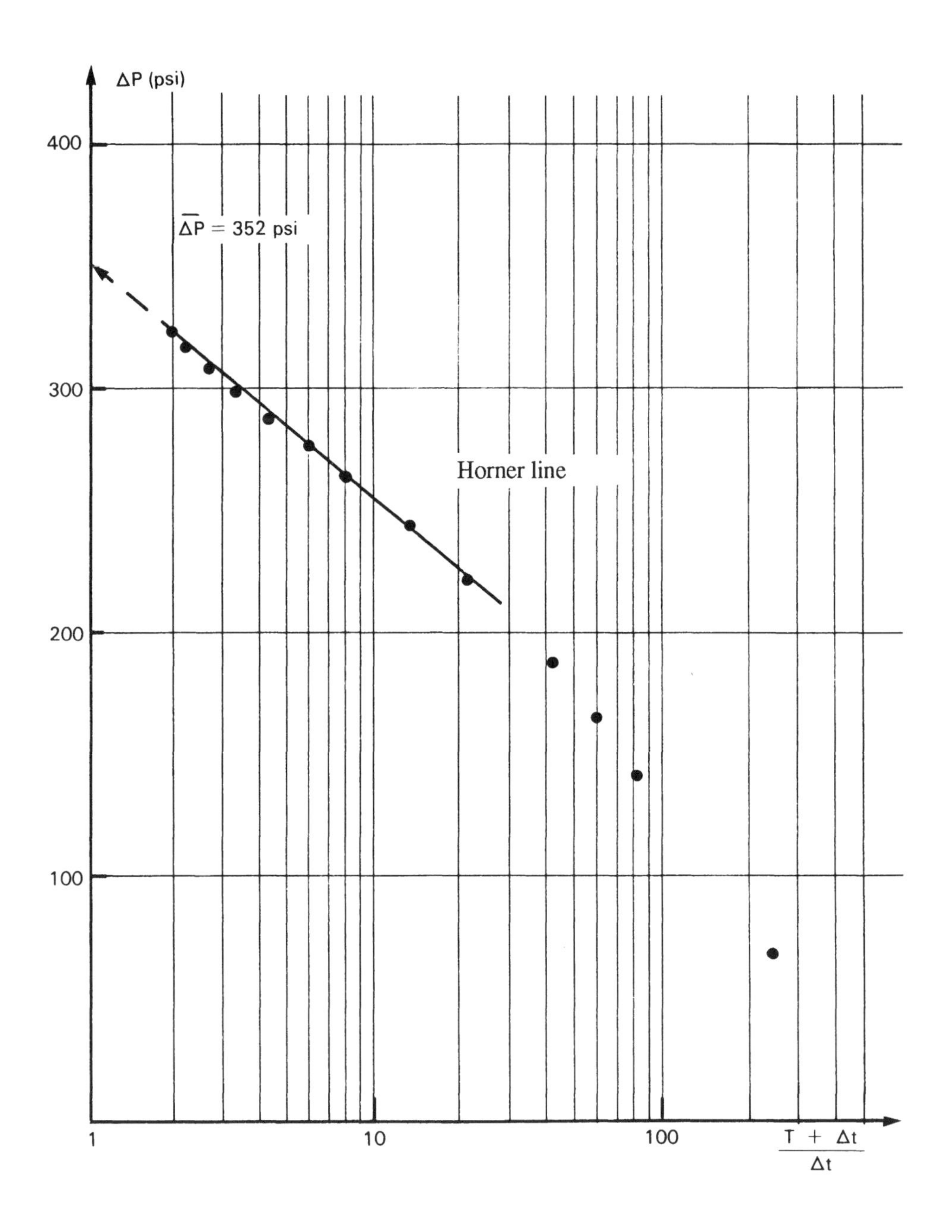
ΔP (psi)
400
300
200
100
$\overline{\Delta P}$ = 352 psi
Horner line
1
10
100
$\frac{T + \Delta t}{\Delta t}$

Fig. 5.25

The result (Fig. 5.25) is the following:

$$\overline{\Delta P} = 352 \text{ psi} \qquad \text{or} \qquad \Delta P_s = 24.3 \text{ bar}$$

Chapter 6

MULTIPHASE FLOW

6.1 GENERAL INTRODUCTION

Flows in hydrocarbon reservoirs are not generally one-phase flows. Yet reservoirs do exist, with oil (above the bubble point) or with gas — dry or wet — without an active aquifer. Other situations are marked by a simultaneous flow of two or even three phases, gas, oil and water, at least in some zones of the reservoir.

What are the different cases of multiphase fluid flow?

These are essentially the following (Figs 6.1 and 6.2):

Oil Reservoir

(a) Change in location of the existing G/O and O/W interfaces (two phases).
(b) Liberation of the gas dissolved in the oil ($P < P_b$): two phases or three phases near the O/W interface.
(c) Injection of gas (two phases) or of water: two phases depending on whether the water is injected in the aquifer or in undersaturated oil, or three phases if the water is injected in saturated oil.

Gas Reservoir

(a) Change in location of existing G/W or G/O interfaces (thin oil ring): two phases.
(b) Condensate gas under retrograde dew-point pressure: two phases, or three phases near a body of water.

Multiphase flow can thus be seen to represent a fairly widespread occurrence. The laws governing these flows are not always satisfactory, particularly for three movable phases.

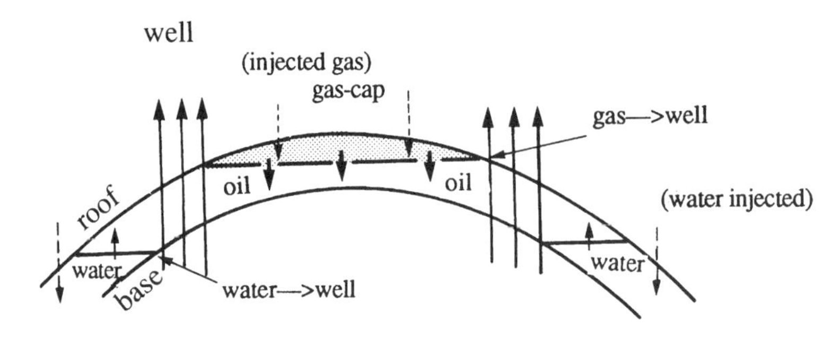

Fig. 6.1

Multiphase flow zones in natural depletion and artificial recovery.

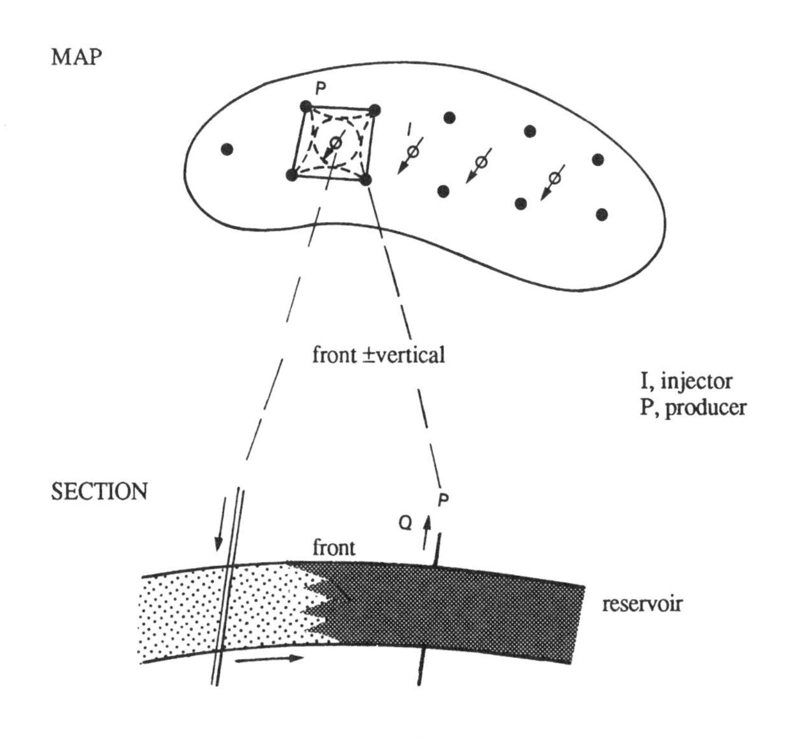

Fig. 6.2

Multiphase flow zones in artificial recovery.

However, the laws governing **two-phase flows** are fairly well represented on the "microscopic" scale (at the pore level), but their integration on the "macroscopic" scale is impossible, because the geometric description of the porous medium is too complex. This means that the laws of two-phase flow are based on experimental considerations of a porous block (a centimeter or decimeter scale) but nevertheless rely on reasoning at the pore level.

The forces acting are those already considered for one-phase flow: viscosity forces (viscous friction) and gravity forces. In addition, however, **capillarity forces** representative of the equilibrium between the fluids and the pore walls are involved, and the concept of fluid **saturation** assumes special importance in this case.

To go from the experimental scale of the block to that of the reservoir, it is naturally essential to consider reservoir geometry and internal architecture, which are often very complex. As a rule, **reservoir simulation models** are used to solve this problem, but a number of simplified cases can nevertheless be dealt with in a simple analytical form, with respect to two-phase flows.

This is described in this chapter, following a definition of the concept of relative permeability which is essential for the analysis of these flows.

6.2 REVIEW OF CAPILLARY MECHANISMS, CONCEPT OF RELATIVE PERMEABILITY

Section 2.3 of Chapter 2 discussed the influence of capillary and gravitational phenomena on the migration of oil and gas, and accordingly on the spatial distribution of saturations.

We shall now discuss the effects of capillary mechanisms during the movement of two fluids in the porous medium.

6.2.1 Capillary Doublet, Development of Drops and Jamin Effect

Let us take a simplified example of two-phase oil and water flow.

Let us consider part of a porous medium with two pores, one narrow and the second wider, in parallel (Figs 6.3a and 6.3b), and let us call this combination a capillary doublet. We shall now examine the general case of a water-wet rock.

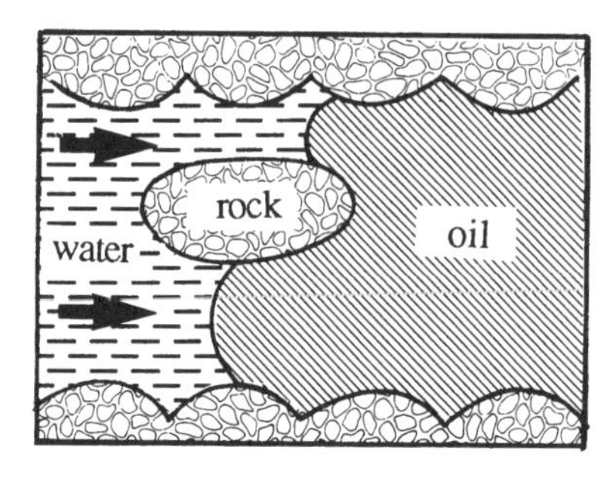

Fig. 6.3a

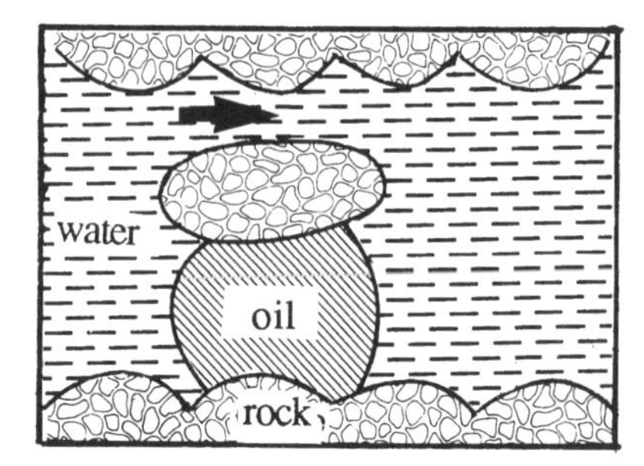

Fig. 6.3b

The (slowly) injected water displaces the oil contained in the pores more easily if the pores are narrower, because the capillarity effect is added to the external forces. The order of magnitude of the speed is 1 ft/day.

The oil/water interface (meniscus) thus advances faster in the narrower pore and reaches the narrow exit of the doublet, which is thus bathed in water at entrance and exit. A second interface thus appears in the large pore, forming a drop of oil which, during its movement, may ultimately be blocked in the pore network, as we shall show concerning the Jamin effect.

Jamin Effect (Fig. 6.3c)

This drop of oil will sooner or later encounter a constriction, and, depending on the pressure gradient associated with the flow, will block the threshold or pass through the constriction.

Since the pressure is the same in the drop along a horizontal plane, we can write:

$$P_{A'} = P_{B'}$$

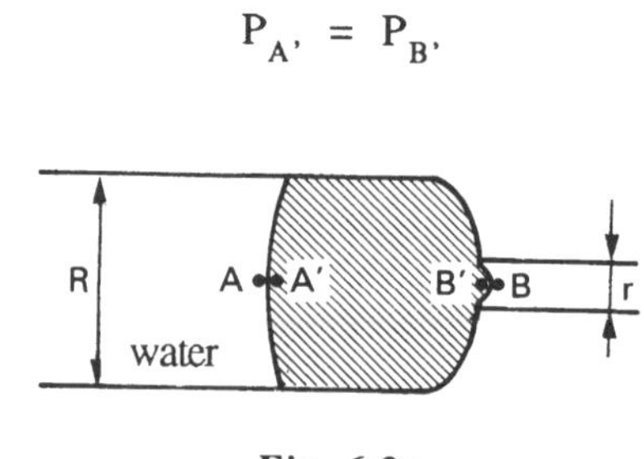

Fig. 6.3c

$$P_{A'} - P_A = \frac{2\ \sigma \cos\ \theta}{R} \qquad P_{B'} - P_B = \frac{2\ \sigma \cos\ \theta}{r}$$

Hence:

$$P_A - P_B = 2\ \sigma \left(\frac{1}{r} - \frac{1}{R}\right) \cos\ \theta$$

This is the pressure difference required for the oil drop to pass the threshold.

For example, if $r = 0.5\ \mu$, $R = 5\ \mu$ and $T\ .\cos\ \theta = 30$ dyn/cm (10^{-3} N/m):

$$P_A - P_B = 1.08\ .\ 10^5\ \text{Pa} \approx 1\ \text{bar}\ (15\ \text{psi})$$

It can be seen that the pressure gradients required to "unblock" certain thresholds may be very high and that, if they are insufficient, the drops of oil remain trapped (Fig. 6.3d). This often occurs, because the pressure gradients associated with production are generally in the range of 1 bar/10 m, and more rarely 1 bar/m. The immobilized drops are called "residual oil" and the corresponding **residual oil saturation** is denoted S_{or}. The average values are often:

$$20\% < S_{or} < 40\%$$

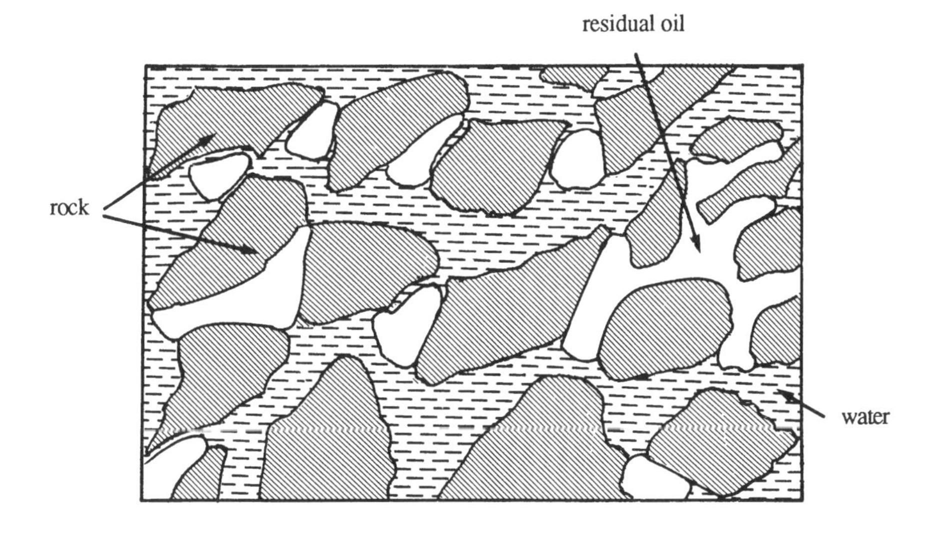

Fig. 6.3d

Residual drops. Drops of oil trapped behind the water front.

Thus only part of the oil (or gas) is displaced, unless both fluids are miscible.

6.2.2 Concept of Relative Permeability

If two fluids flow simultaneously in a sample, this can be observed to reduce the permeability of each fluid.

We can perform the following experiment in the laboratory: two immiscible fluids are injected by pumps into a cylindrical porous sample with cross-section A and length L.

Let Q_1 and Q_2 be the flow rates of these fluids, and ΔP_1 and ΔP_2 the pressure differences corresponding to the length L for each fluid (see diagram below).

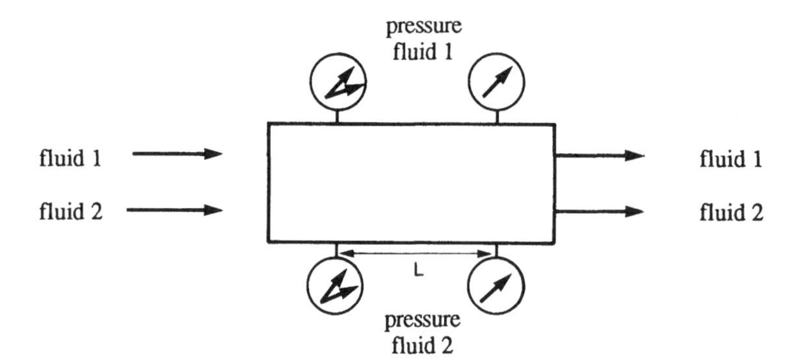

The experiment shows that a Darcy type of equation can be written:

$$Q_1 = \frac{k_1}{\mu_1} A \frac{\Delta P_1}{L} \qquad \text{and} \qquad Q_2 = \frac{k_2}{\mu_2} A \frac{\Delta P_2}{L}$$

where k_1 and k_2 are called the **effective** permeabilities to fluids 1 and 2. They depend on the specific permeability of the medium and on the saturations.

If the injection flow rate of one of the two fluids is changed, changing the average saturation of each of the two fluids (obtained after an injection time interval leading to equilibrium), the experiment also shows that the coefficients k_1 and k_2 are modified, and increase directly with the saturations 1 and 2.

The **relative permeabilities** k_{r1} and k_{r2} are generally introduced, which depend only on the saturation:

$$k_{ri} = \frac{k_i}{k}$$

where k = absolute permeability (one-phase flow).

The following is observed in practice:

Water: $k_{rw} = \frac{k_w}{k}$ Oil: $k_{ro} = \frac{k_o}{k}$ Gas: $k_{rg} = \frac{k_g}{k}$

The relative permeabilities range between 0 and 1.

6.2.3 Variation in Relative Permeability as a Function of Saturation

6.2.3.1 Oil/Water (or Gas/Water) Pair

Let us consider a sample saturated with oil containing pore water ($S_w = S_{wi}$). Let us slowly inject water (wetting fluid). This type of displacement is called "displacement by imbibition".

The following can be observed (Fig. 6.4a):

(a) The permeability to oil decreases steadily. It is not substantially affected by the presence of water, whereas permeability to water is more affected by the presence of oil. This can initially be explained by observing that the water coats the pore walls and fills the small pores, allowing the oil to pass mainly through the centers of the large pores.

(b) The oil stops circulating at a minimum saturation which is the residual oil saturation S_{or}.

(c) The water only circulates above the pore saturation S_{wi}. The per-

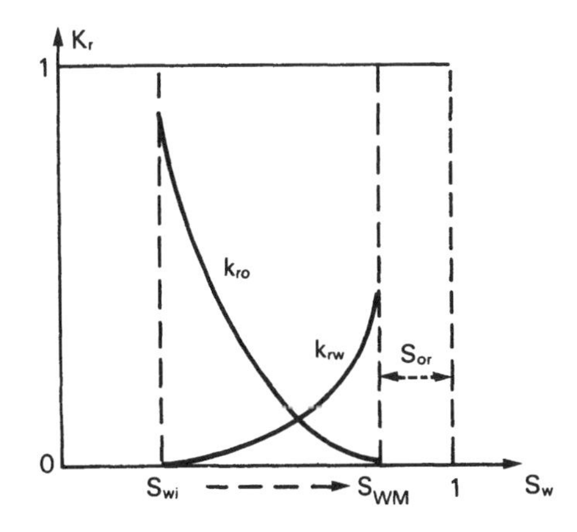

Fig. 6.4a

meability to water then rises steadily up to the maximum water saturation: $S_{WM} = 1 - S_{or}$.

(d) $k_{ro} + k_{rw} < 1$, which shows that both fluids mutually hinder each other during their simultaneous movement: the **total flow capacity is reduced**.

Note:

Since the oil no longer flows at $S_{WM} = 1 - S_{or}$, we can consider that the state reached is that of an oil reservoir swept by water, and we can now examine the displacement of water by oil, called "drainage"[1]. The water saturation varies between S_{WM} and S_{wi}, and the relative permeability curves are not exactly the same as the previous ones, because of hysteresis, especially for k_{ro}.

The experiment shows that the k_r are virtually independent of the flow rate, for the flow rate range corresponding to reservoir production.

6.2.3.2 Oil/Gas Pair

The observations are similar, but with a difference concerning the gas phase. A minimum gas saturation is necessary for the gas to flow, called the **critical gas saturation S_{gc}**. In fact, if we decompress a pressurized sample, so that the pressure falls below the bubble-point pressure, gas bubbles appear. Yet these do not move at the same time as the oil, towards a side of the sample subjected to a lower pressure, until there is sufficient gas saturation to form a continuous gas phase in the porous medium (if not, the gas bubbles will remain blocked by the Jamin effect). The values of S_{gc} are generally a few per cent.

The relative permeability curves are represented by the example shown in Fig. 6.4b.

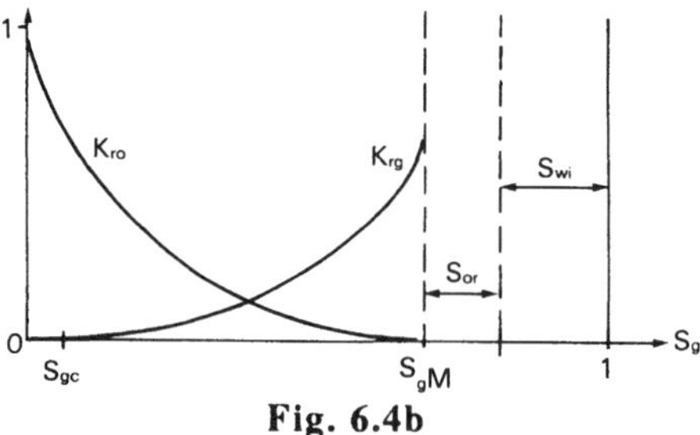

Fig. 6.4b

1. In oil laboratory terms, displacement of a wetting fluid by a non-wetting fluid. Imbibition is the reverse mechanism.

6.2.4 Comments on Relative Permeabilities

The concept of relative permeability is intended to extend the concept of permeability to two-phase flow in a simple way. Thus Darcy's Law can be used directly. A simple equation exists for each of the fluids:

$$Q_1 = k \cdot \frac{k_{r1}(S_1)}{\mu_1} \cdot A \cdot \frac{dP_1}{dl}$$

This concept in fact implies the idea of average saturation at a centimetric or even metric scale of the fluid flow region. This average saturation varies with time. The relative permeabilities provide only a general description of the process, disregarding the detailed physical mechanisms which govern them. Yet they do allow a quantitative formulation of the two-phase flows. Note that the present trend is to use them as matching parameters in mathematical models, because these curves, obtained by a few measurements, may not actually represent the medium.

Let us return to practical applications. **Which curve should be selected,** drainage or imbibition?

(a) If oil is displaced by water (O/W interface and/or water injection), the "imbibition" curve must be used (at least for a water-wet medium). This also applies to a gas reservoir with an active aquifer.
(b) By contrast, if oil is displaced by gas (G/O interface and/or gas injection), since the gas is non-wetting in comparison with the oil, the "drainage" curve must be selected.

6.2.5 Determination of Relative Permeabilities

If possible, it is always necessary to perform laboratory measurements. Sometimes, however, coring is difficult if not impossible (unconsolidated sands, drilling problems, cost, etc.). The modern trend is to take measurements in reservoir conditions (P, T) on large samples (whole cores). Empirical equations are used in the absence of cores.

6.2.5.1 Laboratory Measurements

Several laboratory measurement methods are available. The two main methods are the following:

A. Displacement of One Fluid by Another, "Unsteady-State", WJBN Method

The water-cut f_w formula is used (Section 6.3.2) to calculate k_{r1} and k_{r2} by injecting fluid 1 into a sample saturated with fluid 2 (see diagram below).

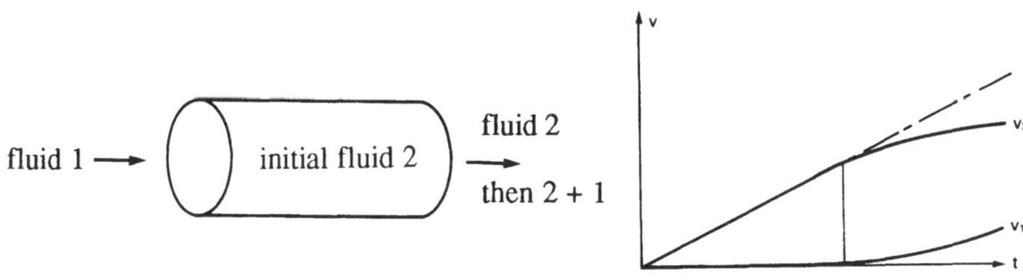

v_1, v_2 = cumulative volumes as a function of time of fluids 1 and 2

The procedure consists in measuring the volumes of fluids 2 and 1 leaving the sample as a function of time. Hence these are essentially volumetric measurements. The calculations are carried out on a microcomputer. For example, the *BEICIP* (*IFP* subsidiary) apparatus is designed for measurements up to 480 bar and 130°C (1990).

B. "Static" Method for Measuring Relative Permeability, "Steady-State" Method

This method is much longer, but is generally considered to be more representative for heterogeneous media.

The **two** fluids are injected into the sample saturated in fluid 1 or 2. They flow through the core until the ratio of the fluids produced is equal to the ratio of the fluids injected. At this time, the system is considered as in steady state flow, and the saturations are considered stable. The ratio of injected flow rates is then changed to cover the range of saturations.

The fluid saturations are determined in one of the following ways:

(a) By volumetric balance of the quantities of fluids injected and produced.
(b) With electrodes giving the resistivity of the sample (if the injected fluid is salt water).
(c) By X-ray absorption.

6.2.5.2 Empirical Equations

Empirical equations have been developed, especially for **unconsolidated sands** which are often difficult, if not impossible to core.

One example for the "gas and oil" pair:

$$k_{ro} = S^{*3} \quad \text{and} \quad k_{rg} = (1 - S^*)^3$$

with

$$S^* = \frac{S_o - S_{or}}{1 - S_{wi} - S_{or}} \quad \text{"reduced" saturation}$$

and for the "oil and water" pair:

$$k_{rw} = S^{+3} \quad \text{and} \quad k_{ro} = (1 - S^+)^3$$

with

$$S+ = \frac{S_w - S_{wi}}{1 - S_{wi} - S_{or}}$$

Important Remark:

The permeability k obtained from **production tests** is actually an **effective** permeability k_o or k_g, generally corresponding to a saturation $S_w = S_{wi}$.

6.2.6 Capillary Imbibition

Let us now examine the influence of capillary mechanisms related to very heterogeneous reservoirs.

Capillary imbibition is the spontaneous displacement of a non-wetting fluid by a wetting fluid. **The typical case of imbibition is the displacement of oil by water**, and it represents a favorable mechanism for oil recovery. It generally corresponds to the rebalancing of the two phases on either side of the borderline zone separating two media, while the initial capillary equilibrium is broken. This applies in particular to two media of highly contrasting permeabilities.

Let us consider a reservoir in which two different porous media coexist. For example:

(a) A **fractured** reservoir (a).
(b) A **stratified** reservoir (b).
(c) A reservoir with more or less compact **nodules** (c) (Fig. 6.5a).

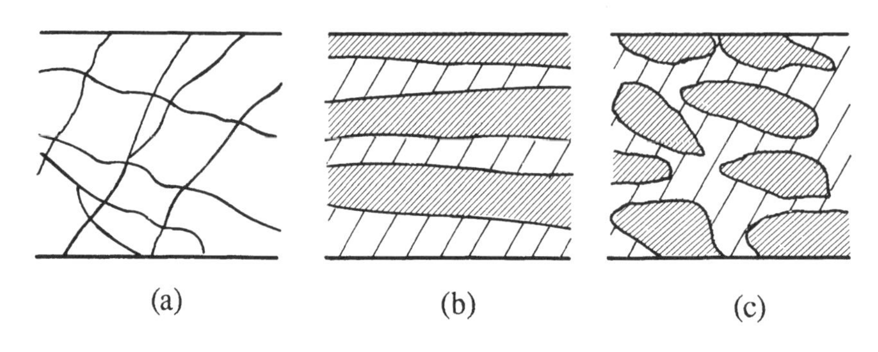

Fig. 6.5a

Let us denote the permeable medium by P and the second, much less permeable medium by I. Zone P has more large pores than zone I. Consequently, for the same saturation S_1, we shall have a capillary pressure $P_cI > P_cP$ for a given S_1.

The injected wetting fluid 1, such as water, occupies the large pores more rapidly (higher permeability), and accordingly we obtain $S_{wP} > S_{wI}$ at a given time (Fig. 6.5b).

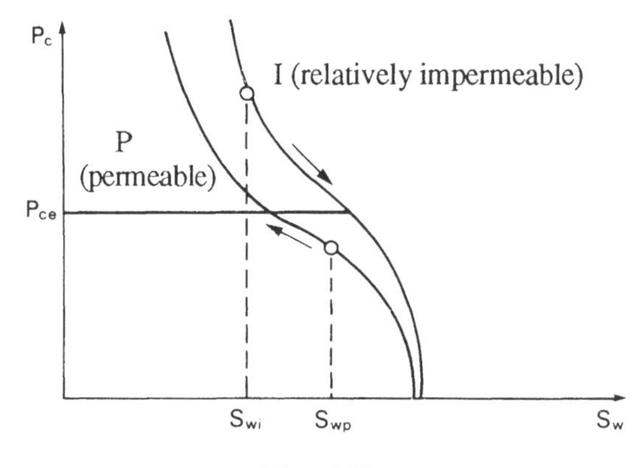

Fig. 6.5b

In these zones, the saturation with displacing fluid S_w thus reaches a value S_{wP}, whereas in the less permeable zones, it only has the value S_{wI}. In fact, the capillary pressures in the less permeable medium, for the same saturation, are higher than those in the more permeable medium, because the pores there are smaller. Yet an equilibrium of the capillary pressures must be established between the two media present, at the **contact of the two zones**. As shown by the figure, this can occur only if the water saturation in zone I rises, while it decreases in zone P until it corresponds to identical pressures P_{ce}.

Thus the low permeability medium tends to be imbibed with wetting fluid 1 (water), at the expense of the high permeability medium, and tends to flush the non-wetting fluid 2 (oil) into the high permeability medium.

This mechanism tends to increase the overall oil (fluid 2) recovery rate.

In practice, effective imbibition is strongly dependent on the **speed** of displacement. In fact, if displacement is very rapid, the less permeable zones can produce oil only through permeable zones already swept by water, where the relative permeability to oil is very low. By contrast, if displacement is slow, the less permeable zones will have the time to produce more oil as the displacement front moves forward.

6.3 THEORY OF FRONTAL DISPLACEMENT

6.3.1 Front Concept

The flow of two immiscible fluids in a **large medium** can be investigated fairly simply if it is **unidirectional**, in other words if the different values such as pressures, saturations, fluid speeds, etc. vary only in a single space direction corresponding to the movement direction. For example, this displacement could correspond to the movement of a G/O or O/W interface during natural depletion, or may occur between two lines of production and injection wells, fairly far from the wells for a thin reservoir.

Let us consider the displacement of a fluid 2 (e.g. oil) by a fluid 1 (e.g. water). The profile of saturations with fluid 1, at a given time, looks like Fig. 6.6 as a function of the displacement direction x. From upstream to downstream, four zones can be distinguished:

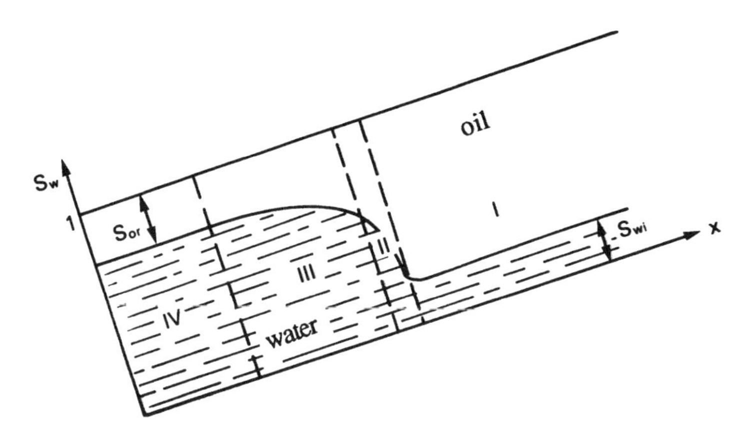

Fig. 6.6

(a) Zone I, not reached by the displacing fluid: only the oil moves.
(b) Zone II, S_w rises sharply: this is the "front".
(c) Zone III, where S_w varies gradually, behind the front.
(d) Zone IV, flooded by water, and where saturation with residual oil S_{or} remains: only the water moves.

This profile moves as a function of time (increasing x), while being deformed, towards the production well.

We shall now make the calculation of the arrival of the front at the producing wells, called "breakthrough", and the rising percentage of water in production. This will enable us to answer the question: what happens to the oil flow rate as a function of time?

6.3.2 Unidirectional Displacement (Capillarity Disregarded) Buckley-Leverett Theory

It is fairly simple to calculate the advance of the front and evaluate the water-cut, for example in a section in which **steady-state flow prevails, by disregarding** the capillary and gravitational forces in the case of unidirectional movement. This applies for example to **horizontal flow** in which the viscosity forces are preponderant (fairly high speed).

6.3.2.1 Water Cut f_w

Let us resume Darcy's equations for water and oil and define the **water cut** f_w:

$$f_w = \frac{q_w}{q_w + q_o} = \frac{\dfrac{k_{rw}}{\mu_w}}{\dfrac{k_{rw}}{\mu_w} + \dfrac{k_{ro}}{\mu_o}}$$

where

$$f_w = \frac{1}{1 + \dfrac{k_{ro}}{k_{rw}} \cdot \dfrac{\mu_w}{\mu_o}}$$

f_w is hence exclusively a function of the saturation S_w through the relative permeabilities. μ_o and μ_w have a given value for the average pressure considered.

A. First Remark

q_w, q_o and f_w are expressed in **reservoir conditions.**

In standard conditions:

$$f_{w\,(C\,.\,S)} = \frac{\dfrac{q_w}{B_w}}{\left(\dfrac{q_w}{B_w}\right) + \left(\dfrac{q_o}{B_o}\right)}$$

with

$$\frac{q_w}{B_w} = Q_w \qquad \text{and} \qquad \frac{q_o}{B_o} = Q_o$$

hence:

$$WOR = \frac{f_w}{1 - f_w} \cdot \frac{B_o}{B_w}$$

Similarly for gas:

$$GOR = R_s + \frac{f_g}{1 - f_g} \cdot \frac{B_o}{B_g}$$

B. Second Remark

By taking **gravity forces** into account, the following is obtained for a **non**-horizontal flow:

$$f_w = \frac{\dfrac{\mu_o}{k_{ro}} - \dfrac{A\,k}{Q_T}\left(\rho_w - \rho_o\right) g \sin\alpha}{\dfrac{\mu_o}{k_{ro}} + \dfrac{\mu_w}{k_{rw}}}$$

or

$$f_w = \frac{1 - 4.9 \,.\, 10^{-4} \,.\, \dfrac{A\,k}{Q_T} \,.\, \dfrac{k_{ro}}{\mu_o} \,.\, \left(\rho_w - \rho_o\right) \sin\alpha}{1 + \dfrac{k_{ro}}{k_{rw}} \,.\, \dfrac{\mu_w}{\mu_o}}$$

in US practical units (square feet, millidarcys, pounds per cubic foot, barrels per day, centipoises), with:

Q_T = $q_w + q_o$ = total flow rate (assumed constant),
A = fluid passage area,
ρ_w, ρ_o = densities,
α = angle of flow to the horizontal.

f_w is accordingly a function of S_w, but also of the flow rate, of k_o, and of the slope of the flow. Note that **gravity has the effect of decreasing f_w**. This effect is at a maximum for interface movement because $\sin\alpha = 1$.

A curve f_w generally has the following shape:

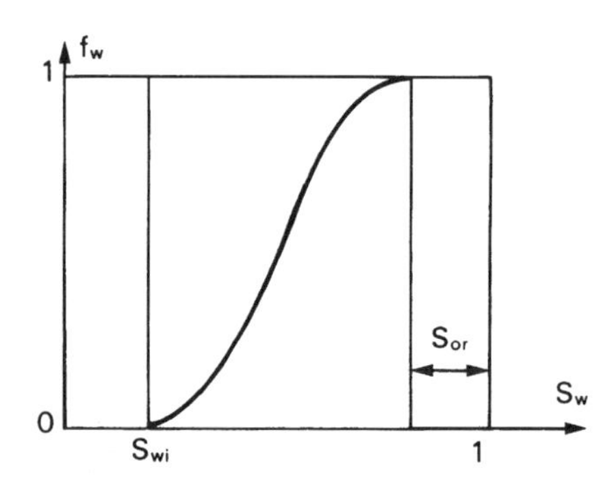

6.3.2.2 Calculation of the Speed of a Section S_w

Let us now calculate the **speed** of a section of fluid of saturation S_w. To do this, let us apply the law of conservation of mass to the water contained in a volume with cross-section A along a section Δx during time Δt. It is assumed that the water is incompressible in the conditions concerned, hence the variation in the volume of water during Δt = the variation in flow rate ΔQ_w:

$$\frac{(\Delta x \, . \, A \, . \, \phi) \, \Delta S_w}{\Delta t} = Q_T \, . \, \Delta f_w$$

Hence:

$$\frac{\Delta x}{\Delta t} = V(S_w) = \frac{Q_T}{A \phi} \, . \, \frac{\Delta f_w}{\Delta S_w}$$

A
Sw
Δx

with

$$V = \left(\frac{dx}{dt}\right)_{S_w}$$

$\frac{df_w}{dS_w}$ = derivative of the function f_w for the value S_w,

ϕ = porosity.

The speed of a "section of given saturation" is **constant** because it depends only on the saturation concerned. The equation is integrated as follows:

$$x\left(S_w\right) - x_i\left(S_w\right) = \frac{Q_T}{\phi A} \, . \, \frac{df_w}{dS_w} \, . \, \left(t - t_i\right) \qquad \text{in US practical units}$$

All the S_w (x,t) profiles are **affine** (assuming that the initial saturation is uniform) and lengthen in proportion to time. The calculation of $f(S_w)$, and hence of df_w/dS_w using the relative permeability curves, yields profiles with the following shape.

Note that part of the curves obtained in Fig. 6.7a has no physical significance. In fact, for the same position x, three values of the saturation are obtained. It is accordingly assumed (Buckley-Leverett hypothesis) that the movement is frontal and exhibits a **saturation discontinuity**: the front is defined as the boundary of the zone into which the injected water has not penetrated. All the injected water is hence located behind the front. This condition is expressed by equating the areas ABCD and DEGBC:

Area ABCD	=	Area DEGBC
(balance of water located behind the front)		(balance of injected water)

What is the value of this saturation discontinuity (S_{wF} - S_{wi})? The Welge method is used.

Let:

$$x_F \, . \, (S_{wF} - S_{wi}) = \int_{S_{wi}}^{S_{wF}} x \, . \, dS_w$$

Since:

$$x_F = V_{(S_{wF})} \, . \, t = 5.615 \, \frac{Q_T}{\phi \, A} \, \frac{df_w}{dS_{w(S_{wF})}} \, . \, t$$

(in US practical units).

We finally obtain:

$$\frac{Q_T}{\phi \, A} \, \frac{df_w}{dS_{w(S_{wF})}} \, . \, (S_{wF} - S_{wi}) \, . \, t = \frac{Q_T}{\phi \, A} \left[f_{w(S_{wF})} - f_{w(S_{wi})} \right] . \, t$$

or

$$\frac{f_{w(S_{wF})} - f_{w(S_{wi})}}{S_{wF} - S_{wi}} = \left(\frac{df_w}{dS_w} \right)_{S_{wF}}$$

This leads to a simple construction to obtain the water saturation behind the front (Fig. 7.7b). Starting with the initial point (assuming that the uniform initial saturation is equal to S_{wi}, though it could have a different value), the **tangent** to the curve $f(S_w)$ is plotted. The contact point gives the desired value S_{wF}, because the slope of the **Welge tangent** at this point is equal to the slope of the chord starting from the initial point. The front is located by the discontinuity of the saturations, with Swi changing abruptly to S_{wF}.

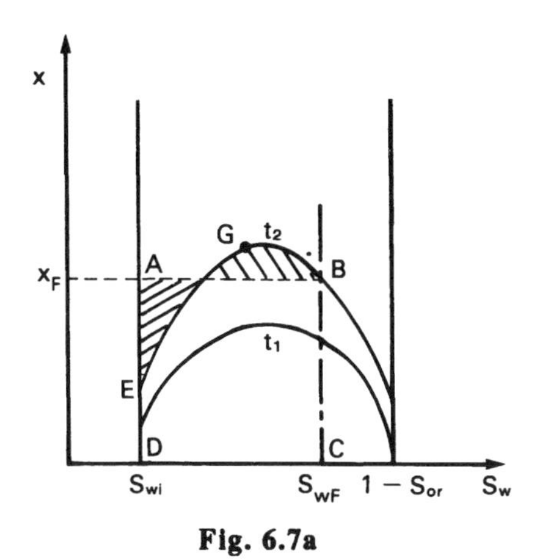

Fig. 6.7a

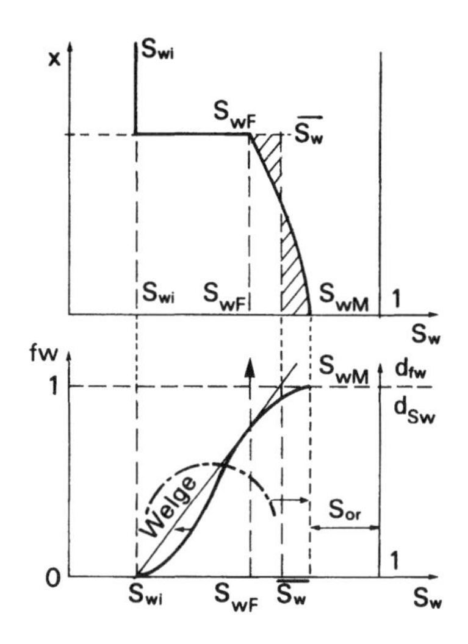

S_{wi} = Pore water saturation (oil zone).
S_{wF} = Water saturation at front.
S_{wM} = Maximum water saturation = 1 − S_{or}.
S_{wm} = Average water saturation behind the front.

Fig. 6.7b

The speed of the front is thus obtained by calculating df_w/dS_w from f_w (Fig. 6.7b). The calculation of the slope of the Welge tangent gives:

$$\left(\frac{df_w}{dS_w}\right)_{S_{wF}} = \frac{1}{S_{wm} - S_{wi}}$$

It can be shown that S_{wm} is the value of the average saturation behind the front.

This leads to the second application of the Welge tangent: the value of the average saturation behind the front S_{wm} is obtained from the intersection of the tangent with f_w = 1. Hence the equation:

$$V_F = \frac{Q_T}{A\ \phi\ \left(S_{wm} - S_{wi}\right)}$$

where V_F = speed at the front.

This equation was foreseeable because:

$$V_F = \frac{dx}{dt},$$

and

$$Q_T \,.\, dt = A\,dx \,.\, \phi\,(S_{wm} - S_{wi})$$

The injected water volume Q_T . dt has altered the water saturation in the pore volume V_p = A dx ϕ, from the initial value S_{wi} to the average value behind the front S_{wm}.

To conclude, the unidirectional displacement of oil by water under the effect of viscosity and gravitational forces can be summarized as follows: since the water inlet flow rate is constant, a saturation front appears ahead of which only oil flows, and behind which water and oil flow simultaneously. The speed of a section of given saturation is constant (depending only on S_w), as is the speed of the front in particular. The saturation profiles over time are thus deduced from one another by an affinity. Since the oil and water flow rates in a section of given saturation depend only on this saturation (through the relative permeability curves), we can calculate the movement of the front as a function of time and consequently deduce the percentage of oil and water at the well after breakthrough, in other words after the arrival of the displacing fluid.

6.3.3 Effect of Capillarity Forces and Influence of Flow Rate

During the displacement of oil by water, the porous medium is the location of sudden variations in saturation, especially at the front, i.e. at the saturation "discontinuity" S_{wF}/S_{wi} (in actual conditions, there is no discontinuity but rather a sharp variation).

The approximation described above, called the Buckley-Leverett approximation, is inadequate whenever these sudden variations in capillary pressure, and hence in saturation (Fig. 6.8) must be considered.

At the front, the capillary forces have the effect of replacing the theoretical sharp profile by a more rounded profile. The higher the total flow rate, the closer the profile obtained by the frontal movement theory approaches the true profile, which was implied by the analysis of the forces in action. Yet it must be noted that, in general, the speeds in reservoirs are fairly low. **In practice, and as a first approximation**, the movement of the front and the water-cut at the well as a function of time are calculated without taking account of capillarity forces.

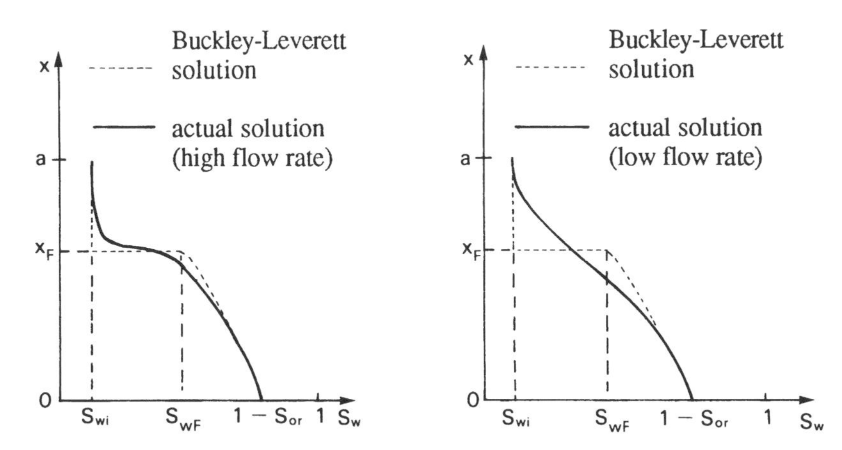

Fig. 6.8 Influence of flow rate on shape of front.

6.3.4 Practical Application

We shall calculate the time taken by the front to reach a producing well, and the change in the percentage of displacing fluid in the production of the well as a function of time. Let us take the following example.

A monoclinal reservoir closed on faults is described schematically as follows: a layer of sand 16 ft (5 m) thick, dip 2 degrees, bounded on the sides by two faults parallel to the dip line and 3300 ft (1000 m) apart.

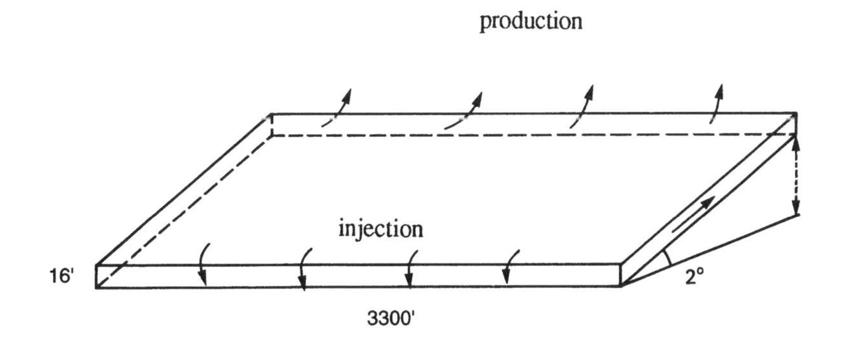

Water is injected through a line of injections wells at a distance of 1650 ft (500 m) from a line of producing wells, at a constant flow rate: Q_T = 1250 bpd (200 m^3/d) (reservoir conditions).

(a) Sand permeability: 500 mD.
(b) Sand porosity: 20%.
(c) Oil density in reservoir conditions: ρ_o = 0.80 g/cm^3.
(d) Water density in reservoir conditions: ρ_w = 1.05 g/cm^3.
(e) Water viscosity in reservoir conditions: 0·5 cP = μw.
(f) Oil viscosity in reservoir conditions: 5 cP = μo.

The relative permeabilities to water and to oil are given in the following table (with S_{wi} = 15% and S_{or} = 20%).

S_w	k_{ro}	k_{rw}	k_{ro}/k_{rw}
0.15	0.96	0.000	∞
0.20	0.69	0.005	138.00
0.30	0.49	0.030	16.33
0.40	0.33	0.070	4.71
0.50	0.20	0.130	1.54
0.60	0.11	0.190	0.58
0.70	0.05	0.260	0.19
0.80	0.00	0.350	0.00

It is first necessary to calculate the values of the function $f_w(S_w)$. With these numerical data, we obtain:

$$f_w = \frac{1 - 0.018\, k_{ro}}{1 + 0.1\, k_{ro}/\, k_{rw}} \approx \frac{1}{1 + 0.1\, k_{ro}/\, k_{rw}}$$

This leads to the following table of results.

S_w	0.15	0.20	0.30	0.40	0.50	0.60	0.70	0.80
f_w	0.00	0.07	0.38	0.68	0.87	0.95	0.98	1.00

Having plotted the curve f_w, the construction of the Welge tangent gives us (Fig. 6.9):

$$S_{wf} = 0.42 \qquad f_{wf} = 0.70 \qquad S_{wm} = 0.52$$

hence the speed of the front:

$$V_f = \frac{5.615\ Q_T}{A\ \phi\ \left(S_{wm} - S_{wi}\right)}$$

$$V_f = \frac{5.615\ .\ 1250}{16\ .\ 3300\ .\ 0.2\ \left(0.52 - 0.15\right)} = 1.8 \text{ feet/day or: } 0.54 \text{ m/day}$$

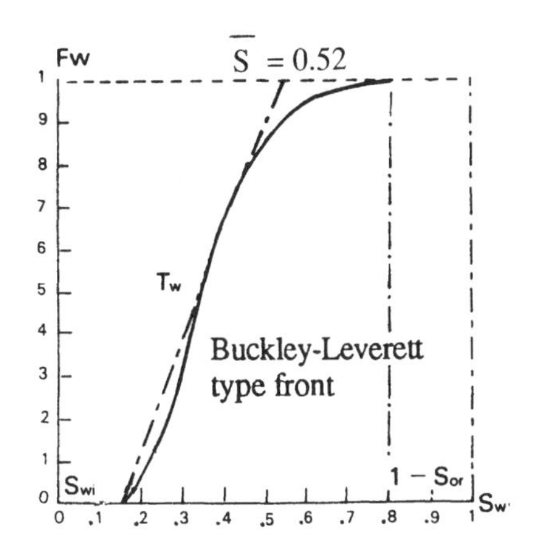

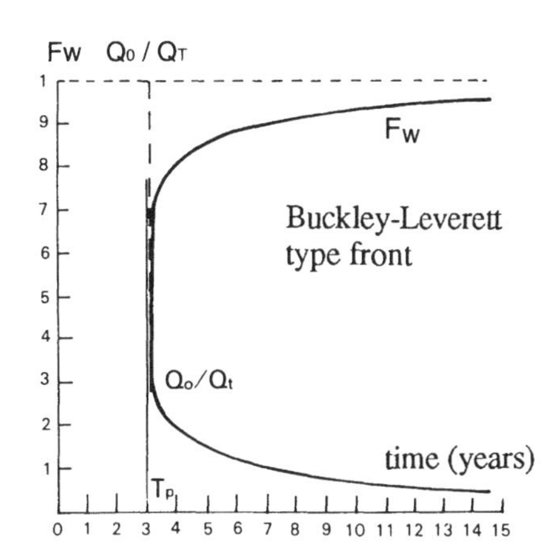

Fig. 6.9

Since the distance between the injection and producing wells is 1650 ft (500 m), the water breakthrough time is:

$$T_{bt} = \frac{1650}{1.8} = 917 \text{ days}$$

After the breakthrough, the speed of water at saturation S_w is:

$$V\left(S_w\right) = \frac{Q_T}{A\ \phi} \cdot \frac{\left(df_w\right)}{\left(dS_w\right)_{S_w}} = \frac{x}{t}$$

x = 1650 ft (500 m)

and

$$5.615 \frac{Q_T}{A\,\phi} = 0.66 \text{ ft/day or: } 0.2 \text{ m/day}$$

hence:

$$t \text{ (days)} \approx \frac{2500}{(d_f/dS)_{S_w}}$$

Thus, for each value of $(df/dS)_{Sw}$, a value of f_w and a time t also correspond. This gives us an equation f_w = f (time).

It suffices to calculate a few values of df/dS. This is done graphically from the curve f_w, by plotting a number of tangents and by calculating their slope for $S_w > S_{wf}$.

For these fixed points, the corresponding times are then calculated and f_w measured on the graph.

S_w	42%	50%	60%
f'_w	2.38	1.32	0.52
f_w	0.70	0.87	0.95
t (days)	917	1894	4808

This gives the curve $f_w(t)$ in Fig. 6.9, and that of $\frac{Q_o}{Q_T}(t)$ which is easily deduced:

$$\frac{Q_o}{Q_T} = 1 - f_w$$

6.4 TWO- AND THREE-DIMENSIONAL TWO-PHASE FLOW

While interface movement can very approximately be analyzed by a calculation of the unidirectional frontal movement, this does not apply to most other cases of two-phase flow. Figure 6.2 in Section 1 of this chapter showed us the case of a cylindrical radial type of displacement around an injection well with secondary recovery. The analytical expression of the flows now becomes much more complex. As a rule, the geometry and internal architecture of the reservoirs require dealing with these problems by simulation on grid pattern numerical **models**. However, this calculation process is more expensive and, to begin with, an attempt is sometimes made to reduce the type of flow to a case of simple geometry and to treat it, for example, by the frontal movement theory. This is valid in particular for a **thin inclined layer that can be treated like a monocline**.

6.4.1 Encroachment, Instability Mechanisms, Definition of the Mobility Ratio

If the reservoir is thicker, the vertical component of the velocities cannot be ignored, and the analysis of the forces acting in the porous medium shows that the interfaces and "fronts" are generally distorted (encroachment).

These encroachments occur on the scale of the front, called the **tongue** phenomenon, but also on a smaller scale (meter or decameter), called **fingering**. Near the producing well, the mechanism is called **coning**.

These encroachments are governed by conditions of stability or instability. A **movement** is said to be **stable** if a small change in the initial conditions of the movement (initial coordinates, initial speeds) causes a variation in the movement that remains small over time, of the same order as the initial disturbance. A movement is said to be unstable in the opposite case.

The many experiments conducted in the oil and gas industry show that these instabilities depend in particular on the mobility ratio.

Mobility ratio is the following equation:

$$M = \frac{k_{rD}^*}{\mu_D} \Big/ \frac{k_{ro}^*}{\mu_o}$$

Since the mobility of the displaced oil or gas depends[1] on k_{ro}/μ_o (for oil) and that of the **displacing** fluid on k_{rD}/μ_D, the relative permeabilities k_{rD}^*

1. The mobility of a fluid i is written: $M_i = k \cdot k_{ri} / \mu_i$.

and k_{ro}* are defined **at two distinct locations in the reservoir:**

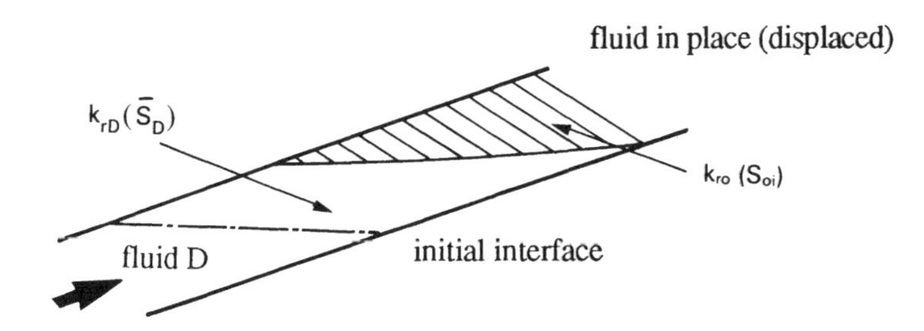

k_{ro}* is the relative permeability to the displaced fluid in the zone saturated by this fluid and not reached by the water, i.e. ahead of the front, or in practice k_{ro} (S_{wi}). As to k_{rD}*, this is the relative permeability to the displacing fluid behind the front and for the **average value of the saturation** with displacing fluid in the invaded zone S_{Dm}.

Hence:

$$M = \frac{k_{rD}(S_{Dm})}{k_{ro}(S_{wi})} \cdot \frac{\mu_o}{\mu_D}$$

N.B.:

If the **viscosity contrast is not too great** between the displacing fluid and the displaced fluid (such as a light oil and water), the saturation S_{Dm} is close to $S_{DM} = 1 - S_{or}$.

The analysis of displacement instabilities, supported by many experimental correlations, shows that **generally the lower the mobility ratio the better the displacement stability**. More precisely, instabilities (tongues, fingering) are more likely to appear if the mobility ratio M is higher than 1[1]. This is why a mobility ratio above 1 is said to be "unfavorable", and a ratio below 1 is said to be "favorable".

1. 1 to 2 in practice: stabilizing influence of capillary forces.

Some **examples** of mobility ratios are given in the table below:

Displaced fluid	**Displacing fluid**	k_{ro}^*	k_{rw}^*	k_{rg}^*	μ_o (cP)	μ_w (cP)	μ_g (cP)	M
Oil:								
. Light	Water	0.9	0.3		0.5	0.6		0.28
. Medium	Water	0.9	0.3		5.0	0.6		2.80
Oil:								
. Light	Gas	0.9		0.5	0.5		0.02	13.90
. Medium	Gas	0.9		0.5	5.0		0.02	139.00
Gas	Water		0.3	0.9		0.6	0.02	0.01

The mobility ratio is favorable (M < 1) only for gas or light oil displaced by water. Note that displacement by gas always yields a mobility ratio M > 1 which is unfavorable (since the viscosity of the gas is very low).

6.4.2 Concept of Critical Speed, Formation of a Tongue

The analysis of the forces in action (viscosity and gravity), disregarding capillary forces, serves to identify a **critical filtration speed** U_c above which the tongue phenomenon occurs (Fig. 6.10):

$$U_c = \frac{k\left(\frac{k_{rD}}{\mu_D}\right) . \left(\rho_D - \rho_o\right) g \sin \alpha}{M - 1}$$

A numerical coefficient is obtained equal to $8.4 \cdot 10^{-4}$ in metric practical units (meters per day, millidarcys, centipoises, grams per cubic centimeter).

Note that the filtration speed U = Q/A (hence the transition from U_c to Q_c).

This is the **Dietz** formula. However, experience shows that the actual values of the critical speeds observed in fields are significantly different from those obtained by this formula, and are generally higher (stabilizing effect of capillary pressures).

Instability mechanisms occur for **M > 1 and for $Q_0 > Q_c$**. These two conditions are necessary for the formation of a tongue. It stretches out as a function of time, and accordingly causes a premature breakthrough of the displacing fluid, which has a **detrimental effect on production**.

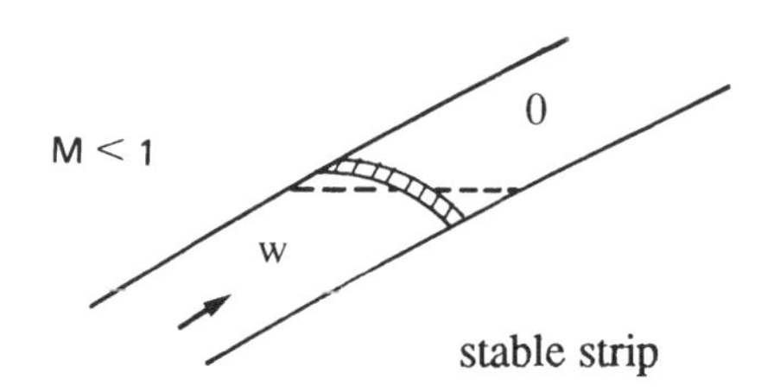

M > 1

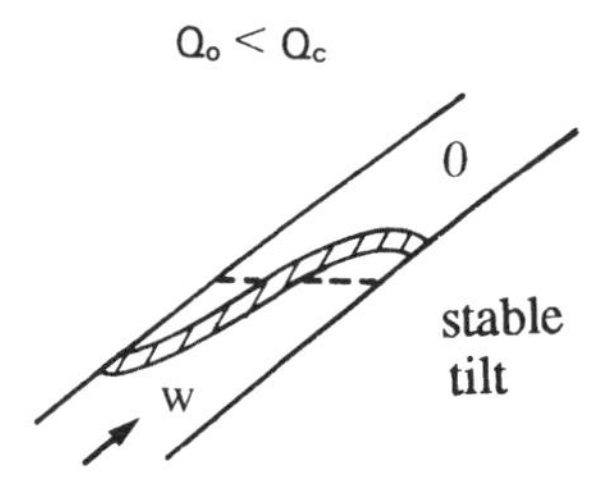

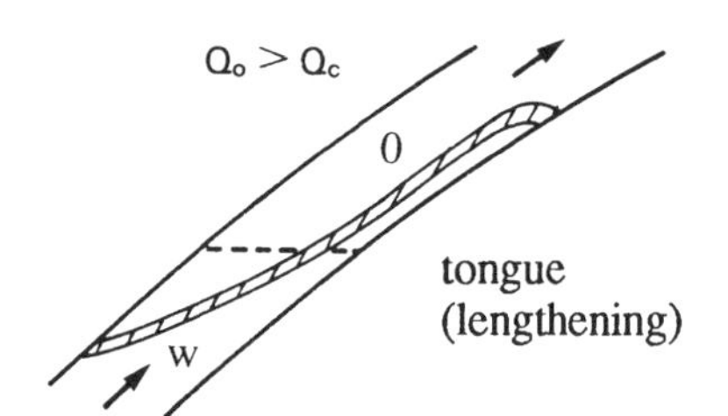

Fig. 6.10

6.4.3 Fingering

Fingering, which forms initially on a small scale, is made possible by the heterogeneity of the rock (variations in permeability) and grows on a metric or decametric scale if the mobility ratio $M > 1$. By contrast, if $M < 1$, incipient fingering is resorbed.

fluid 2

fluid 1

The higher M and the more pronounced the reservoir heterogeneities, the more likely fingering is to develop. Fingering is superimposed on the tongue phenomenon described above.

6.4.4 Coning

One important example of interface encroachment consists of a **local deformation of the interface (G/O or O/W) near a producing well.** Draw off is related to a pressure difference between the well and the interface, and the interface tends to be distorted and to approach the well.

If the well is in production, the flow displays symmetry of revolution if the medium is homogeneous and isotropic, but this flow is not exactly radial cylindrical if the well is not perforated along the entire thickness of the bed, because the stream lines of the lower part of the bed straighten up to reach the base of the well (O/W interface). This upward flow of the oil and water distorts the water/oil contact surface, which remains a surface of revolution and assumes a roughly conical shape.

Several types of coning can be distinguished according to the input and distribution conditions of the fluids. Figures 6.11 and 6.12 illustrate the two main types.

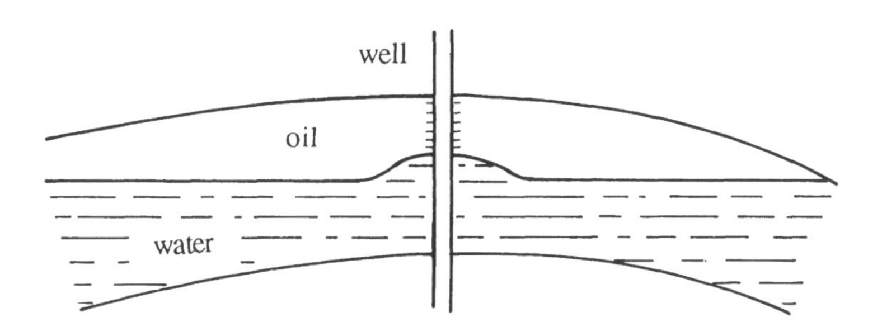

Fig. 6.11
Bottom coning (O/W or G/O).

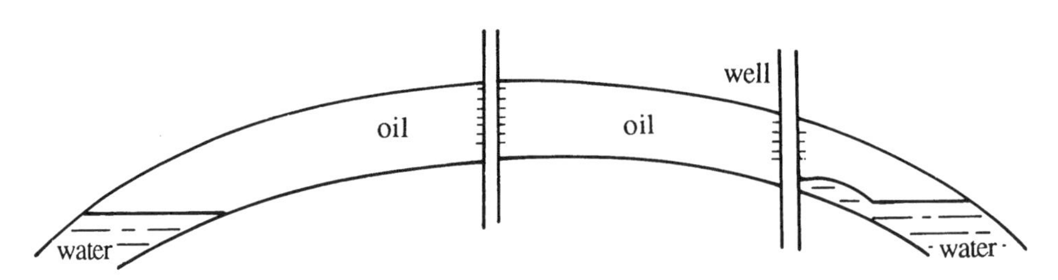

Fig. 6.12
Edge coning (O/W or G/O).

Figure 6.13 shows a borehole drilled to depth h_p in a layer of thickness H impregnated with oil to height h_o and by water to height $(H - h_o)$. The ratio h_p/h_o is the **penetration** of the borehole. The more the draw off effect of the well makes itself felt, i.e. the greater the penetration and flow rate, the higher the "cone" is. Conversely the gravitational forces exert a stabilizing effect.

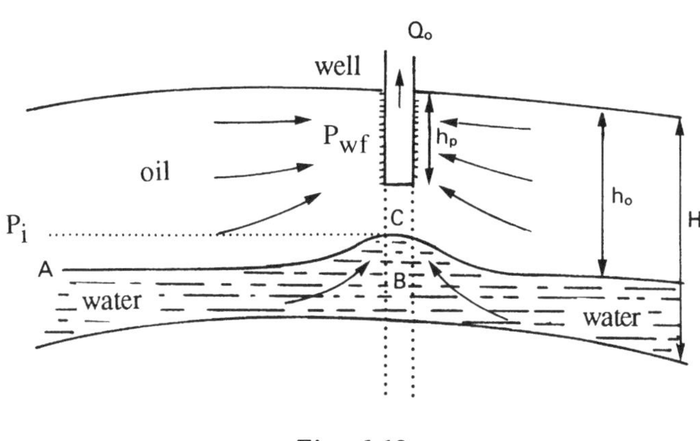

Fig. 6.13
Stable cone.

6.4.4.1 Infra-Critical Flow

For a given value of these parameters, and if, in particular, the flow rate is sufficiently low with respect to penetration, the cone forms a fixed surface in space. It is stable and does not reach the base of the perforations. No water encroachment occurs in the well (Fig. 6.13).

6.4.4.2 Supercritical Flow

For flow rates above the **critical flow rate Q_c characteristic of the well,** the cone reaches the perforations and water encroachment occurs in the well. This is the supercritical condition. Naturally, this may also involve a gas cone with oil or a water cone with gas (Figs 6.14 and 6.15).

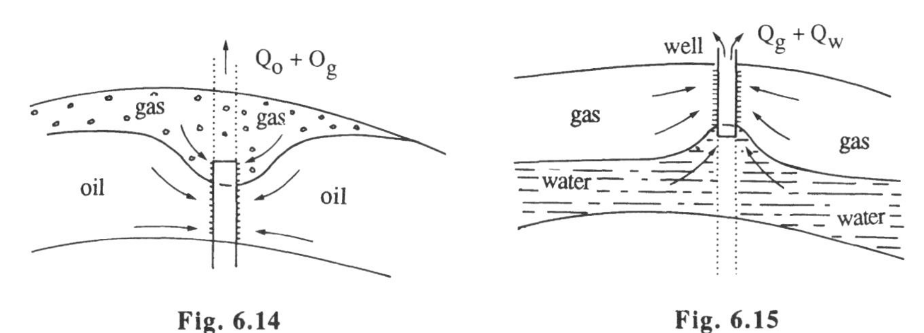

Fig. 6.14
Gas encroachment in well.

Fig. 6.15
Water encroachment in well.

6.4.4.3 Value of the Critical Flow Rate

Investigations conducted by Bournazel and Jeanson *(IFP)* clarify the correlations established by Sobocinsky *(Esso)* concerning bottom coning which is the only one we shall discuss. Edge coning requires very complex analysis and is studied by simulation in practice. The critical flow rate is given by the following equation, in US practical units (barrels per day, pounds per cubic foot, millidarcys, feet, centipoises):

$$Q_c = \alpha \cdot \frac{(\rho_w - \rho_o)\, k_h \cdot h_o (h_o - h_p)}{\mu_o B_o}$$

($\alpha = 1.41.10^{-5}$ for Sobocinsky and $\alpha = 1.15.10^{-5}$ for Bournazel: results of correlations),
($\alpha = 1.52.10^{-3}$ and $\alpha = 1.24.10^{-3}$ respectively in metric p.u.).

This formula is derived simply from the expression of the forces in action (Fig. 6.13).

Let:

$$h_c = \text{cone height}$$

This gives:

$P_S = P_A - \rho_o g\,hc$ $\quad$ $P_{wf} \approx P_C = P_B - \rho_w g\,hc$ $\quad$ $P_A = P_B$

hence:

$$P_S - P_{wf} \approx (\rho_w - \rho_o)\,g\,hc$$

Darcy:

$$P_S - P_{wf} = \frac{\mu_o Q_o B_o}{2\pi h_o k_o} \text{Ln} \frac{R}{r_w}$$

If $h_c \rightarrow (h_o - h_p)$, by definition $Q_o \rightarrow Q_C$ and:

$$Q_c = \frac{2\pi g}{\text{Ln}\dfrac{R}{r_w}} \left(\rho_w - \rho_o\right) \frac{k_o h_o \left(h_o - h_p\right)}{\mu_o B_o}$$

with

$$h_o - h_p = \text{water blanket}$$

The coefficient α is obtained from an average correlation including Ln R/r_w and the partial penetration effect.

Note that the critical flow rate depends on the horizontal permeability k_h.

Note also that the well results often give a critical flow rate higher than the one given by this equation. This stems partly from the fact that the formation of the cone is considered "statically", whereas the vertical movement of water also occurs, and also from the fact that little is known about the permeabilities k_h and k_v near the well.

Bournazel's study also serves to **calculate the breakthrough time** and the change in the water/oil ratio (WOR) after breakthrough. This study also shows that the WOR reaches a limit WOR for horizontal input and in steady-state flow.

6.4.4.4 Production Aspect, Coning Parameters

Although coning has been investigated methodically since the 1950's, it is so complex and unstable that two basic **practical** questions remain unanswered for the producer:

(a) **How to drill a well subject to coning?**
(b) **What is the right production rate?**

Controversies exist on the subject of these two questions. **The cautious solution** consists in perforating over a short stretch of pay zone and adopting a low flow rate to delay the arrival of the undesirable fluid at the well for as long as possible. It was, for many years, and is still, adopted if Q_o is not much higher than Q_c (see (1) in diagram below).

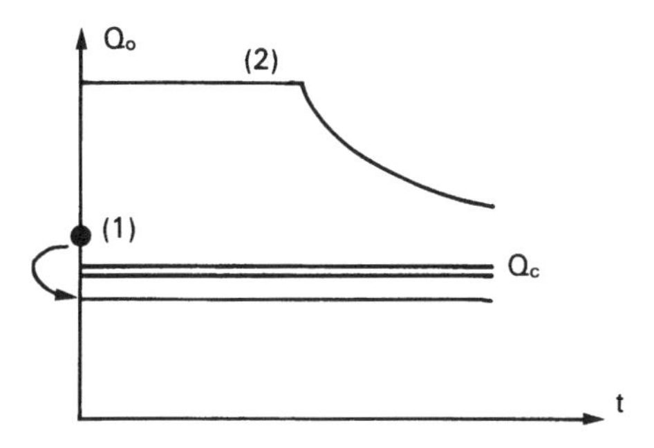

From this standpoint, it would be better to avoid drilling a well crossing the O/W, G/O or G/W interfaces with a large thickness of undesirable fluid.

The more recent **"full pot" solution**, on the contrary, consists in perforating widely and withdrawing at maximum flow rate (see (2) in diagram above), hence with considerable production of the second fluid, mainly if the second fluid is water. In fact, in largely supercritical conditions, the final recovery, which depends on production, can be improved with high flow rates. And intermediate solutions are available, giving the engineer a choice according to the economic data.

It should be added that **empirical laws** serve to make production forecasts for coning in supercritical conditions **after** breakthrough. Included are the methods of **Hutchinson** and **Henley**, who established correlations between recovery and WOR as a function of the mobility ratio M, of k_v/k_h, of the estimated recovery area and of the penetration or of the flow rate.

However, the numerical models adapted to coning (RZ) are widely employed to simulate these problems. A vertical subdivision helps to represent the horizontal and vertical permeability variations of the reservoir, providing a better approximation of the potential results. Here also, various assumptions on heterogeneities serve to achieve some sensitivity to the probable results selected.

Studies are also under way to prevent water influxes by the injection of **polymers** (designed to plug the water influx zones).

6.5 CONCLUSIONS

Two-phase (or three-phase) flows are difficult to deal with analytically, and we have merely described the most important aspects in simple qualitative form. Only unidimensional movement has been dealt with quantitatively, for a two-phase flow, disregarding capillarity forces.

The main conclusions to be drawn are the following:

(a) In general, the more complex cases are dealt with by simulation on models.
(b) In simple (or simplified) cases, the front concept is introduced, allowing an approximate calculation of the advance of the displacing fluid, the change in the water-cut or in the GOR at the well, and also of the oil flow rate as a function of time, giving the recovery: the **Buckley-Leverett** analysis.
(c) Injected fluids that are more mobile than oil ($M > 1$ to 2) can cause harmful instabilities (tongue and fingering) and a critical speed or critical flow rate is calculated to set the production conditions in infra- or supercritical conditions.
(d) Coning (encroachment of the interface near a well) is highly unfavorable and difficult to control.
(e) Capillary imbibition (spontaneous displacement by water) is, by contrast, a favorable regulating mechanism in heterogeneous media.

Chapter 7

PRIMARY RECOVERY, ESTIMATION OF RESERVES

7.1 DRIVE MECHANISMS

7.1.1 General Introduction

In earlier days, it sufficed in a preliminary period (and this still sometimes happens today) to open the wells and to allow the reservoirs to decompress: hence the term **primary recovery or natural depletion**. When the reservoir pressure and hence the well flow rates became too low, an attempt was sometimes made to attenuate or offset this decompression of the reservoir by the injection of water or gas, called **secondary recovery**.

Reservoir engineering is far more advanced today, and the production plan established includes secondary recovery methods (often very soon), which have been vastly improved.

This being said, the question arises as to how, based on an assessment of the hydrocarbons in place (seismic maps, core analyses, etc.) it is possible to determine a recovery ratio R% such that:

Hydrocarbons in place x R% = reserves

We shall examine only natural drive mechanisms here, and our analysis will lead us to identify them. We shall deal with the most important mechanisms with the goal of calculating the reserves.

7.1.2 Reserves

The term **"reserves"** concerns the **estimated recoverable volumes in place** (to be produced). For the total volumes already produced and to be produced in the future, the term of **"initial reserves"** applies.

The reserves obtained by primary recovery depend on the following:

(a) Amount of oil and gas in place and their distribution.
(b) Characteristics of the fluids and of the rock.
(c) Existing drive mechanisms and production rate.
(d) Economic factors.

The first two points are easily understood and have already been discussed. It therefore remains to determine the drive mechanisms and to show how the production rate can influence recovery, with respect to the technical aspect (although the production rate is linked to economic policy).

Classification of reserves is both technical and economic.

These terms always apply to **recoverable reserves** and not to reserves in place:

Proven: Discovered reserves that can reasonably be expected to be produced in present economic and technical conditions.

Probable: Discovered reserves which have a reasonable probability of production with technology and profitability close to those that exist today.

Possible: Reserves not yet discovered, but whose existence is presumed with a reasonable degree of probability.

Ultimate: Proved + probable + possible.

7.1.3 Drive and Recovery Mechanisms

These are the mechanisms that allow production. We shall list them rapidly and then examine them in greater detail (Fig. 7.1):

(a) **One-phase expansion**: gas or undersaturated oil reservoir. Considerable for gases, it allows the recovery of only a few per cent for oil (wide difference between the compressibilities of gas and oil).

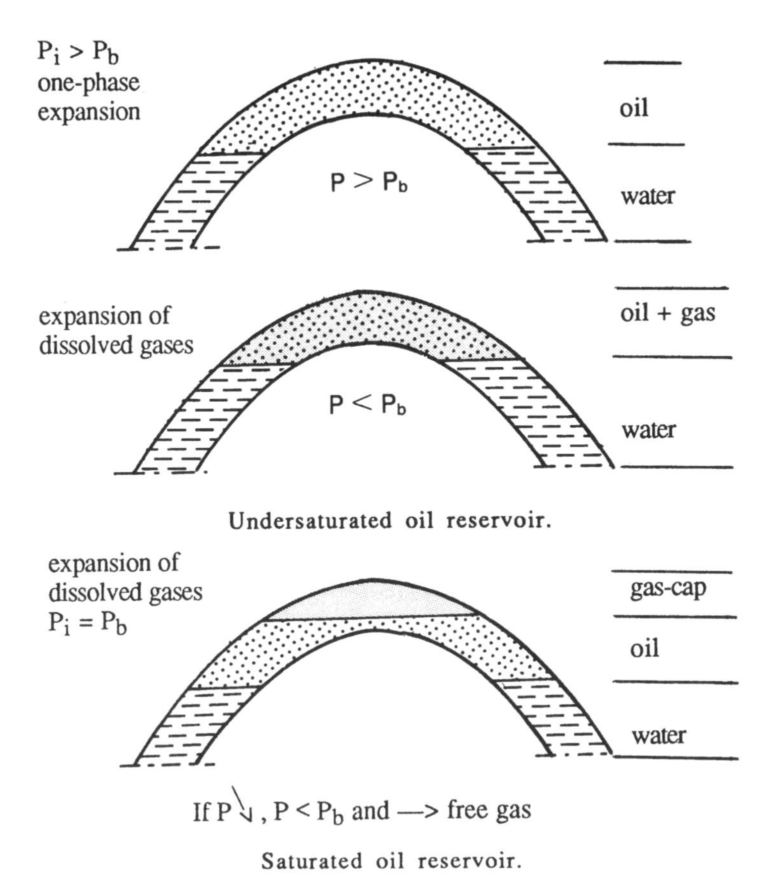

Undersaturated oil reservoir.

Saturated oil reservoir.

Fig. 7.1

Undersaturated and saturated oil reservoirs.

(b) **Expansion of gases coming out of solution** (at lower pressure than the bubble-point pressure), called **dissolved gas drive or solution gas drive**. As we have shown, since the production GOR rises rapidly, each volume of oil becomes increasingly expensive in terms of reservoir energy.

(c) **Expansion of the water of an aquifer** associated with the reservoir, which limits the pressure drop of an oil reservoir and drains it, called **water drive**. For a gas reservoir, however, this mechanism could be harmful, by trapping high-pressure gases behind the gas/water interface thus flooding the wells.

(d) **Expansion of a gas cap** overlying the oil (saturated oil reservoir), called **gas-cap drive**.

(e) **Imbibition,** important for very heterogeneous reservoirs. This is a very slow mechanism.
(f) **Gravitational forces** causing the segregation of fluids, particularly between gas and oil.
(g) **Rock compressibility**: relatively important for one-phase oil.

7.1.4 Influence of the Production Rate

The production rate has a real influence in many situations. Let us consider the following three examples.

7.1.4.1 Oil Reservoir without Aquifer

For a reservoir without a gas cap, recovery is independent of the flow rate. Hence it can be produced rapidly if necessary. This may also apply to a reservoir with a gas cap. In this case, however, it is important to consider harmful gas coning that is related to well flow rates (Section 7.1.6).

7.1.4.2 Oil Reservoir Associated with an Aquifer with Mediocre Petrophysical Characteristics

(a) Production at a **rapid rate**: the aquifer does not have the time to react as a whole. The predominant drive mechanisms are successively:

- one-phase expansion of the oil (to which the expansion of the pore water and compression of the rock are added),
- solution gas drive ($P < P_b$).

Recovery is typically about 20 to 30%.

(b) Extremely **slow production**: the aquifer has the time to react as a whole (preponderant mechanism: water drive).
Recovery is typically about 40%.

7.1.4.3 Fractured Oil Reservoir Associated with a Large Aquifer with Good Characteristics

(a) Production at a **rapid rate**: the aquifer acts, but drains only the fractures in which circulation is easiest. In fact, the fracture porosity accounts for only a small portion of the total porosity. The oil contained in the matrix can only move slowly into the fractures already flooded by water, and may still remain trapped when the field is abandoned.

(b) **Extremely slow** production: imbibition has the time to act. The rising water thus simultaneously drains the fractures and the matrix. Recovery is improved.

In actual fact, **economic factors** are inherent in every situation, together with reservoir conservation policy, which means that the final solution lies between the above two. To produce fast, secondary recovery must be resorted to.

7.1.5 Compressibility Coefficients, Fluid Expansion

The wide differences observed in these recovery rates are primarily explained by the compressibility coefficient of the different fluids. It is well known that a gas is much more compressible than a liquid or, if one prefers, its **expansion** is much greater as the reservoir pressure decreases. Note that a compressibility coefficient is defined by:

$$C = -\frac{1}{V} \cdot \frac{dV}{dp}$$

The orders of magnitude for oil, water and the porous medium are as follows:

$C_o = 7$ to $20 \cdot 10^{-6}\ psi^{-1}$ $(1$ to $3 \cdot 10^{-4}\ bar^{-1})$.
$C_w = 3$ to $5 \cdot 10^{-6}\ psi^{-1}$ $(0.4$ to $0.6 \cdot 10^{-4}\ bar^{-1})$.
$C_p = 2$ to $10 \cdot 10^{-6}\ psi^{-1}$ $(0.3$ to $1.5 \cdot 10^{-4}\ bar^{-1})$.

As to the compressibility of gas, this is of the form:

$$C_g \sim 1/P$$

In fact, PV ~ constant for gases (if Z ~ constant) and by deriving:

$$\frac{dP}{P} + \frac{dV}{V} \sim 0 \qquad \text{and} \qquad C_g \sim \frac{1}{P}$$

hence the orders of magnitude of C_g:

P	1000	3000	5000	7000 psi	
C_g	[1000	333	200	143]	$10^{-6}\ psi^{-1}$

One can immediately see the high values and wide variations of C_g. Since gases expand much more than liquids, it is easy to explain that the recovery of gas and that of oil with a gas cap are much higher than that of one-phase oil.

7.1.6 Multiphase Flow, Reservoir Heterogeneities

Two or three fluids may be present in the reservoir rock in proportions that vary with production because, as the pressure falls, an increasingly large gas phase appears (case of oils at pressures lower than the bubble-point pressure) or a hydrocarbon liquid phase (case of retrograde condensate gas), or even an aqueous liquid phase originating in an underlying aquifer (Fig. 7.2).

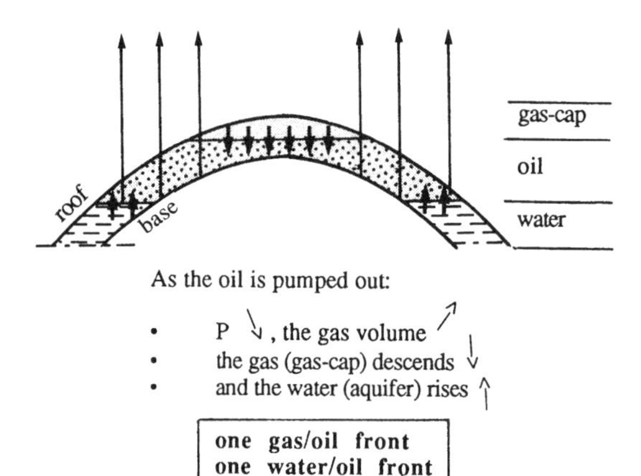

Fig. 7.2

A reservoir with natural depletion.

The presence of these different phases means that, since each fluid occupies only part of the pores, each flow rate can no longer be calculated from the **permeability** of the rock previously defined. It is accordingly necessary to introduce the concepts of **effective and relative permeabilities** of a fluid. This means that the fluids are "mutually hindered", their flow is accordingly slower, especially that of the oil (or of the gas if it is the main reservoir fluid): see Chapter 6 related to these flows.

In addition, interface distortions may occur and cause the premature arrival of an undesirable fluid in the well (water or gas), so that **oil production will decrease**.

These processes do not always reduce recovery, but often require the drilling of a larger number of wells to reach a given production rate.

Reservoir **heterogeneities** also cause a decrease in recovery, in comparison with an ideal homogeneous medium. Consider, for example, a level consisting of two media of very different permeabilities (kI > kII) as shown in Fig. 7.3:

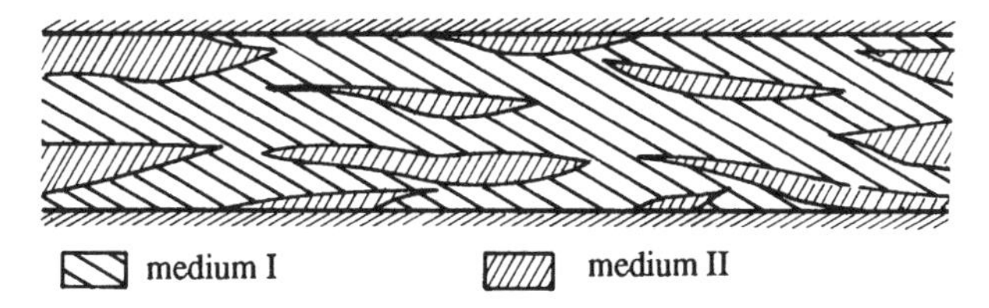

Fig. 7.3

The hydrocarbons will be drained easily from medium I, which is permeable, and which is also continuous, whereas those in medium II will move much more slowly and will depend on the imbibition associated with this heterogeneity (Chapter 6). On the whole, recovery will be more limited. Figure 7.4, in its own way, illustrates the difficulty of interpolating the values of the parameters identified exclusively at the wells.

Fig. 7.4

Drawing by Mick With the kind authorization of *SNEA(P)*.

7.1.7 Simplified Calculation Methods

After having estimated these drive mechanisms, how can one calculate the recovery ratio and the **size of the reserves**?

Three types of calculation are employed:

(a) Material balance (single-cell model).
(b) **Numerical models.**
(c) Decline laws.

The models are dealt with separately in view of their considerable importance (Chapter 9).

As to the **material balance** and **decline laws**, they form part of simplified calculation methods. They are often necessary at the start and end of production respectively. These methods are described in Sections 7.3 and 7.4.

Figure 7.5 shows the periods of use and the relative importance of the different methods employed.

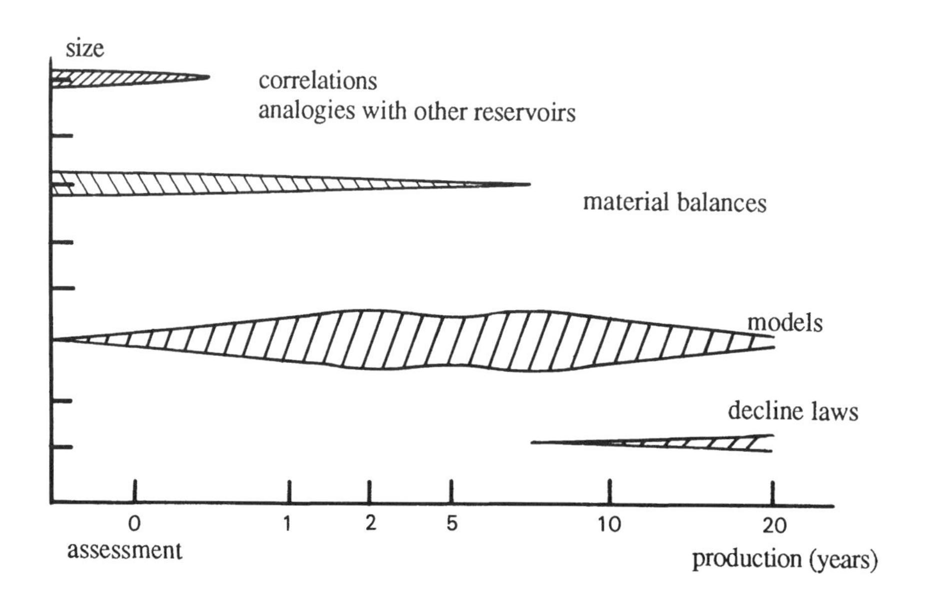

Fig. 7.5

Estimation of reserves and production forecasts.

7.2 RECOVERY STATISTICS

7.2.1 Recovery as a Function of the Type of Reservoir

Most reservoirs already produced have yielded recovery figures (reserves recovered in comparison with volumes in place) shown in the table below. The recovery ranges for each type of reservoir result from the properties of the fluids, the thermodynamic conditions, the petrophysical properties, and from the variations due to the architecture and the heterogeneities of the reservoir, and to the production rate.

Reservoir type	Recovery	Remarks
One-phase oil	< 10%	Pb < Pa (abandonment)
Oil with dissolved gas drive	5 to 25%	Pa < Pb
Oil with gas cap	10 to 40%	
Oil with aquifer	10 to 60%	Aquifer ± active
Gas	60 to 95%	
Average oil ≈ 30% and average gas ≈ 75%.		

7.2.2 World Reserves

The following figures indicate the **proven** reserves as of 1 January 1990, according to the present methods of primary and secondary recovery (Figs 7.6 and 7.7).

Oil production is also shown in Fig. 7.8.

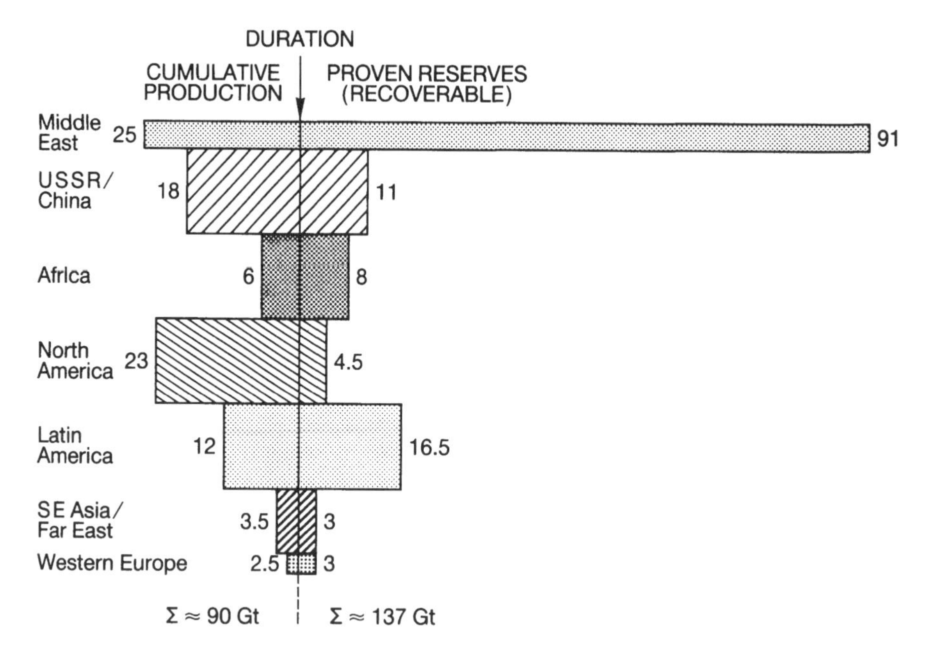

Fig. 7.6

Cumulative production and oil reserves (in billions of tons), 1990.

Oil production of the 20 leading countries in 1990 (Mt)

1	USSR	569	11	Canada	91
2	United States	410	12	Norway	81
3	Saudi Arabia	336	13	Nigeria	79
4	Iran	157	14	Indonesia	72
5	Mexico	147	15	Libya	71
6	China	138	16	Algeria	58
7	Venezuela	114	17	Kuwait	(58)
8	United Arab Emirates	110	18	Egypt	46
9	Iraq	(100)	19	Oman	33
10	United Kingdom	93	20	Brazil	32

(According to *CPDP*, 22 February 1991)

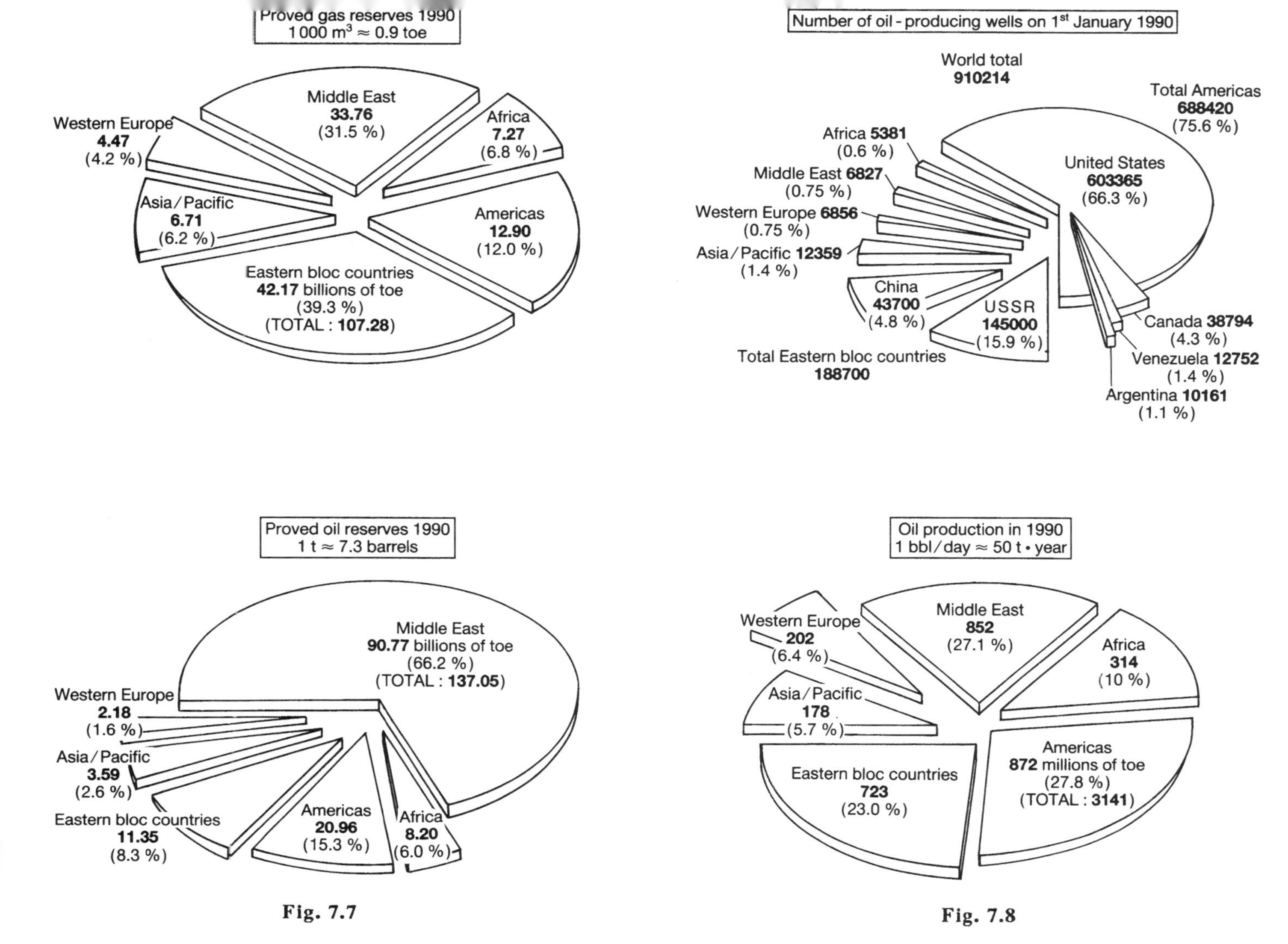
Proved gas reserves 1990
1 000 m³ ≈ 0.9 toe
Middle East
33.76
(31.5 %)
Africa
7.27
(6.8 %)
Western Europe
4.47
(4.2 %)
Asia/Pacific
6.71
(6.2 %)
Americas
12.90
(12.0 %)
Eastern bloc countries
42.17 billions of toe
(39.3 %)
(TOTAL : 107.28)
Number of oil - producing wells on 1st January 1990
World total
910214
Total Americas
688420
(75.6 %)
United States
603365
(66.3 %)
Africa 5381
(0.6 %)
Middle East 6827
(0.75 %)
Western Europe 6856
(0.75 %)
Asia/Pacific 12359
(1.4 %)
China
43700
(4.8 %)
USSR
145000
(15.9 %)
Total Eastern bloc countries
188700
Canada 38794
(4.3 %)
Venezuela 12752
(1.4 %)
Argentina 10161
(1.1 %)
Proved oil reserves 1990
1 t ≈ 7.3 barrels
Middle East
90.77 billions of toe
(66.2 %)
(TOTAL : 137.05)
Western Europe
2.18
(1.6 %)
Asia/Pacific
3.59
(2.6 %)
Eastern bloc countries
11.35
(8.3 %)
Americas
20.96
(15.3 %)
Africa
8.20
(6.0 %)
Oil production in 1990
1 bbl/day ≈ 50 t • year
Western Europe
202
(6.4 %)
Middle East
852
(27.1 %)
Africa
314
(10 %)
Asia/Pacific
178
(5.7 %)
Eastern bloc countries
723
(23.0 %)
Americas
872 millions of toe
(27.8 %)
(TOTAL : 3141)

Fig. 7.7

Fig. 7.8

PROVEN RESERVES

Region	Oil (10^6 bbl)	Gas (bcf)
Western Europe		
Denmark	799	4,488
France	185	1,324
(West) Germany	425	12,400
Italy	693	11,624
Netherlands	157	60,900
Norway	7,609	60,674
United Kingdom	3,825	19,775
Total.	**14,476**	**175,265**
Middle East		
Abu Dhabi	92,200	182,800
Saudi Arabia	257,504	180,355
Iran	92,850	600,350
Iraq	100,000	95,000
Kuwait	94,525	48,600
Oman	4,300	7,200
Qatar	4,500	163,200
Total.	**662,598**	**1,324,265**
Africa		
Algeria	9,200	114,700
Angola	2,074	1,800
Cameroon	400	3,880
Congo	830	2,580
Egypt	4,500	12,400
Gabon	730	490
Libya	22,800	43,000
Nigeria	17,100	87,400
Tunisia	1,700	3,000
Total.	**59,892**	**285,143**
Americas		
Canada	5,783	97,589
United States	26,177	166,208
Mexico	51,983	72,744
Argentina	2,280	27,000
Brazil	2,840	4,045
Colombia	2,000	4,500
Ecuador	1,420	3,950
Venezuela	59,040	105,688
Total.	**153,050**	**506,080**

PROVEN RESERVES (CONTINUED)

Region	Oil (10^6 bbl)	Gas (bcf)
Asia		
Australia	1,566	15,433
Brunei	1,350	11,200
India	7,997	25,050
Indonesia	11,050	91,450
Malaysia	2,900	56,900
Total.	**26,242**	**263,260**
Eastern Bloc Countries		
China	24,000	35,300
USSR	57,000	1,600,000
Other	1,855	29,000
Total.	**82,855**	**1,664,300**
Total World.	**999,113**	**4,208,315**

Note: 10^3 m^3 gas ~ 0.9 toe.

(*OGJ*, December 1990).

7.3 MATERIAL BALANCE

7.3.1 The Material Balance and its Use

Material balances **equate the volume of the fluids contained in a reservoir and the volume of the pores of the reservoir at any given time.**

The behavior of a reservoir can only be known from observations made in the wells, essentially flow rate and pressure measurements and fluid analyses.

These observations, interpreted according to a number of guidelines in the "material balances", provide the data used to analyze and predict the behavior of the reservoirs for which the available information is still limited (one or a few wells).

It may be observed that a material balance basically represents only the **equation of continuity** for the overall reservoir (or of one of its compartments) for a finite time interval, thus entailing the consideration of finite variations in masses and pressures. It is also the **simplest reservoir simulation model**, and represents a reservoir as a **single cell**.

If the reservoirs are better known, a number of wells having already been drilled, it is necessary to use grid pattern simulation models, commonly called "models". In these models, the reservoir is subdivided into cells, each one containing quantities of fluids subject to the laws of fluid mechanics. Their essential advantage is their ability to represent the variations in the reservoir fluid characteristics, the well flow rates, and the pressure in space.

In concrete terms, they are programs run on scientific computers.

A material balance always has two possible uses:

(a) **Production forecasts** N_p, G_p, (W_p), variation in pressure, and in production GOR and WOR.
(b) Calculation of **volumes in place** N, G, (W), production already having begun (production case history).

These values are compared with those obtained previously by volumetric methods.

This leads to the following table, with AIME notations:

	Oil	**Gas**	**Water**
Volumes in place (1)	N	G	W
Cumulative production (2)	N_p	G_p	W_p
Recovery $= \frac{(2)}{(1)}$	n_p	g_p	–

These volumes are expressed in **standard conditions**.

In practice, N and N_p are expressed in 10^6 bbl (or 10^6 m^3), or metric tons, and G and G_p in 10^9 cu.ft (or 10^9 m^3).

7.3.2 Undersaturated Oil Reservoir

If the compressibility of the rock and of the pore water is disregarded, it can be written that the reservoir volume occupied by hydrocarbons remains cons-tant:

$$(N - N_p) B_o = N B_{oi}$$

Remaining volume of oil = Initial volume of oil

Since B_o is a function of P, the relationship between the reservoir pressure and production is thus defined.

However, it is necessary to introduce the concept of compressibility:

$$C = -\frac{1}{V} \cdot \frac{dV}{dp}$$

This serves to account for the compressibility of the formation C_p and of the pore water C_w.

For a pressure drop ΔP:

(a) The oil volume $(V_p S_o)$ increases by $V_p S_o C_o \Delta P$.
(b) The water volume $(V_p S_{wi})$ increases by $V_p S_{wi} C_w \Delta P$.
(c) The pore volume (V_p) decreases by $V_p . 1 . C_p \Delta P$,

and the oil production (reservoir volume) $N_p B_o$ is the sum of these three terms:

$$N_p B_o = V_p \Delta P \left(C_o S_o + C_w S_{wi} + C_p\right) = V_p S_o \Delta P \left(C_o + C_w \frac{S_{wi}}{S_o} + \frac{C_p}{S_o}\right)$$

Everything proceeds as if, since the rock and the water are incompressible, the oil has an equivalent apparent compressibility C_e:

$$C_e = C_o + C_w \frac{S_{wi}}{S_o} + \frac{C_p}{S_o}$$

hence:

$$N_p B_o = V_p S_o C_e \Delta P \qquad \text{or} \qquad = N B_{oi} C_e \Delta P$$

and

$$\boxed{N_p = N . \frac{B_{oi}}{B_o} . C_e \Delta P}$$

with $V \sim N B_{oi}$ and $dV \sim N_p B_o$.

Numerical example:

$N = 100 \cdot 10^6$ bbl $\qquad C_o = 14 \cdot 10^{-6}\ psi^{-1}$

$S_{wi} = 25\%$ $\qquad C_w = C_p = 4 \cdot 10^{-6}\ psi^{-1}$

For P_i (initial) = 4300 psi, $B_{oi} = 1.50$. For P = 2900 psi, $B_o = 1.53$.[1]

Calculation of C_e:

$$C_e = \left(14 + \frac{4 \cdot 0.25 + 4}{0.75}\right) \cdot 10^{-6}\ psi^{-1}$$

$$C_e = 20.7 \cdot 10^{-6}\ psi^{-1}$$

hence:

$$N_p = 100 \cdot 10^6 \cdot \frac{1.50}{1.53} \cdot 20.7 \cdot 10^{-6} \cdot 1400$$

$$N_p = 2.9 \cdot 10^6\ bl$$

and

$$n_p = 2.9\% \text{ for a considerable depletion of 1400 psi}$$

Note that if C_p and C_w were disregarded, a very large relative error would have been made (use of C_o instead of C_e).

7.3.3 Dissolved Gas Drive

Initial conditions: $P_i = P_b$ (and subscript i ≈ b). The "initial" pressure, in the following calculations, corresponds to the bubble-point pressure.

The rock and water compressibility, which is generally low compared with that of free gases, is disregarded. It can then be written that the volume of hydrocarbons remaining in place at pressure P is equal to the initial volume of hydrocarbons, leading to the following table:

1. Since $B_o = B_{oi}\,(1 + C_o\,(P_i - P))$.

Pressure	Volumes	Reservoir volumes
P_i	Oil N (dissolved gas $N\ R_{si}$)	$N\ B_{oi}$
P	Oil $(N - N_p)$ remaining dissolved gas: $(N - N_p)\ R_s$ Free gas: $N\ R_{si} - G_p - (N - N_p)\ R_s$	$(N - N_p)\ B_o$ $[N\ R_{si} - G_p - (N - N_p)\ R_s]\ B_g$

We therefore have:

$$(N - N_p)\ B_o + [N\ R_{si} - G_p - (N - N_p)\ R_s]\ B_g = N\ B_{oi}$$

This equation is sufficient to calculate the oil in place N. However, it is inadequate for production forecasts N_p. This is because $\Delta G_p = \overline{R} \cdot \Delta N_p$ with $\overline{R}$ = average GOR for the calculation step corresponding to a selected pressure drop ΔP. Hence the calculation of N_p depends on the change in the GOR.

But GOR = $f(P, S_o)$ with S_o = oil saturation, because:

$$R = R_s + \frac{k_g}{k_o} \cdot \frac{\mu_o\ B_o}{\mu_g\ B_g} \qquad \text{and} \qquad \frac{k_g}{k_o} = \varphi\left(S_o\right)$$ [1]

The reservoir volume of initial oil $V_p\ S_{oi} = N\ B_{oi}$ becomes at pressure P:

$$V_p\ S_o = (N - N_p)\ B_o$$

hence the equation:

$$S_o = S_{oi}\left(1 - \frac{N_p}{N}\right)\frac{B_o}{B_{oi}}$$

1. Note the role of the critical gas saturation S_{gc}. In fact, $k_g = 0$ for $S_g \leq S_{gc}$ (and $R = R_s$ even for $P < P_b$).

Thus G_p depends on R, which is a function of S_o, which varies with N_p, unknown. A **computation** must be performed by **iterations** as follows: assumption on R for a given ΔP, computation of N_p in the general equation, hence computation of S_o and k_g / k_o, and subsequently computation of R. Compare with the assumption on R.

The accuracy required is often 1%. These computations are carried out on desktop microcomputers.

Computation of $k_{rg}/k_{ro} = \varphi (S_o)$ according to production case history:

For each calculation step (ΔP), krg/kro is obtained from $R - R_s$, as well as S_o (foregoing equations), giving the curve that can often be extrapolated linearly on a graph:

$$\log \frac{k_{rg}}{k_{ro}} = f\left(S_o\right)$$

7.3.4 Oil Reservoir Associated with an Aquifer

The expansion due to an aquifer maybe considerable. This is because, while C_w and C_p are small, the volume of the aquifer may be very large, such as a regional aquifer. The activity of this aquifer also depends on its properties, particularly its permeability.

The problem to be solved consists in determining the **"water inflow"** W_e into the reservoir as a function of time. These water inflows are more specifically cumulative water inflows as a function of time and of the drop in pressure in the "oil + aquifer".

7.3.4.1 Aquifer Extension, Bottom Aquifer and Edge Aquifer

Some reservoir beds are continuous over very long distances, indeed on the scale of a basin. An oil reservoir accordingly represents an extremely small volume of fluid compared with the volume of the aquifer with which it is connected. Thus the withdrawal of hydrocarbons, oil or gas, can cause the gradual decompression of a large volume of water which, with the water inflows, tends to offset or practically offset the pressure drop that should have occurred inside the reservoir.

This leads to the following results:

(a) The water inflows, a consequence of the gradual expansion of the aquifer, occur in **transient conditions** over a rather long time interval.
(b) The subsequent decline in pressure in the reservoir depends on the **size** of the **aquifer** and on its **characteristics**.

A very extensive, very continuous and highly permeable aquifer guarantees virtually perfect pressure maintenance. This is termed a **perfect "water drive"**.

A relatively small, noncontinuous or mediocre-permeability aquifer can guarantee only limited pressure maintenance. This is termed **partial "water drive"**.

If the reservoir is in contact with an aquifer over its entire extent, the water inflows occur vertically. This is a "**bottom**-water drive" or **bottom aquifer**. The main mechanism in this case is **coning** (Fig. 7.9a).

If the aquifer surrounds the reservoir, the water inflows occur laterally. This is called "**edge**-water drive" or **edge aquifer** (Fig. 7.9b).

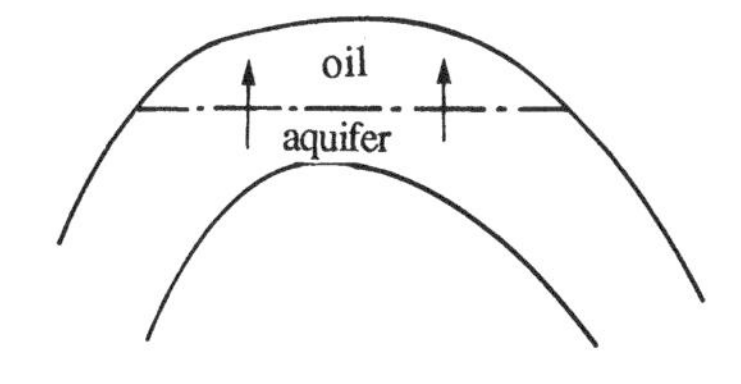

Fig. 7.9a

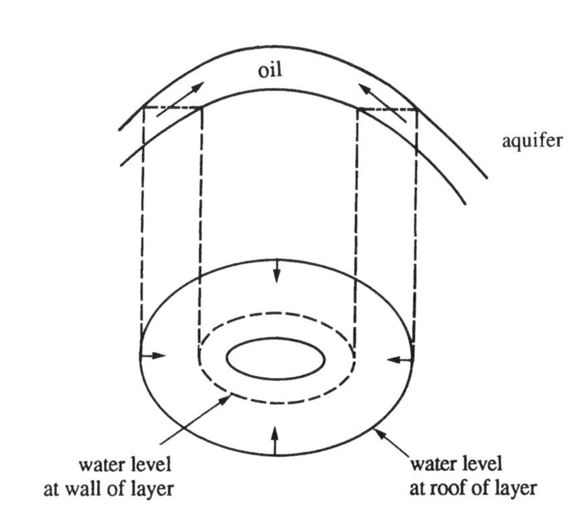

Fig. 7.9b

With the bottom aquifer, the water inflows involve the **vertical** permeabilities of the reservoir, and in the edge aquifer, the **horizontal** permeabilities.

The second case is more generally encountered.

7.3.4.2 Aquifer Functions

With respect to the aquifer, an oil reservoir can be considered to behave as a large-diameter well. This is the **Van Everdingen and Hurst method.** The water influx, which corresponds to the cumulative production of water in the "big well", is thus defined by an aquifer function, similar to the well function relating the pressures and the flow rate as a function of time for a producing well.

The contours of the O/W boundaries and of the aquifer are treated as circles (radial cylindrical flow) with R_i = inside radius and R_e = outside radius (Fig. 7.10).

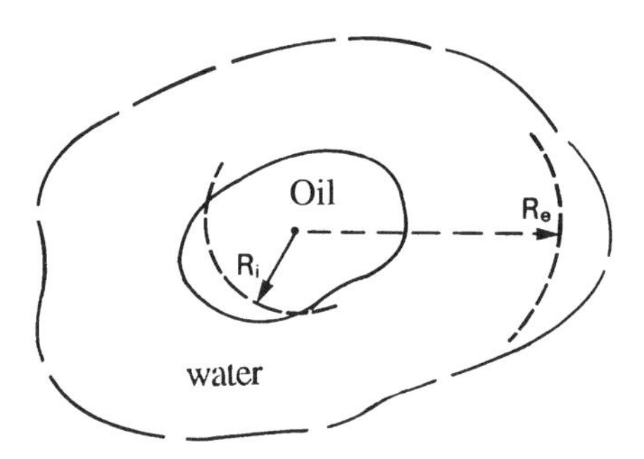

Fig. 7.10

It is necessary to calculate the water inflows W_E (P, t). The water flow rate entering the oil zone is written:

$$Q_e = S \frac{k_e}{\mu_e} \left(\frac{dP}{dr}\right)_{r = R_i}$$

with

$$S = 2 \pi R_i h \qquad \text{and} \qquad W_e (P, t) = \int_0^t Q_e \, dt$$

In this geometry, the horizontal O/W contact surface is replaced by a virtual vertical surface in accordance with the amount of oil in place.

The water inflows resulting from gradual expansion of the aquifer continue in **transient conditions** over a relatively long period. Since the pressure at the oil/water interface ("big well") drops as a function of time, the superimposition theorem is used after subdividing the pressure curve into n successive increments and the following is obtained:

$$W_e (P, t) = B \sum_{i=0}^{n} \overline{C} (t_D - t_{Di}) \Delta P_i$$

with

B = constant = $2 \pi R_i^2 h \phi (C_w + C_p)$,

$\overline{C}$ = tabulated aquifer function (from time t considered to time t_i): a reduced time is used, $t_D = K . t / R_i^2$

ΔP_i = half-pressure drop at the interface, from time (i – 1) to (i + 1), for the time interval from i to i + 1 (see Fig. 7.11).

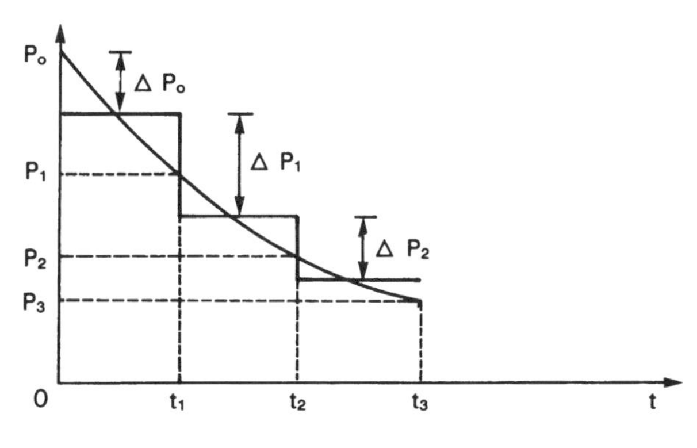

Fig. 7.11

The following two $\overline{C}$ functions are used:

$\overline{C}_3$ = function defined for a **closed aquifer** bounded at distance R_e,
$\overline{C}_5$ = function defined for an **"infinite" aquifer** (large size).

Each of these functions exists in two distinct forms according to whether the supply and flow conditions are radial cylindrical or linear.

The functions $\overline{C}_3$ and $\overline{C}_5$ in radial cylindrical conditions are shown schematically in Fig. 7.12 (the functions $\overline{C}_3$ are generally indicated for radius ratios

$r_{eD} = R_e / R_i$ ranging from 1.5 to 10).

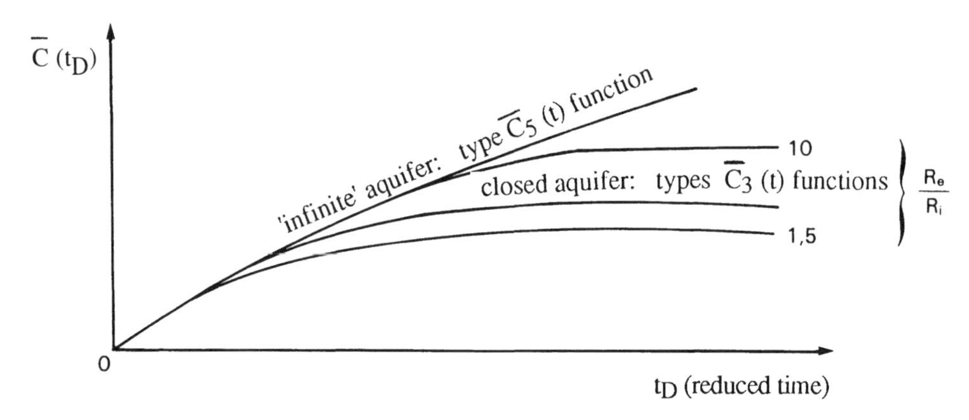

Fig. 7.12

Some forms of aquifer functions.

For a bounded aquifer, the function $\overline{C}_3\,(t_D)$ approaches an asymptotic value of:

$$\overline{C}_3 = \frac{\left(R_e/R_i\right)^2 - 1}{2}$$

since the water flow is semi steady state for large time values.

7.3.4.3 Calculation of Water Inflows

A. At Start of Production

Based on the aquifer characteristics (generally hypothetical), it is possible to calculate the water inflows as a function of pressure drop. The reservoir is treated as a "big well": Van Everdingen and Hurst method of which the principle is described above.

B. With Production Records: Determination of the Water Influx Law

The water inflows are estimated from a material balance. Consider for example a one-phase oil reservoir.

The material balance is written:

$$N_p B_o = N B_{oi} C_e (P_i - P) + W_e$$

or for a given time t_j:

$$N_{pj} B_{oj} = N B_{oi} C_e (P_i - P_j) + W_{ej}$$

Hence the numerical values of $W_e(t)$.

With its values during the production history, the aquifer function is determined up to the present time, enabling its extrapolation for production forecasts.

Also depending on the regional geological data, optimization methods serve to determine the most probable values of the size and permeability of the aquifer.

These analyses are of fundamental importance, because, depending on the predicted activity of the aquifer, recovery may vary within a considerable range (by a factor of 1 to 10).

Example:

This can be illustrated by resuming the example shown in Section 7.3.2. For a pressure drop of 1400 psi, one-phase recovery was only 3%.

Let us assume that a vast aquifer exists, but one whose activity is unknown before the start of production.

What is the potential recovery range for this reservoir?

Let us still consider a pressure drop of 1400 psi. The maximum theoretical recovery corresponds to the decompression of the entire aquifer, with semi steady-state conditions from the beginning (favorable assumption of very high permeability). This leads to a maximum theoretical water influx of W_{eM}.

In these conditions:

$$W_{eM} = V_w (C_w + C_p) (P_i - P)$$

with the **estimated** volume of aquifer water $V_w = 500\ Mm^3$ (~ 30 N):

$$W_{eM} = 0.5 \,.\, 10^9 \,.\, 8 \,.\, 10^{-6} \,.\, 1400 = 5.6\ Mm^3$$

$$R\ \% = \frac{N_p}{N} = \frac{B_{oi}}{B_o} C_e (P_i - P) + \frac{W_{eM}}{N B_o}$$

Hence $N = 100 \cdot 10^6$ bl $\rightarrow 16 \cdot 10^6$ m^3:

$$\frac{W_{eM}}{N\,B_o} = \frac{5.6 \cdot 10^6}{16 \cdot 1.53 \cdot 10^6} \approx 0.23$$

This gives an additional maximum theoretical recovery of 23%, and a (theoretical) range of 3 to 26%.

Thus knowing the activity of the aquifer is vitally important. Yet it can be known accurately only after the first few years of production.

7.3.5 Generalized Material Balance for a Saturated Oil Reservoir with Gas Cap and Aquifer

7.3.5.1 Gas-Cap Drive

We introduce the ratio:

$$m = \frac{\text{volume of gas cap}}{\text{volume of oil}} \qquad \text{(reservoir conditions)}$$

Since the oil volume is $N\,B_{oi}$, the volume of the gas cap in reservoir conditions is written $G\,B_{gi} = m\,N\,B_{oi}$. Hence the gas cap contains a quantity of gas (expressed in standard conditions) $G = m\,N\,B_{oi}/B_{gi}$. At pressure P, this gas occupies a volume of:

$$\frac{m\,N\,B_{oi}}{B_{gi}}\,B_g \qquad \text{in reservoir conditions}$$

The expansion of the gas cap, expressed in volume at pressure P, is written:

$$G\,B_g - G\,B_{gi}$$

or

$$\frac{m\,N\,B_{oi}}{B_{gi}}\,B_g - m\,N\,B_{oi} = m\,N\,B_{oi}\left(\frac{B_g}{B_{gi}} - 1\right)$$

7.3.5.2 Generalized Material Balance

Material balance of the oil zone without gas cap (Section 7.3.3):

$$N B_{oi} = (N - N_p) B_o + [(N R_{si} - G_p) - (N - N_p) R_s] B_g$$

with $G_p = R_p B_g$.

Gas resulting from expansion of the gas cap:

$$m N B_{oi} \left(\frac{B_g}{B_{gi}} - 1 \right)$$

Net water influx:

$$W_e - W_p \quad \text{(water influx — produced water)}$$

It is assumed that the gas from the gas cap is not produced, hence:

$$N B_{oi} = (N - N_p) B_o + [(N R_{si} - G_p) - (N - N_p) R_s] B_g + m N B_{oi} \left(\frac{B_g}{B_{gi}} - 1 \right) + W_e - W_p$$

or

$$\boxed{N_p [B_o + (R_p - R_s) B_g] = N (B_o - B_{oi} + (R_{si} - R_s) B_g) + m N B_{oi} \left(\frac{B_g}{B_{gi}} - 1 \right) + W_e - W_p}$$

The calculation of the forecast N_p is the same as that discussed for dissolved gases in Section 7.3.3.

7.3.5.3 Efficiency of Various Drive Mechanisms

The volume of the hydrocarbon produced at any pressure P is:

$$N_p B_p \quad \text{in reservoir conditions}$$

with $B_p = B_o + (R_p - R_s) B_g$.

The water that has invaded the reservoir has displaced an equivalent volume of oil. The ratio:

$$(W_e - W_p) / N_p B_p$$

thus represents the per cent input of the **water drive** in production.

The ratio:

$$\frac{\text{Expansion of gas cap}}{N_p\ B_p} = \frac{m\,N\,B_{oi}}{N_p\ B_p} \cdot \left(\frac{B_g}{B_{gi}} - 1\right)$$

similarly expresses the input of the **gas-cap drive**.

The input due to **dissolved gas drive** can be similarly calculated.

7.3.5.4 Segregation

In the foregoing discussion, it is assumed that the gas released does not reach the top of the layer by segregation. Segregation is the mechanism due to gravitational forces, since gas is much lighter than oil. Thus part of the gas is produced and part either joins the existing gas cap, or forms a "secondary" gas cap in the case of an initially undersaturated oil. The extent of this segregation depends on the pressure field prevailing in the reservoir and on the anisotropy ratio of the permeabilities k_v/k_h. Simultaneously, oil replaces the released gas. This is called countercurrent segregation (analyzed by simulation models).

7.3.5.5 Variation in Production Data According to Drive Mechanisms

Figures 7.13 and 7.14 show the typical changes in pressure drop and GOR for reservoirs with and without a gas cap, and with an active aquifer.

7.3.6 Specific Case of Volatile Oil

Volatile oils, or high-contraction oils, constitute an intermediate case between light oil and condensate gas.

Volatile oil is characterized by:

(a) A high reservoir temperature (often close to the critical temperature).
(b) A high percentage of C_2 to C_{10} (hence their volatility).
(c) An abnormally high solution GOR (more than 200 m^3/m^3).
(d) A high formation volume factor B_o (over 2).
(e) A very light stock tank oil (often more than 45°API).

For ordinary oil, the material balance was somewhat simplified by assuming (a fact not confirmed for rather heavy oil) that the reservoir gas phase

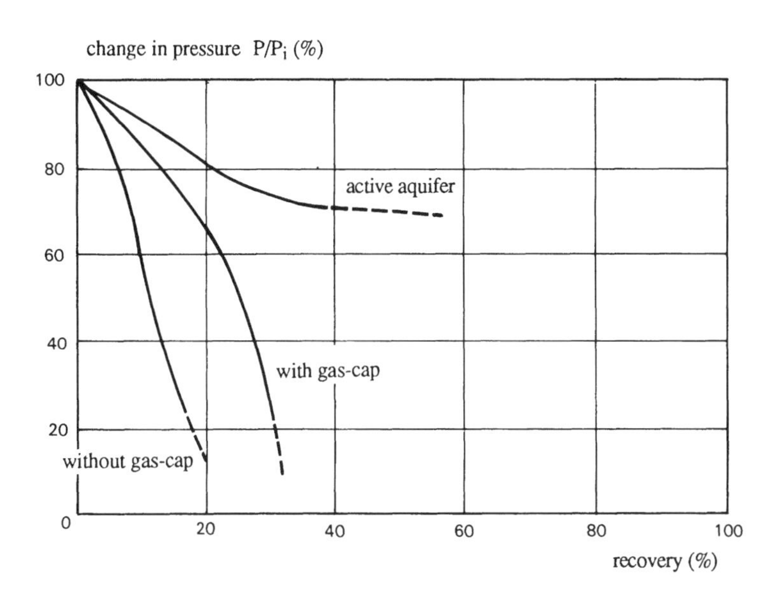

Fig. 7.13 Typical developments.

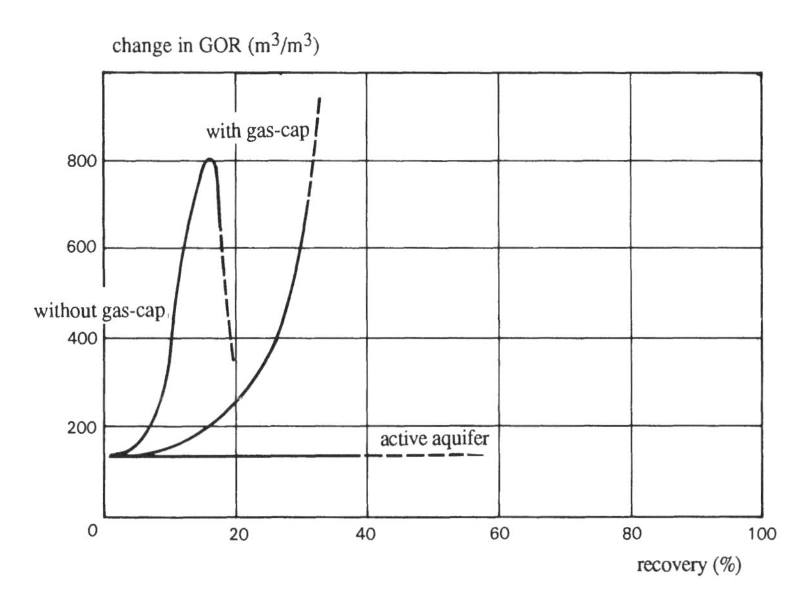

Fig. 7.14 Typical developments.

did not contribute in any way to the production of liquid at the surface. With volatile oil, a large share of the stock tank oil is transported to the well in the gas phase. The equations used for ordinary oil are inadequate and lead to false predictions.

Two methods are available.

7.3.6.1 Jacoby and Berry Method ("Volumetric" Method)

This is a step-by-step method. For each pressure drop ΔP, the following are calculated in succession:

(a) Changes in composition of the oil and gas in place (flash calculation).
(b) Total volume of fluid produced in reservoir conditions (volumetric material balance).
(c) Respective volumes of oil and gas produced in reservoir conditions (iterative method to solve the material balance and the relative permeability equation simultaneously).
(d) Production of stock tank oil and production GOR by a flash calculation.

7.3.6.2 Compositional Material Balance Program

Its validity depends on the method for calculating the thermodynamic equilibrium and the accuracy of the values of the equilibrium coefficients K_i employed (possible need of "matching" with a minimum of laboratory experiments).

7.3.7 Dry (or Wet) Gas Reservoir: Not Giving Rise to Retrograde Condensation in the Reservoir

7.3.7.1 Material Balance (without Water Influx)

Remaining volume of gas = initial volume:

$$(G - G_p)\, B_g = G\, B_{gi}$$

This equation is similar to that for undersaturated oil reservoirs and is treated similarly (C_w and C_p are disregarded). However, knowing that:

$$B_g = Z \frac{P_s}{P} \cdot \frac{T}{T_s} \qquad (P_s \text{ and } T_s \text{ in standard conditions})$$

We can also write:

$$G_p = G\left(1 - \frac{Z_i}{P_i} \cdot \frac{P}{Z}\right)$$

In principle, the function P/Z is hence linear with G_p, but this may not apply to reservoirs with abnormal (high) pressure, due to possible compaction of the rock, or expulsion of the water from the shales.

7.3.7.2 Calculation of Gas in Place

It is easy to find G, which is represented by the constant term of the linear equation (Fig. 7.15).

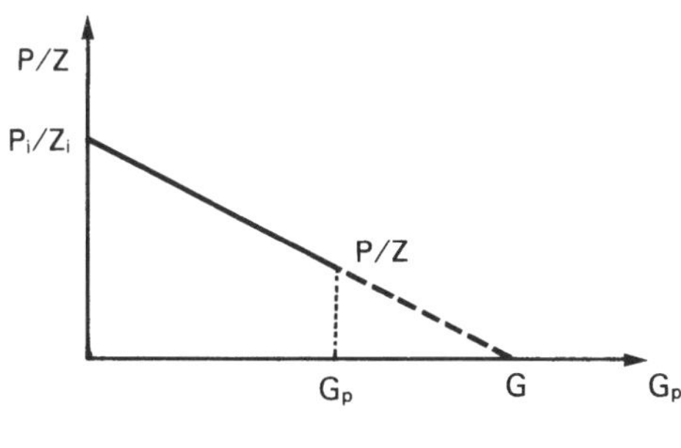

Fig. 7.15

7.3.7.3 Recovery with Water Influx

$$G_p\, B_g = G\,(B_g - B_{gi}) + W_e - W_p$$

From which it is written:

$$g_p = \frac{G_p}{G} = \left(1 - \frac{Z_i}{P_i} \cdot \frac{P}{Z}\right) + \frac{W_e - W_p}{G\, B_g}$$

Note that:

(a) A moderate water influx increases recovery.
(b) A **large water influx decreases** recovery because a large amount of gas is trapped by the rise of the G/W interface, and the water reaching the

wells means a higher abandonment pressure P_a (high water-cut), and hence a shorter production period (Fig. 7.16).

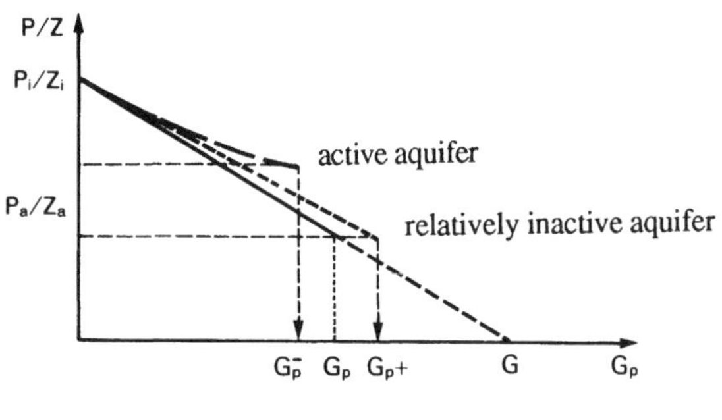

Fig. 7.16

What can be done in this case? The activity of the aquifer can be adjusted by depleting the reservoir faster or more slowly.

This means that if water inflows are expected, and better recovery is desired, it may be necessary **to produce the reservoir very rapidly, not leaving the aquifer the time to expand and to sweep the reservoir.**

Whereas, on the contrary, if the reservoir is produced very slowly at a very low production rate, the aquifer performs its role fully and may succeed in practically perfect pressure maintenance, but with a detrimental effect on the recovery of gas. This means that a decision must be taken in the case of an active aquifer associated with a gas reservoir (everything concerning dry gas is also valid for condensate gas, since the water inflows do not depend on the type of fluid in place).

For this type of reservoir, it is essential to determine the precise value of the saturation with trapped gas S_{grw} (core analyses and logs).

7.3.7.4 Example of Dry Gas Material Balance (without Water Influx)

A gas reservoir has the following gas in place:

$$G = 7.4 \;.\; 10^{11} \text{ std. cu.ft.}$$

The initial pressure is:

$$P_i = 3800 \text{ psi,} \quad \text{and} \quad Z_i = 0.94$$

What is the final recovery for an abandonment pressure estimated at:

$$P_a = 600 \text{ psi}$$

with $Z_a = 0.96$.

This gives:

$$G_p = G\left(1 - \frac{Z_i}{P_i} \cdot \frac{P_a}{Z_a}\right)$$

$$G_p = 7.4 \cdot 10^{11}\left(1 - \frac{0.94}{3800} \cdot \frac{600}{0.96}\right) \text{ std m}^3$$

$$G_p = 2.26 \cdot 10^{11} \text{ std. cu.ft.}$$

and the recovery ratio:

$$g_p = \frac{G_p}{G} \approx 85\%$$

7.3.8 Condensate Gas Reservoirs

A condensate gas reservoir is one containing a complex hydrocarbon mixture such that, by lowering the pressure, a liquid part called condensate is deposited in the reservoir. The production of gas is itself accompanied by a high proportion of condensate, which is a **light oil with strong market demand**. These reservoirs are generally rather deep ($Z > 2000$ m). The material balance is compiled from a detailed PVT analysis (Standing, Lindblad, Parsons method).

The most general case is that in which the in situ saturation of condensate remains lower than the critical saturation, i.e. in which the liquid phase deposited never becomes movable (the other alternative is the one in which this saturation becomes higher than the critical saturation, leading to simultaneous mobility of both phases). Differential analysis is used in this general case, proceeding by decompression steps of 300 to 400 psi, with the elimination of the volume of surplus gas in each case.

Note that, in the case of a water influx, the volume of the reservoir gas does not remain constant. In practice, however, it is assumed that this does not alter the results obtained by the differential PVT analysis carried out at constant volume.

Another remark: although the condensate is immobile within a reservoir, this does not apply around the well, where the condensate saturation is always higher, and where the condensate is always susceptible to entrainment by the gas. However, although the condensate also causes a **reduction** in the relative permeability to gas and in the **productivity of the well**, this mechanism, from the standpoint of the analysis of **reservoir** behavior, concerns only a very small volume, and the effect is consequently assumed to be unimportant for the material balance.

7.3.9 General Remarks

As we have already pointed out, the material balance represents the simplest simulation tool. But the geometry, the internal architecture of the reservoir, and its heterogeneities do not appear in it. The reservoir is considered as a sort of **large bubble** containing gas, oil and water (with a single pressure).

This means that, both for calculating the oil and gas in place and for production forecasts, the results given by this type of calculation must be interpreted with caution. Nevertheless, they provide very rapid **semiquantitative indications which can guide further studies**. The material balance can be used advantageously for relatively homogeneous reservoirs with a simple structure, which are rare, and especially for **reservoirs at the outset of production** (lack of data).

A specific remark is necessary concerning the calculation of the oil and gas in place. If this is done after a fairly short production history, the figure obtained may be lower than the real value (approximated by area/depth method).

This is because the volumes obtained by the material balance represent **"active" reserves** that have effectively participated in production. Certain parts of the reservoir, however, which are little drained, if at all, may still be in the initial state (P ~ initial P) and the corresponding volumes in place may be virtually inactive.

The **drive mechanisms** must be determined as early as possible. Is there a gas cap in an oil reservoir? A comparison between the initial pressure and the bubble-point pressure (PVT) can give the answer if no well has crossed the gas/oil interface. Yet caution is necessary, since some reservoirs have a highly variable bubble point.

An **aquifer's characteristics must be optimized (R_e/R_i and k)** as soon as permitted by the production history. At the very start of production, the uncertainty may be significant and recovery could vary considerably depending on the size of the aquifer. Avoid analogies with neighboring fields, which may spring unpleasant surprises.

7.4 STATISTICAL CORRELATIONS AND DECLINE LAWS

Statistical correlations and decline laws are empirical methods that were widely used some decades ago, and which may still be valuable, at least with respect to decline laws.

7.4.1 Statistical Correlations

These correlations have been obtained from the production data of a large number of **depleted** reservoirs. For example, those of ARPS (API) result from the analysis of 312 American oil reservoirs. Two classes of reservoirs have been distinguished: those subject to an active aquifer, and those without an active aquifer (of dissolved gas drive). The correlations have the following form:

$$R = a \left(\frac{\phi \left(1 - S_w\right)}{B_o} \right)^{\alpha} \cdot \left(\frac{k}{\mu_o} \right)^{\beta} \cdot S_w^{\ \delta} \cdot \left(\frac{P_i}{P_a} \right)^{\gamma}$$

with

R = recovery,

P_a = abandonment pressure,

$a, \alpha, \beta, \delta, \gamma$ = constants,

$\frac{\phi \left(1 - S_w\right)}{B_o}$ = volume of oil per unit volume of impregnated rock,

$\frac{k}{\mu_o}$ = oil mobility.

These correlations can be seen to disregard the geometry and heterogeneity. Charts are available for using these formulas immediately. But they are unreliable in the sense that a new reservoir treated in this way may deviate significantly from this mean statistical correlation. It could possibly be used after the discovery well has been drilled.

7.4.2 Decline Laws

These are **methods for extrapolating the parameters of a well**, particularly the **flow rate**, at the end of production of the field. These laws are essentially used for small reservoirs, or for very complex reservoirs that are difficult to model.

Two laws are routinely used for hydrated wells.

7.4.2.1 Exponential Decline of Flow Rate

$$Q_o = Q_{oi} \; e^{-\alpha t}$$

with a decline rate α.

It can be shown that the water-cut f_w is a linear function of the cumulative oil production N_p.

If this corresponds to the case of a well, it is easy to extrapolate f_w and the flow rate Q_o (t).

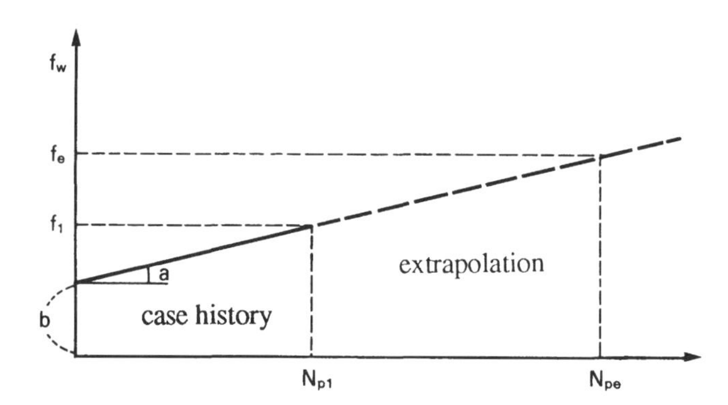

7.4.2.2 Hyperbolic (and Harmonic) Decline

This corresponds to log $(1 - f_w) = a\,N_p + b$.

Similarly, $(1 - f_w)$ and the flow rate Q_o (t) are extrapolated.

These declines at the wells must be checked continuously.

7.5 PRODUCTION IN FRACTURED FORMATIONS

7.5.1 General Introduction

Fractured formations generally exhibit a dual feature: the wells are good or very good producers in the fractured zones, and often mediocre elsewhere. The amount of oil and gas in place may also be fairly small.

The internal architecture of fractured reservoirs is more complex than that of matrix reservoirs. This stems precisely from the presence of an additional network of fractures in the porous medium, which results from tectonic forces which have "broken" the rock.

The fractures themselves have allowed water flows that have caused the deposition of watertight materials such as calcite, as well as the local dissolution of the matrix with the formation of cavities. Moreover, in the matrix blocks, these tectonic stresses give rise to another network of deposits of chemical residues, or stylolites, which form virtual barriers.

Thus three discontinuous units are superimposed on the matrix, which is of variable permeability (hard and brittle rock because fractured). These units are especially difficult to describe (Fig. 7.17):

(a) The fracture network.
(b) Channels opened in these fractures and cavities.
(c) The network of stylolites.

The image of a fractured formation is more complex than a simple superimposition: blocks that are slightly (or not) permeable which contain volumes in place + network of open and/or closed fractures causing most of the flows towards the wells.

Yet a distinction is drawn between "porous fractured formations" and "non-porous fractured formations". The latter can be expected to contain very small volumes in place as a rule.

The order of magnitude of the opening of useful fractures is variable, often **a few dozen microns**. Analyses of these formations show that the thickness of the water film coating the fracture walls does not exceed 0.02 μm.

One important consequence is that, in the fractures, S_w is negligible. The same applies to the capillary pressure P_c, given the relatively large size of

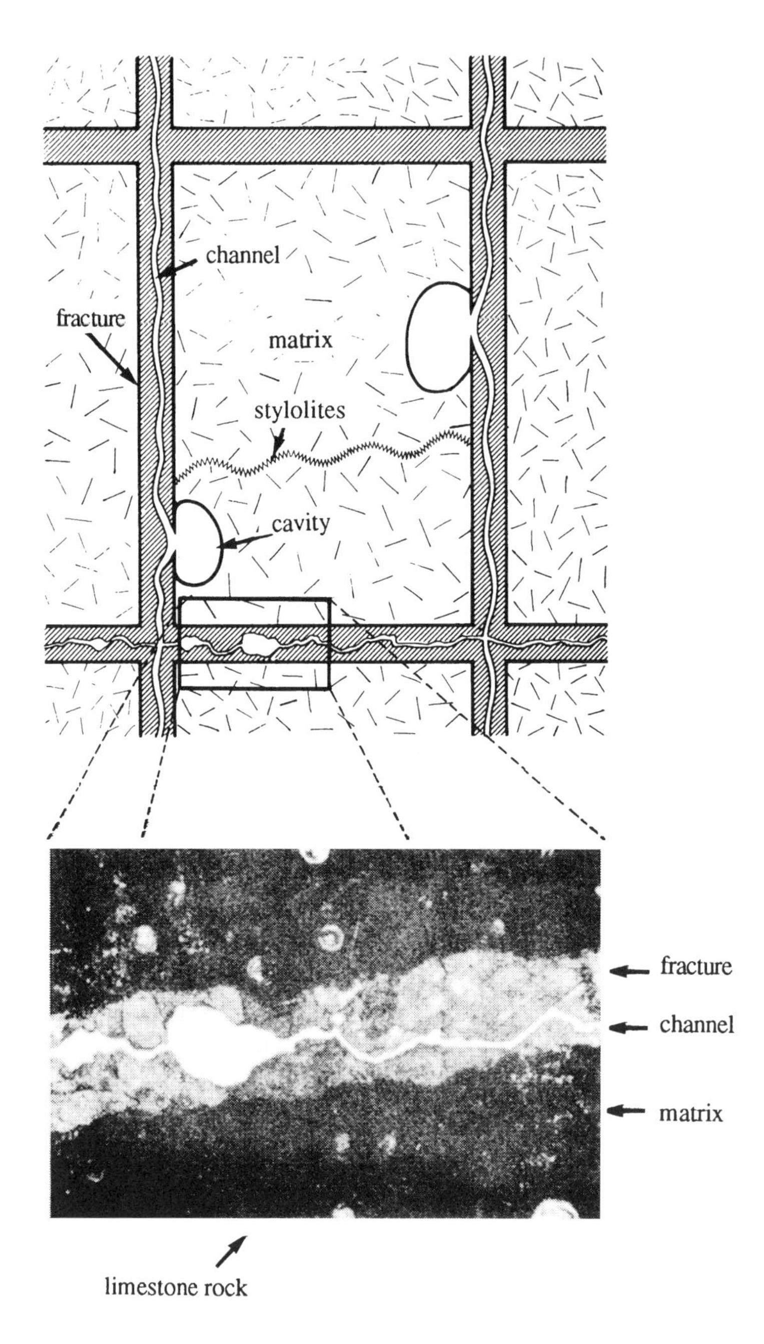

Fig. 7.17 Diagram of a piece of matrix and illustration of secondary porosity within a fracture.

the fractures in comparison with the pores in the blocks. Thus, as a rule, the fractures are not the seat of capillary mechanisms.

As to the geometry of the fracture network, the following can be distinguished (Fig. 7.18):

(a) "Matches": system with three main fracture planes.
(b) "Sheets": system with a single series of parallel planes.
(c) Composite systems.

Some orders of magnitude concerning fractured formations:

(a) Block dimensions: from a few centimeters to a few meters.
(b) Opening of useful fractures: normally a few dozen microns, sometimes a few millimeters (but combined with very high permeabilities).
(c) Fracture permeability: from a dozen millidarcys to a dozen darcys.
(d) Matrix permeability: often low (one millidarcy or less).
(e) Fracture porosity: generally about 0.01 to 1% (related to the total volume).
(f) Vug porosity (karsts): about 1%.
(g) Matrix porosity: variable, sometimes nil (karsts).

One of the major difficulties in characterizing the system (as for a conventional porous medium) stems from the lack of information, since most of the data are obtained from wells, namely from a strictly localized system.

Thus the reservoir specialist is faced with the following problem:

How to describe the internal architecture of the reservoir comprising the matrix unit and the discontinuous units? How to describe the fluid flow between the matrix blocks and the fractures on the one hand, and between the fractures and wells on the other?

7.5.2 Geological Aspect

7.5.2.1 Description of Cores

Fractures are characterized by their opening, filling, length, dip and azimuth of the fracture plane and distance between two consecutive fractures. Parameters concerning stylolites are also measured (Fig. 7.18).

Data obtained by the visual observation of cores are correlated from well to well, using observations of any outcrops that exist, and of rock mechanics models.

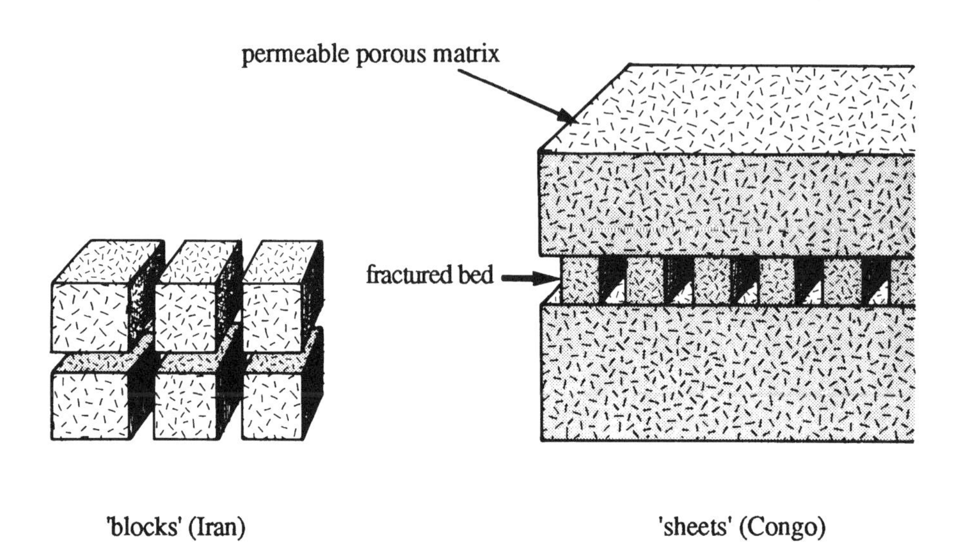

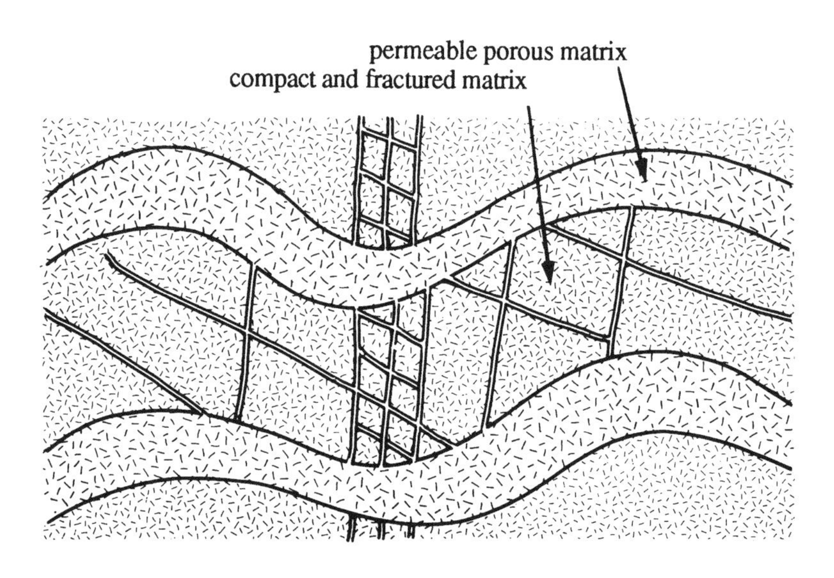

Fig. 7.18

Typical fractured reservoirs.

7.5.2.2 Observation of Outcrops

It is necessary to have "fresh" and fairly extensive outcrops for the scale effect to give any overall statistical value. The ideal outcrop is a mined quarry. The most important measurements concern the fracture density per unit length or area, and their trend.

7.5.2.3 Rock Mechanics Model

Information on the stress (geostatic pressure, fluid pressure, tectonic forces associated with the distortion of the unit) and on an experimental criterion of rock fracture, plus a number of general hypotheses, serve to describe a probable state of fracturing of the rock, which helps to improve the image of the fracture distribution within the reservoir.

7.5.2.4 Visual Analysis in the Borehole

Exceptionally, because a very light mud is needed (hence a pressure ~ hydrostatic pressure), a televised recording helps to observe the rock in situ. A fine example is the "downhole camera", a film shot by *SNEA(P)* on an Emeraude well offshore Congo (Fig. 7.19).

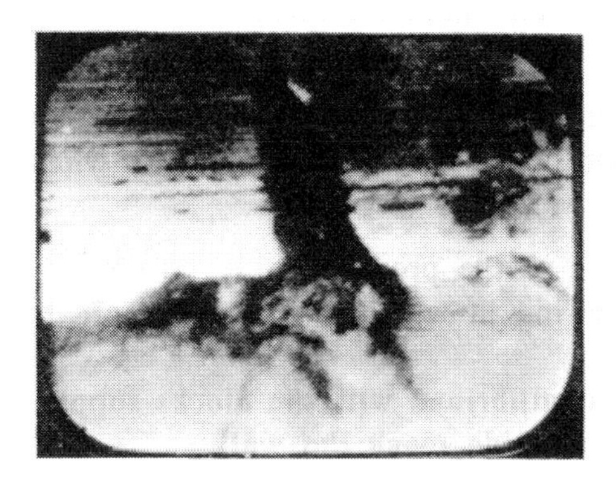

C13 - 275.20 m. Vertical fracture intersected, with a change in facies.

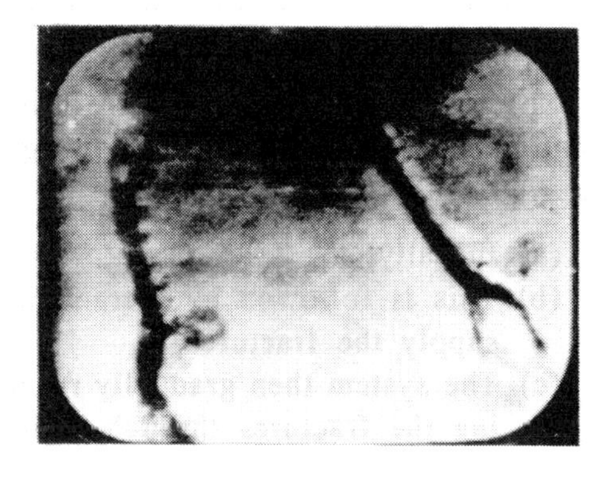

C10 - 276.65 m. Fractured limestone bed.

Fig. 7.19

Television photographs, Emeraude Field (Congo) (enlargement: about 1).

7.5.2.5 Reservoir Seismic Shooting

A significant **decrease** in the quality and continuity of **markers** is sometimes observed in areas of intense fracturing. So it is possible to survey the extension of these zones, after calibration by one or more wells (cores).

7.5.3 Data Obtained from Logs and Production Tests

The essential information includes the permeability k_f, and, to a lesser degree, the porosity ϕ_f of the fracture system.

7.5.3.1 Logs

Two types of instruments can provide essential data: production instruments (flow metering + temperature measurement), and acoustic ones with analysis of wave attenuation (including *SNEA(P)*'s EVA).

Seismic instruments in particular can locate the fracture zones, and help to supplement the core analysis (recovery < 100%, coring very expensive offshore).

7.5.3.2 Production Tests

Initial tests showing very high production indexes materialize the presence of fractured zones. For oil (and for gas), the relationship between the flow rate and ΔP is not linear, since high flow rates generate turbulence near the hole.

The curve of flowing pressure drop, or of shut in pressure build-up, does not display a simple line in semi-log coordinates. **What in fact occurs when a well starts flowing?**

(a) Initially, the fractures provide most of the production.
(b) This is followed by a transition period in which the blocks begin to supply the fractures.
(c) The system then gradually reaches equilibrium, with the blocks supplying the fractures through which the fluids reach the well.

This gives a first line with slope $h\,k_f$, then a connecting curve, and the second line with the same slope (since the fractures still supply the well). This is the "bayonet" curve. The first part is often marked by the wellbore storage and skin effect, and the third is not visible over short-term tests. The interpretation is hence generally very difficult. This is carried out mainly from type-curves.

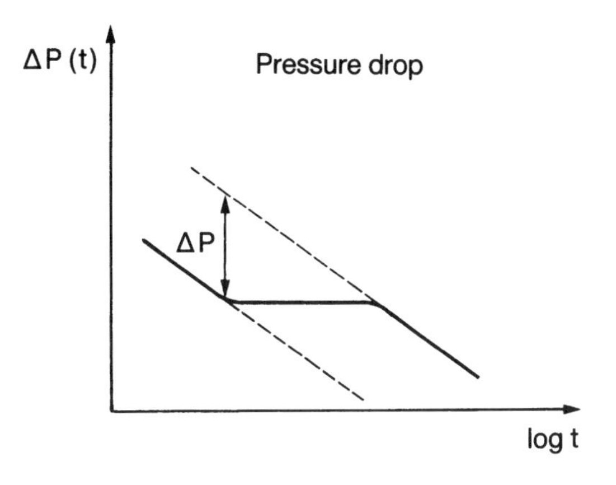

Fig. 7.20

7.5.4 Drive Mechanisms

The superimposition of a fracture network on a porous matrix represents the specificity of the system. The analysis of drive mechanisms thus con-cerns the supply to the fractures by the blocks, in one-phase or multiphase flow.

Two key mechanisms are involved, expansion and exudation.

7.5.4.1 One-Phase Flow, Expansion

The expansion is due to the total compressibility of the block + fracture system, which is not very different from that of the same formation without fractures.

7.5.4.2 Two-Phase Flow

A. Expansion

Oil/water: identical for expansion, but with the water rising very fast through the fractures, which can therefore flood the well very rapidly, hence the **detrimental** aspect for production.

Gas/oil: since the essential compressibility is that of the gas, there is not much difference from a nonfractured formation, apart from the rapid formation of a secondary gas cap due to the fractures.

B. Exudation

The combined action of capillarity forces (static imbibition) and gravitational forces cause the expulsion of the hydrocarbon from the block into the fractures. This is **exudation.**

Oil/water and water-wet rock: the water tends to penetrate spontaneously into the block by imbibition, and also by gravity from the top downward. Both processes are cumulative.

Oil/water and oil-wet rock (some limestones): the capillary forces oppose the entry of the water into the block. Exudation is possible only if the gravitational forces prevail, and hence if the blocks are large in size (Fig. 7.21):

$$q_i = \frac{k_o}{\mu_o} \cdot \frac{h_b (\rho_w - \rho_o) g - P_d}{h_b}$$

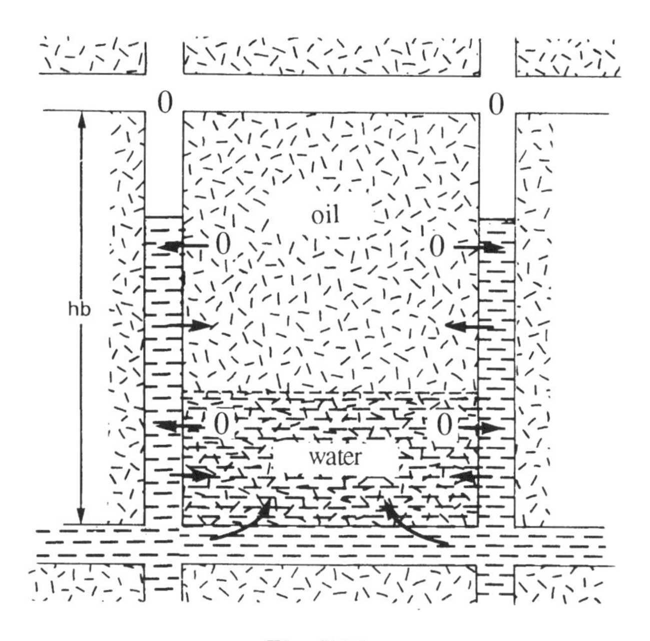

Fig. 7.21

Schematic exudation from a block.

with:

q_i = flow rate of oil leaving the block initially per unit area,
P_d = displacement pressure,
h_b = block height.

The O/W boundary is clear in the fractures ($P_c \cong 0$) and, by contrast, gradual variations of S_w occur in the blocks.

Gas/oil: the gas is the non-wetting fluid. Hence opposition between the gravity and capillarity terms. The same applies as in the previous case. Exudation can only occur with very large blocks.

Thus exudation is significant in the oil/water case with a water-wet rock, and only for large blocks (decimeter or meter dimensions) in other cases.

Figure 7.22 summarizes the drive mechanisms in fractured reservoirs (oil with gas cap and water level here).

Note also that in fractured limestones with large blocks (quite commonly found), it may be advantageous to use exudation caused by gas, hence the idea of reinjecting gas into the upper parts.

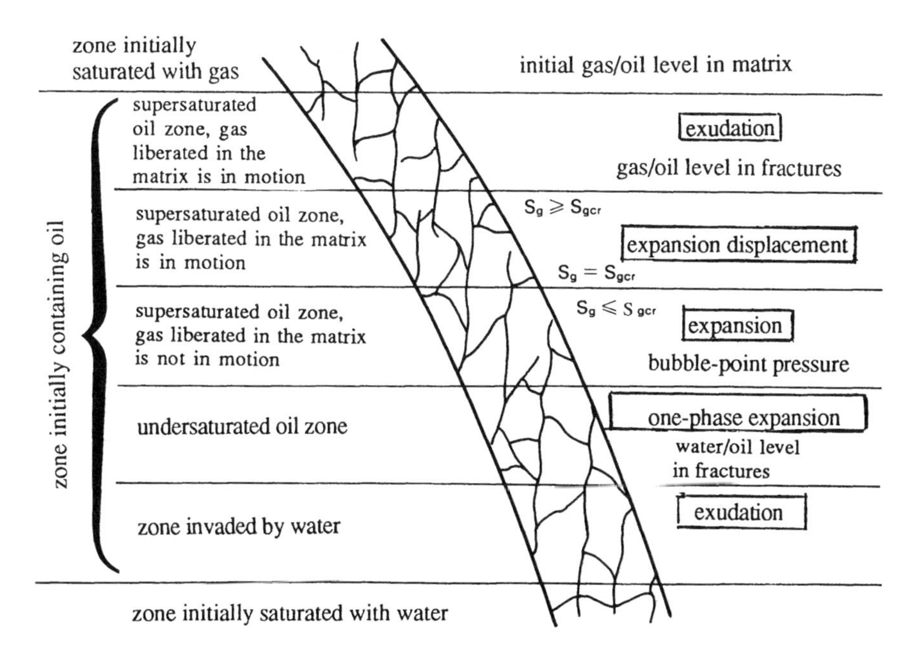

Fig. 7.22

7.5.4.3 Transfer Function (Exudation)

This represents the quantity of oil leaving the block as a function of time. It can be obtained by numerical and computer methods, by a simplified calculation or by an experimental study in the laboratory. Research is under way, especially by numerical methods, and each case is nearly a specific one. Figure 7.23 illustrates a transfer function.

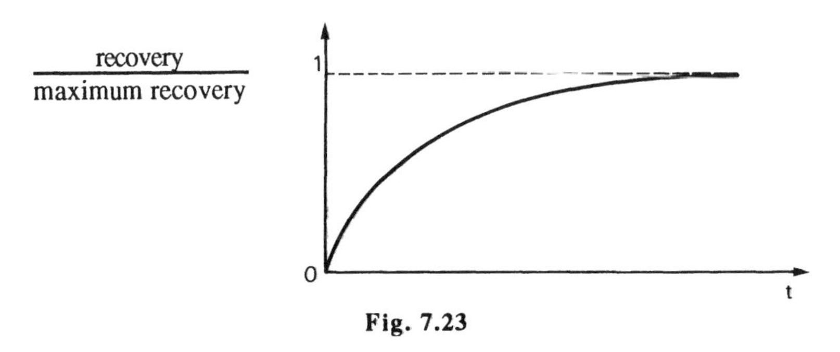

Fig. 7.23

Note that, for **karsts** (nonporous fractured reservoirs), recovery is high (60 to 80%), because displacement by water or gas is excellent. There are practically no capillary mechanisms, and the fractures and vugs are swept.

Note: The figures in **Section 7.5** are taken from the reference "Reservoir Engineering Aspects of Fractured Formations" by L.H. Reiss *(SNEA(P))*, course at the *ENSPM*, *Editions Technip*, 1980.

7.6 A SPECIFIC PRODUCTION TECHNIQUE, THE HORIZONTAL DRAIN HOLE

The horizontal well technique was recently developed, and only thanks to progress achieved in the accurate monitoring of the well bore trajectory.

7.6.1 What are the Cases for Application of the Horizontal Drain Hole (in Terms of Reservoir Engineering)? What are the Advantages?

It is well known that the **productivity** of a well is proportional to h . k. It is therefore **reduced** if a reservoir is **thin**. Hence the already old idea of the horizontal drain hole. With this method, productivity rises with the penetration distance into the reservoir, but slower because the pressure/distance relationship is also involved in logarithmic form:

$$P\,I_h \approx \frac{2\,\pi\,k_a\,\dfrac{L}{\mu\,B}}{\dfrac{L}{h}\,\mathrm{Ln}\left(\dfrac{4\,R_e}{L}\right) + \mathrm{Ln}\left(\dfrac{h}{2\,\pi\,r_w}\right)}$$

with

L = drain hole length,
h = reservoir thickness,
R_e = (equivalent) drainage radius,
r_w = well radius,
k_a = $\sqrt{k_h \cdot k_v}$
PI_h = production index (horizontal oil well).

Note that, for a gas well, the productivity can rise even more due to the reduction in the flow speed (length L), hence a decrease in turbulence and quadratic pressure drops.

The gain in productivity over a vertical well is commonly multiplied by a factor of 3 to 5. The cost of drilling and completion, which is variable, lies within a range of 1.5 to 2 (Figs 7.24 and 7.25).

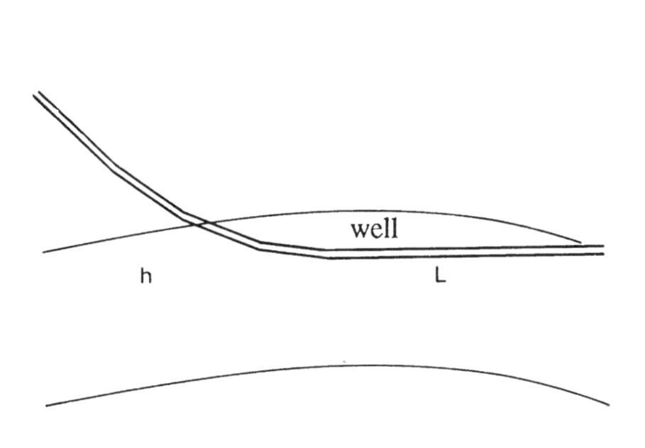

Fig. 7.24

Horizontal drain hole.

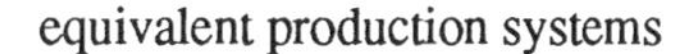

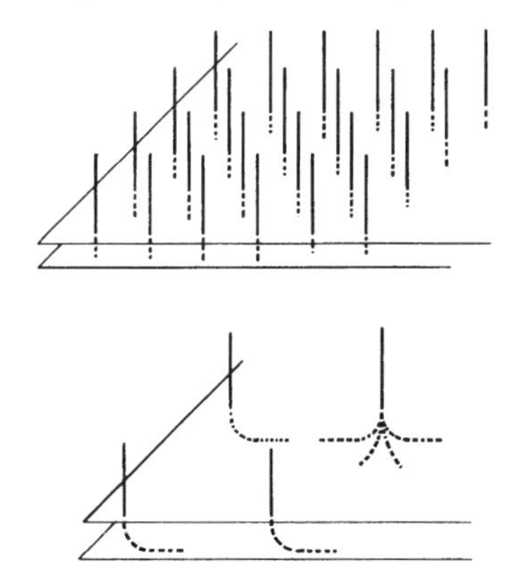

Fig. 7.25

Reduction in number of wells by use of horizontal boreholes.

Figures 7.26 and 7.27 show the history of the first horizontal wells and the example of a number of wells drilled in France, a country that has pioneered this system.

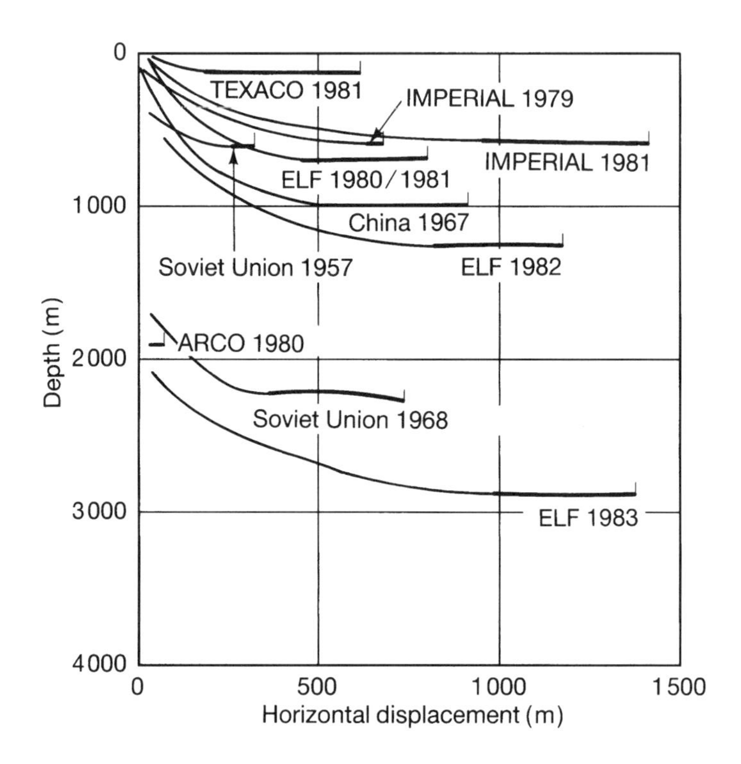

Fig. 7.26

World panorama of existing horizontal wells (1984).

7.6.2 Ideal Situations

7.6.2.1 Fractured Reservoirs

A horizontal drain hole, positioned perpendicular to a subvertical fracture network, intersects a large number of fractures and can achieve very high productivity gains (Fig. 7.28).

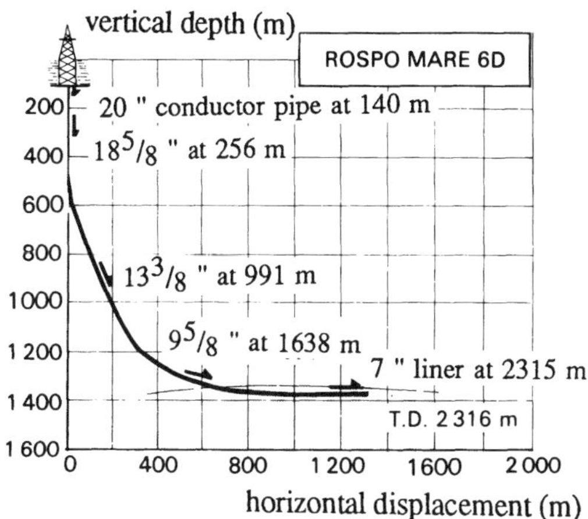

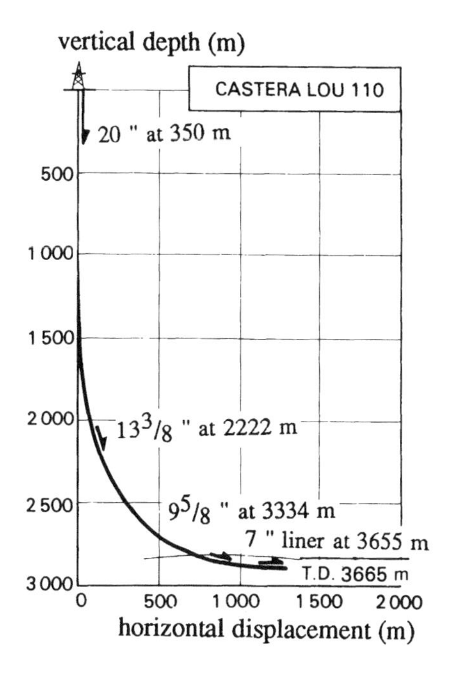

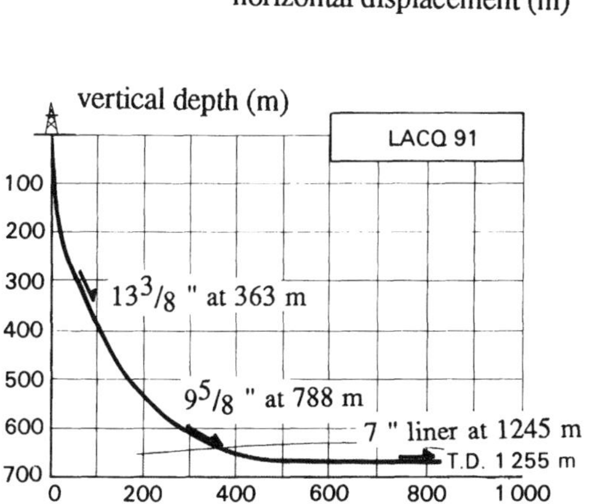

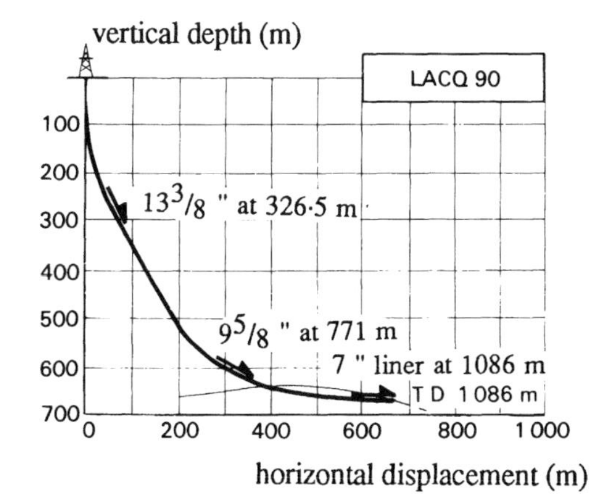

Fig. 7.27

Profiles of Elf horizontal wells.

7.6.2.2 Karst Reservoirs (Nonporous Fractured)

These are extremely heterogeneous reservoirs. A horizontal drain hole has better chances of intersecting high productivity zones. The example of the Italian Rospo Mare Field, produced by *Elf Italiana*, is significant in this respect: for this reservoir, the productivity can be multiplied by a factor ranging from four to ten times that of a vertical well.

This reservoir will be developed completely using horizontal wells drilled in a "star" pattern.

7.6.2.3 Reservoirs with Aquifers

The horizontal drain hole has two additional assets: it can be placed at the top of the reservoir to obtain a sufficient water blanket. In addition, the great length of the drain hole stresses the aquifer less, per unit length of drain hole. Furthermore, the vertical sweep is more effective and recovery is higher. In some cases, production may remain anhydrous for a much longer time (Fig. 7.29).

This is also true for protection against gas influxes from a gas cap, by placing the drain hole in the lower part of the oil zone (if there is no aquifer).

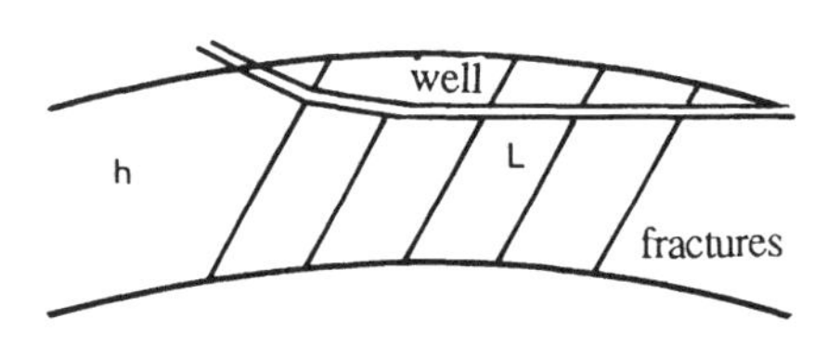

Fig. 7.28

The horizontal drain hole intersects many more factures.

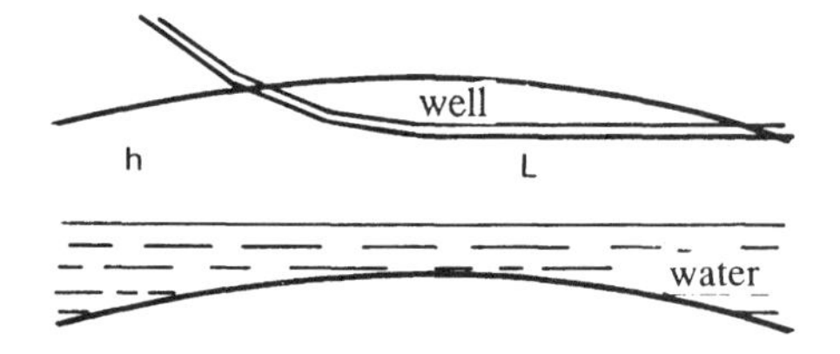

Fig. 7.29

The cone is replaced by a water crest.

7.6.2.4 Enhanced Recovery

The horizontal drain hole seems advantageous for **steam** injection, the injection tube being stretched over a much greater length.

7.6.2.5 Value for Reservoir Characterization

Thanks to the performance of horizontal coring and logging, horizontal drain holes have proved to be a good way for determining the lateral changes in facies and the changes in **fluids in porous media**. The drilling of horizontal wells in a "star" pattern may prove extremely fruitful in this respect.

7.6.2.6 Unfavorable Case

In **stratified** reservoirs, with very low or zero vertical permeabilities, recovery is mediocre, since only one (or a few) beds can be drained, and the "average" permeability is lower:

$$k_a = \sqrt{k_h \cdot k_v}$$

In this case, it is recommended to drill sharply inclined wells, but not horizontal, in order to drain the different levels.

It must also be pointed out that this type of recovery adds a further complication in reservoir engineering studies. Thus every project must be the subject of a thorough and cautious investigation.

Chapter 8

SECONDARY AND ENHANCED OIL RECOVERY

8.1 GENERAL INTRODUCTION

Recovery by natural drive mechanisms rarely exceeds **30 to 40%**, and is often lower for **oil** reservoirs.

This is why the need soon appeared **to inject energy into these reservoirs to achieve better recovery**. The first processes employed (injection of water or gas) were employed in a second phase, after the decompression of the reservoir, hence the name secondary recovery.

Today, these injections are sometimes used from the start of the life of the reservoir. Before undertaking one of these processes, it is still necessary to make sure of the inadequacy of natural mechanisms, and this is difficult to determine at the start of field production. It is also generally necessary to have a minimum of production data (one to two years of production).

Other more elaborate techniques have been developed and employed on oil fields in recent decades, and their use is justified by the determination to achieve higher recovery. These "improved" or tertiary methods are termed **enhanced oil recovery** (EOR).

A large group of these methods, thermal methods, are applied to heavy oils. Improved production results not only from the displacement of one fluid by another, but also, and above all, by an input of heat.

The new artificial recovery processes could raise world recovery levels to between 30 and 40% in the near future. This is based on the assumption that the price of crude oil will rise rather than fall in the long term.

The challenge is a major one, because an increase of 1% in the recovery ratio means the production of an additional six billion tons of oil, i.e. the equivalent of two years of present world production.

Figure 8.1 illustrates this potential improvement in billions of tons.

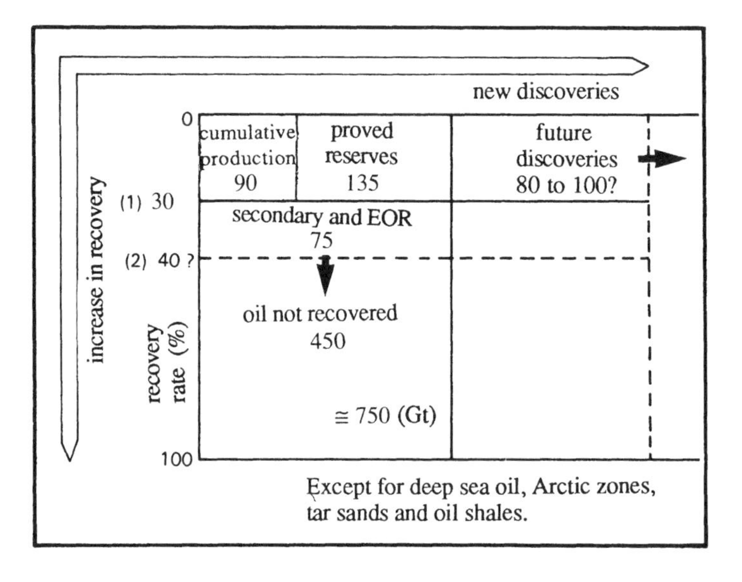

(1) Present average recovery rate.
(2) Foreseeable recovery rate.

Fig. 8.1

Oil reserves and recovery.
(*Source: IFP*, 1990).

Typical Applications

(a) Reservoirs with low natural energy: undersaturated oil, very small or low-permeability aquifer.
(b) Low-permeability or large oil reservoirs (wide pressure differences between producing wells and aquifer or gas cap).

(c) Heterogeneities: unfavorable spatial distribution of natural inflows of water or gas, causing delays in the sweeping of certain parts of the reservoir (example, local permeability barriers).
(d) Condensate gas reservoirs.
(e) Mediocre conventional secondary recovery. Search for an improvement in recovery by EOR processes: injection of water with chemical additives, of miscible fluids, steam, air, etc.

Note:

Secondary recovery and EOR do not concern dry and wet gas reservoirs. For condensate gas reservoirs, the goal is to recover more natural gasoline by cycling dry gas. **These methods accordingly concern additional recovery of liquid hydrocarbons.**

The different methods can be summarized by the table below:

Secondary Recovery and EOR

- **Secondary recovery ≈ "Conventional" artificial recovery**
  - Water flood
  - Flooding by (immiscible) hydrocarbon gases

  (condensate gas reservoirs: cycling of dry gas)

- **Enhanced oil recovery ≈ "Improved" or "tertiary" recovery**
  - Miscible methods (CO_2, CH_4+, etc.)
  - Chemical methods (polymers, microemulsions, etc.)
  - Thermal methods: heavy oils (steam, in situ combustion)

An **economic constraint** associated with injections lies heavily on all these methods. Contrary to a production well which, especially onshore, produces rapidly implying a fairly rapid "payback", an injection well achieves additional production for the reservoir only after a certain time, and the pay-back time is therefore longer.

Given the prices of crude oil at the present time (1990), conventional processes (water flood and gas injection) in general, and sometimes steam and miscible methods, are economically advantageous. The other processes are nevertheless undergoing intensive research and are being tested in experimental pilot schemes.

8.2 FACTORS INFLUENCING RECOVERY

The different types of secondary recovery reveal that all these oil production processes involve sweeping the reservoir between injection wells and production wells. They always imply fluid flow, and accordingly a number of reservoir rock characteristics have a strong influence on recovery, together with the types of fluids in place and injected fluids.

We shall discuss the extent and the way in which these characteristics tend to act on sweep efficiency, and shall then analyze the importance of the volume injected and the injection pattern.

8.2.1 Reservoir and Fluid Characteristics

8.2.1.1 Reservoir Geology

Since sweep is the result of flows between injection and production wells, one of the conditions for success is the absence of any impermeable barrier opposing this circulation. While this principle is obvious, the detection of impermeable barriers in the reservoir is often much more haphazard. Some formations, such as shale/sandstone or carbonate formations, exhibit a sedimentation of permeable materials and close-packed materials, creating interlocking masses that make fluid circulation extremely difficult.

A small fault that is difficult to identify could also prevent any sweep in a specific zone. The detailed analysis of **cores** and **logs**, as well as **interference tests**, are essential for understanding interwell communication.

8.2.1.2 Permeability

A. Absolute value

Good permeability is favorable for the following reasons:

It indicates the existence of large-diameter pores, in which the initial oil saturation is high. Moreover, the thresholds are less constricted than in compact media, and desaturation by displacement is more complete, because of the action of capillary forces. This twofold reason results in a larger swept oil volume, for the same volume of displacement fluid injected.

Good permeability also provides a high flow rate, thus increasing the well spacing and decreasing the flooding pressure required.

B. Heterogeneities

Let us consider the schematic reservoir model as a regular stratification of beds of different permeabilities. Let us also assume that there is no possible communication between these beds (Fig. 8.2).

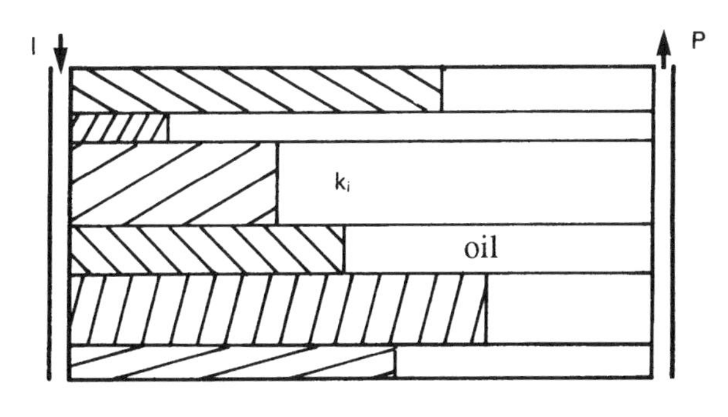

Fig. 8.2

The movement of the displacement front is faster in the more permeable beds than in the other beds. Since in actual fact the displacement fluid, whether water or gas, usually has a lower viscosity than the oil, this imbalance between the movement of the front in the different layers increases with time. Note however that, if communication exists between the layers, imbibition reduces the impact of this mechanism.

If the differences in permeability are wide, this means that the displacing fluid breaks through into the production well via preferential paths. An extreme case would be that of a fracture that places the injection and production wells in direct communication. All the matrix oil would remain trapped. Thus **great heterogeneity in permeabilities is an unfavorable factor for injection.**

Other properties have a lesser influence on the validity of injection: capillary pressures, the dip of the beds, and the depth of the reservoir in particular.

8.2.1.3 Viscosity of Fluids and Mobility Ratio

Darcy's Law immediately shows that, in one-phase flow, the oil flow rate is lower and hence recovery is economically poorer for viscous oils.

Moreover, for a given water cut (or a given GOR), the average saturation of injected fluid behind the front is lower, and hence the amount of oil trapped is greater for a more viscous oil: an example is shown in Fig. 8.3.

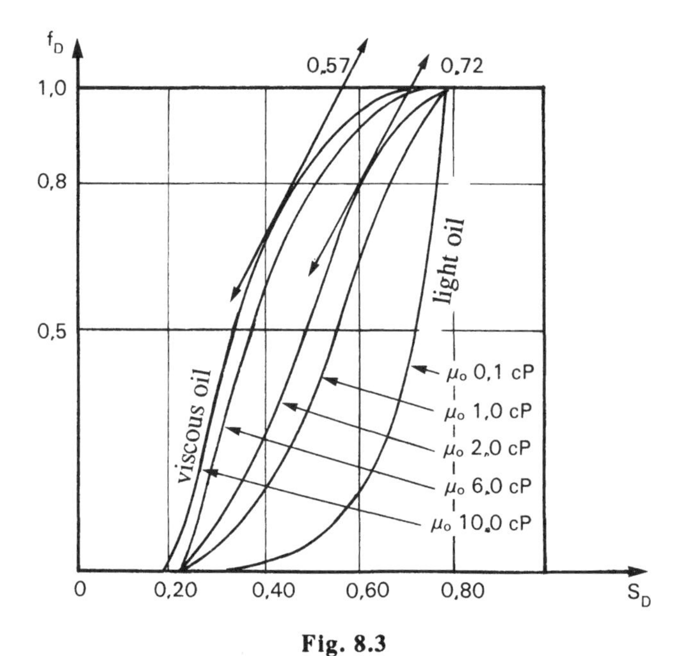

Fig. 8.3

To sum up the influence of the reservoir and fluid characteristics, **recovery** is **higher** if there are:

(a) Few or no barriers.
(b) Good or high k.
(c) Narrow range of heterogeneities.
(d) High-angle dip.
(e) Low viscosity: light oil.
(f) High viscosity of injected fluid (clear advantage of water over gas).

8.2.2 Injection Characteristics

Injection is essentially governed by the volumes of fluid available, the type of fluid injected, and the injection pattern.

8.2.2.1 Volume of Injected Fluid

If this volume, in reservoir conditions, is equivalent to the volumes of oil, gas and water produced, the **pressure is maintained**. If the injected volume is higher, recompression occurs: in the case of a highly depleted reservoir. Yet the injected volume primarily depends on the possible injection fluid sources (aquifer levels, associated gas or gas-bearing levels), and it is not always possible to maintain the pressure.

The better injection is distributed to sweep a maximum amount of oil, the more efficient it is. So efficiency also depends on the number of wells. The technico-economic optimum indicates the number of wells to be drilled.

8.2.2.2 Type of Fluid

As we have already pointed out, injection is more effective if the injected fluid is more viscous and hence the mobility ratio M lower. Water is a good vector for light oils, but not as effective for more viscous oils. In any case, injected gas poorly sweeps the oil in place, since its viscosity is very low. This is discussed in the section on "Analysis of Efficiencies".

8.2.2.3 Injection Patterns

The relative layout of the injection and production wells depends on the geology of the reservoir, on its fluid content, and on the volume of impregnated rock that must be swept.

Two injection patterns can be distinguished and may be used jointly on a number of reservoirs:

(a) Grouped flood, in which the injection wells are grouped locally.
(b) Dispersed flood, in which the injection and production wells are in an alternate arrangement.

A. Grouped flood

In a fairly high-dip reservoir, an attempt is made to locate the injection wells so that gravitational forces make displacement as regular as possible. Especially if a reservoir has a gas cap and/or an aquifer, it is advisable to inject either gas into the gas cap, or water into the aquifer close to the water/oil interface (**peripheral** flood).

This is because the interfaces move "piston-like" over considerable areas, they advance slowly and uniformly, and only break through to the well later, which is desirable. This is termed grouped flood (Fig. 8.4). Note also that the injectivity is maximum because of the absence of relative permeabilities.

B. Dispersed Flood

On the contrary, if the reservoir is virtually horizontal and extensive, gravity cannot be exploited as above, and only a limited zone is effectively flooded, particularly in a **low-permeability** reservoir.

The production and injection wells are laid out in a fairly regular pattern: this is termed **dispersed** flood in the oil zone (Fig. 8.5).

A number of patterns are employed: the wells are laid out in a line or alternately with a **five-spot**, seven-spot or nine-spot pattern (i.e. five, seven or nine wells).

These three patterns involve a ratio of the number of injection to production wells of 1/1, 1/2 and 1/3 respectively. The choice of a pattern is determined by calculation or by simulation on a model.

However, this is theoretical. In actual conditions, it is necessary to evaluate the **various heterogeneities**. They are identified only gradually, by successive boreholes. This means an adjustment of the actual pattern with respect to the basic forecast pattern, involving studies on models. Sweep problems associated with possible faults must be considered very closely (Fig. 8.6).

Example of well sites on a field (Fig. 8.7)

A large carbonate field in Abu Dhabi is characterized by a wide variation in petrophysical properties from the south to the north of the structure. The south is characterized by good permeability: hk = 90 m . 400 mD = 36,000 mD/m, and the north, by contrast, is mediocre: hk = 30 m . 50 mD = 1500 mD/m.

Consequently, a five-spot pattern was selected in the north, while peripheral flood was selected for the south.

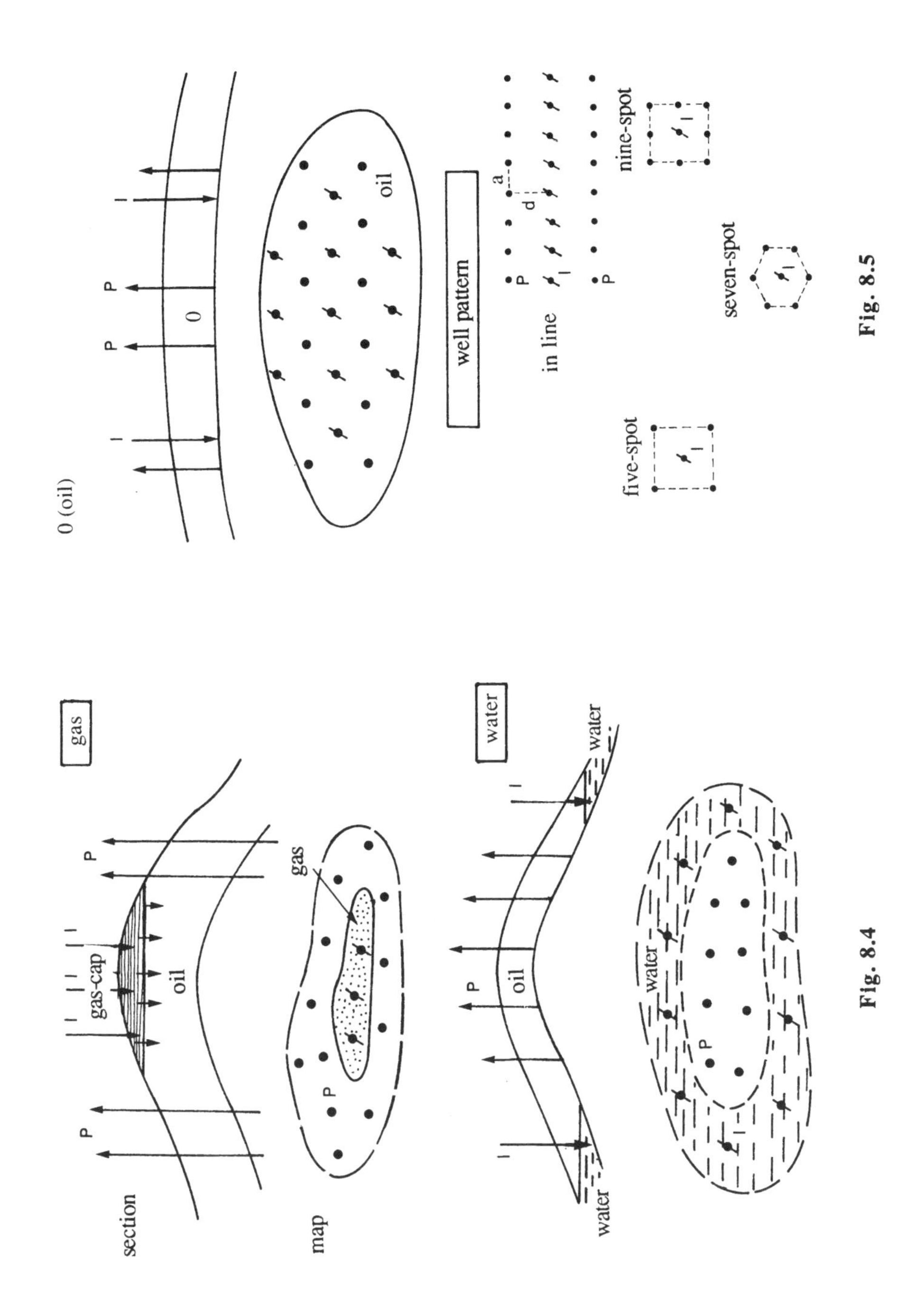
gas
section
gas-cap
oil
P
I
map
gas
water
oil
water
water
0 (oil)
0
oil
well pattern
in line
a
d
five-spot
seven-spot
nine-spot

Fig. 8.4

Fig. 8.5

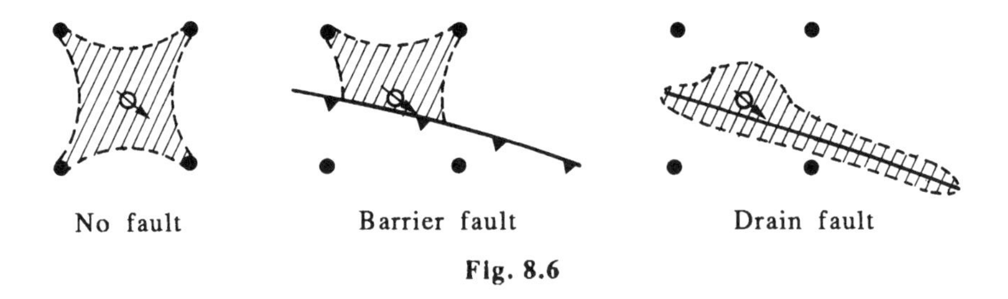
No fault
Barrier fault
Drain fault

Fig. 8.6

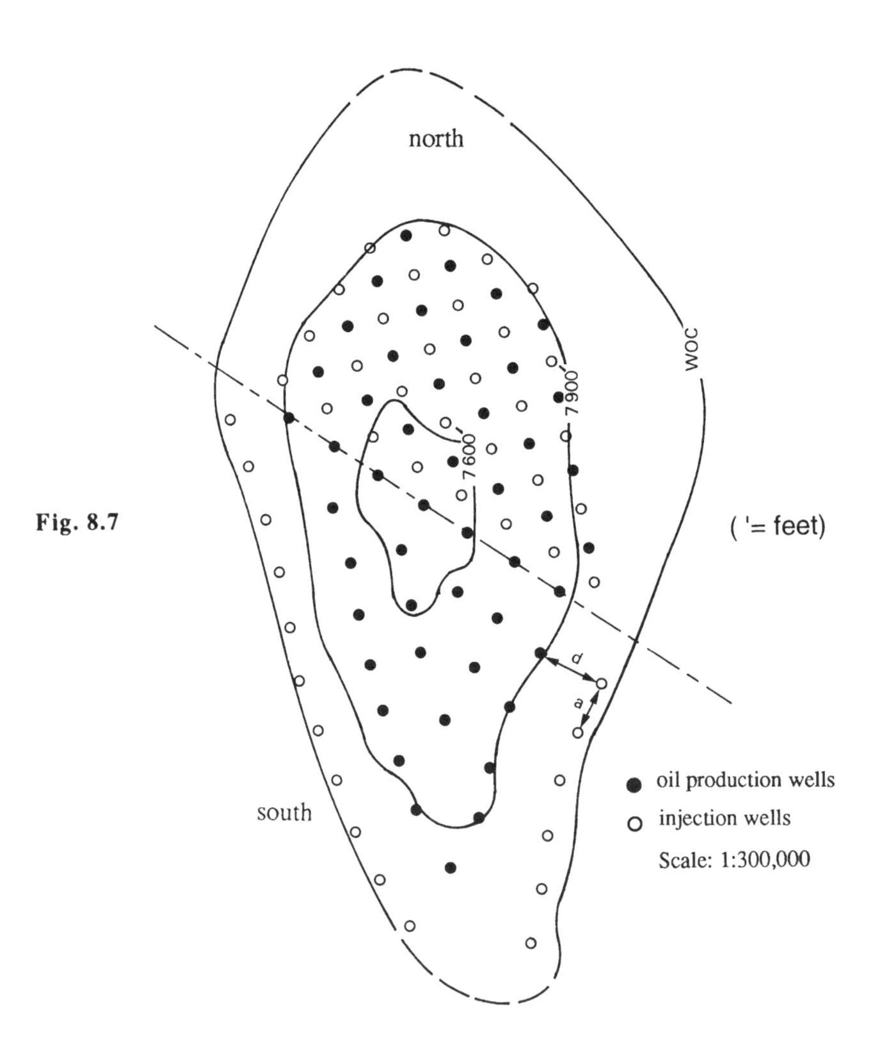
north
WOC
7900
7600
('= feet)
d
a
south
oil production wells
injection wells
Scale: 1:300,000

Fig. 8.7

Injection Well Pattern (Summary)

a. Grouped or Local Flood

Peripheral flood (water) or central flood (gas). Predominant gravity and low pressure gradients (high k).

Choice of well site depending on type of fluid injected.

Advantages:

(a) Front with wide area, hence not advancing too fast, in relation to the volume injected.
(b) Countercurrent gravity vector.
(c) No k_r, hence maximum injectivity.

b. Dispersed Flood in Oil

Valid for mediocre k. Used offshore (shorter interwell distances).

Regular geometric patterns:

(a) In line: straight, **staggered**.
(b) In five-, nine-, seven-spot patterns (reverse patterns):

$$\left(\frac{\text{Number of injection wells}}{\text{Number of production wells}} : 1, \frac{1}{3}, \frac{1}{2}\right)$$

In general:

(a) Five-spot for water or staggered lines.
(b) Nine-spot for gas or staggered lines.

8.3 ANALYSIS OF EFFICIENCIES

Recovery is generally analyzed on models. However, a **simplified calculation**, already mentioned in the study of natural drive mechanisms, allows a **necessary approach** to any more intensive study, and/or any study requiring further information about the reservoir. We shall accordingly define the concept of efficiency in flooding.

8.3.1 Injection Efficiency and Definition

The total efficiency is the recovery factor (for the zone subjected to flooding) in reservoir conditions:

$$E = \frac{N_p \, B_o}{V_p \, S_{oi}}$$

with S_{oi} at the **start** of flooding.

The total efficiency E of flooding can be defined as the product of the following three efficiencies (Fig. 8.8a):

$$E = E_A \, . \, E_V \, . \, E_D$$

with

E_A = areal sweep efficiency (in the same phase as the bed),
E_V = vertical or invasion efficiency (in vertical cross-section),
E_D = displacement efficiency, at the scale of the pores (miscroscopic efficiency).

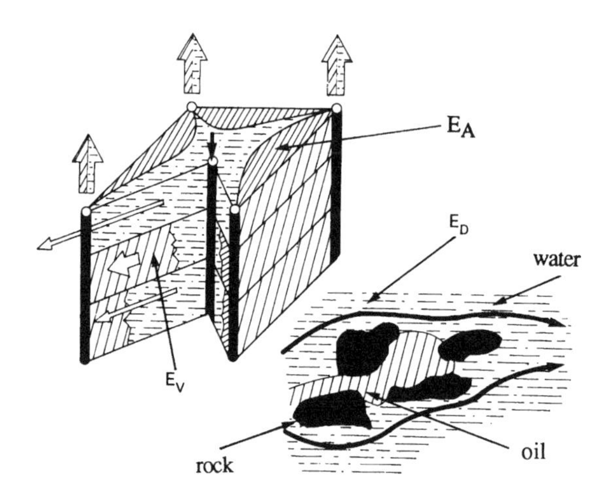

Fig. 8.8a

8.3.2 Areal Sweep Efficiency E_A

$$E_A \text{ (areal sweep efficiency)} = \frac{\text{Area swept by the front}}{\text{Total area}}$$

The surfaces are viewed in a horizontal plane (or bedding plane).

For a five-spot pattern, for example, the diagrams in Fig. 8.8b show the "front" at three times, for a homogeneous medium and draw off at four equal wells.

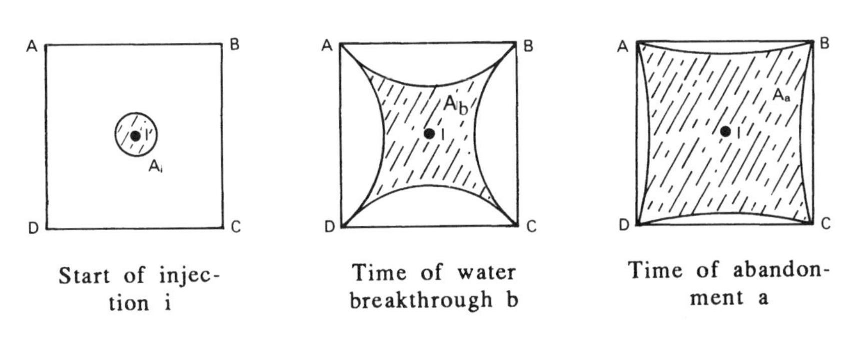

Fig. 8.8b

The area sweep efficiency is written:

$$E_{Ai} = \frac{A_i}{\text{Area ABCD}} \qquad E_{Ab} = \frac{A_b}{\text{Area ABCD}} \qquad E_{Aa} = \frac{A_a}{\text{Area ABCD}}$$

E_A depends on the time (volume injected), the well pattern, and also on the **mobility ratio M**. Let us consider the direct streamline between an injection and a production well, and assume that the flood fluid is more mobile than the displaced fluid (M > 1).

Since the total pressure drop over the streamlines is constant, the pressure gradient is higher (shorter distance), and the speed of the front is greater on the direct line. Hence, if a bulge is formed in the front, it tends to lengthen and the flood fluid is produced before the zone between the injection and production wells is thoroughly swept.

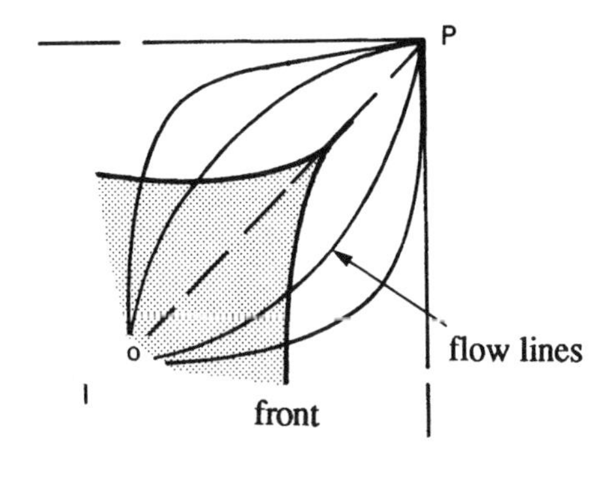

Fig. 8.8c

The areal sweep efficiency decreases with rising M. A number of authors have determined this variation by calculation or simulation. Caudle and Witte obtained the following results (Fig. 8.9), as a function of the fractional flow rate of displacing fluid f_D, for a five-spot pattern.

Note that, at the breakthrough (arrival of the displacing fluid at the well), i.e. $f_D = 0$ in Fig. 8.9, we have $E_A = 70\%$ for $M = 1$ which approaches $E_A = 40\%$ when M is very large.

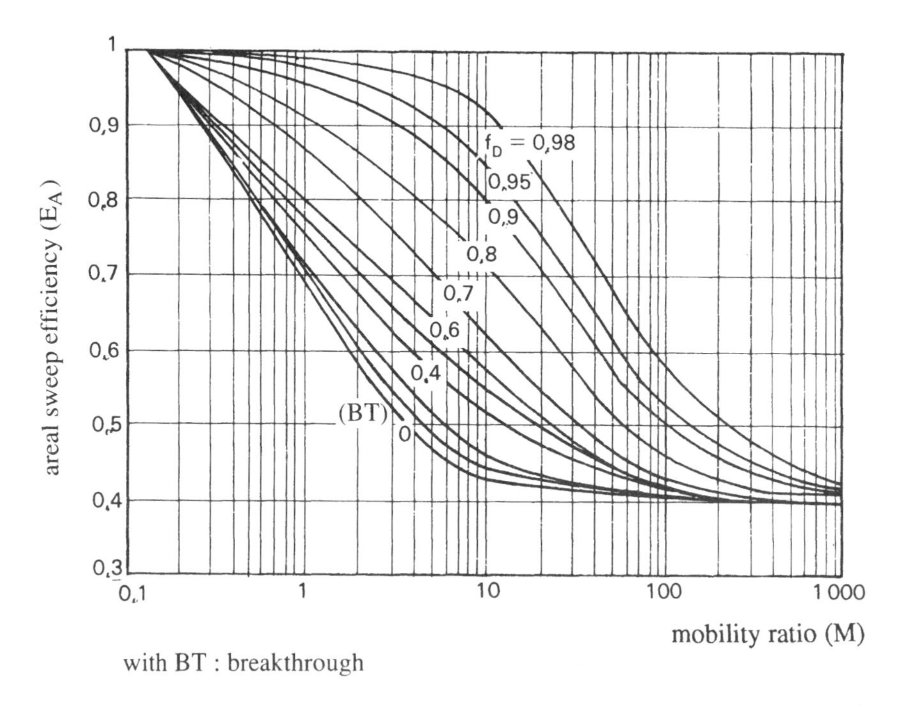

Fig. 8.9

8.3.3 Vertical (or Invasion) Efficiency E_V

An efficiency is also defined in the vertical direction of the swept layers. In analyzing two-phase flows, we showed that the front could become distorted (fingering when $M > 1$). The reservoir heterogeneities (different permeabilities, strata, drains, fractures) considerably hinder the regular movement of the front and are detrimental to sweep. This vertical efficiency is defined as the ratio of the area swept to the total area, for a vertical cross-section (Fig. 8.10):

$$E_V = \text{Vertical efficiency} = \frac{\text{Area swept by the front}}{\text{Area A'B'C'D'}}$$

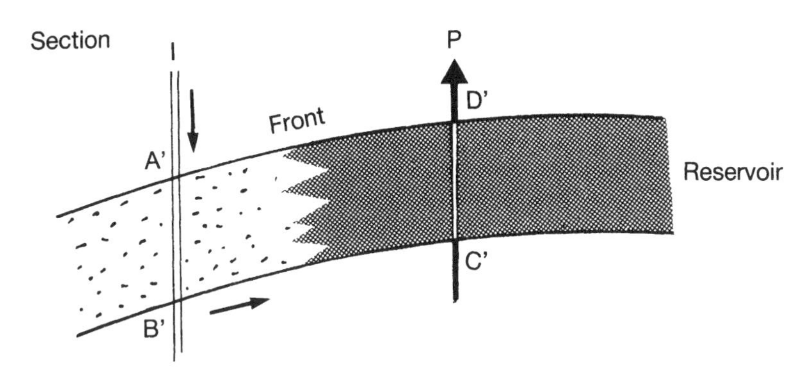

Fig. 8.10

The vertical efficiency can be determined by calculation, by subdividing the reservoir vertically into uniform sections. The simplified methods most widely employed are those of Stiles and of Dyskra-Parsons. It is also analyzed by simulation in vertical cross-section models (xz).

Depending on the forces in action and the heterogeneities, highly variable values of E_V are obtained, both at breakthrough and at abandonment. As for E_A, it is clear that E_V decreases with rising M.

The product of these two efficiencies $E_A \cdot E_V$ is called the sweep (or volumetric) efficiency.

8.3.4 Displacement Efficiency E_D

The ratio of "volume of oil displaced" to volume of initial oil has already been discussed in describing two-phase flow and concerns the effectively swept zones.

This efficiency E_D is defined by:

$$E_D = \frac{S_{oi} - S_{om}}{S_{oi}}$$

with

$$S_{om} = 1 - S_{Dm} \quad \text{and} \quad S_{oi} = 1 - S_{wi} \text{ (one-phase oil)}$$

S_{Dm}, the mean saturation of displacing fluid **behind the front**, is determined by the **Welge tangent**. In the "piston" simplification, S_{Dm} is replaced by $(1 - S_{or})$, a value that is approximately the final S_{Dm} at the end of sweeping. Obviously, this efficiency E_D increases as a function of time (volume injected), but does not depend on the mobility ratio M. It is essentially governed by the irreducible water and oil (or gas) saturations:

$$E_{D\ (final)} = \frac{1 - S_{wi} - S_{or}}{1 - S_{wi}}$$

8.3.5 Conclusion

To simplify matters, it can be stated that:

(a) E_A and E_V decrease with rising M.
(b) $E_D \leq (1 - S_{or} - S_{wi}) / (1 - S_{wi})$.
(c) These three efficiencies increase as a function of time (injected volume).
(d) The total efficiency, a product of $E_A \cdot E_V \cdot E_D$, represents the oil recovery for zones subjected to flooding.

Numerical Example

Recovery by water flood in a five-spot pattern (pressure maintenance):

Area = 250 acres Thickness = 30 ft $\phi = 0.25$

$S_{wi} = 0.20$ $S_{or} = 0.30$ $\mu o = 2.5$ cP $\mu w = 0.6$ cP

$k_{ro} = 0.85$ $k_{rw} = 0.35$ Injection: $Q_{wi} = 3000$ bwd

$Bo = 1.33$ $B_{wi} = 1$

— **Volume of oil in place**

$N = V_R \cdot \phi\ (1 - S_{wi}) / B_o$ with $\alpha = 7{,}758$ US p.u

(1 acre-foot = 7,758 STB)

$$N = \frac{7{,}758 \ . \ 250 \ . \ 30 \ . \ 0.25 \ . \ 0.8}{1.33} \text{ STB}$$

$$N \approx 8.75 \text{ MMSTB } (1.4 \ . \ 10^6 \text{ m}^3)$$

— **Calculation of E_A**

$$M = \frac{k_{rw} \ . \ \mu_o}{k_{ro} \ . \ \mu_w} = \frac{0.35 \ . \ 2.5}{0.85 \ . \ 0.6} = 1.715$$

Figure 8.9 in Section 8.3.2 indicates that $E_A = 0.6$ at breakthrough.

— **Calculation of E_V**

This demands a knowledge of the vertical permeability distribution. For this example, we can consider that, in the absence of such knowledge, $E_V \sim E_A$.

— **Calculation of E_D**

It is assumed that, at breakthrough $S_{wm} \sim 1 - S_{or}$:

$$E_D = \frac{1 - S_{or} - S_{wi}}{1 - S_{wi}} = \frac{1 - 0.3 - 0.2}{1 - 0.2} = 0.625$$

hence:

$$E = (0.6)^2 \ . \ 0.625 = 0.225$$

$$Np = E \ . \ N = 1.97 \text{ MMSTB } (3.150 \ . \ 10^5 \text{ m}^3) \textbf{ at breakthrough}$$

— **Time of breakthrough**

With a constant flooding rate, $N_p \, B_o = Q_{wi} \, B_{wi} \, t_b$, hence:

$$t_b = \frac{N_p \, B_o}{Q_{wi} \, B_{wi}} = \frac{1.97 \ . \ 10^6 \ . \ 1.33}{3000 \ . \ 1} \sim 900 \text{ days}$$

The production of oil is then calculated after breakthrough as a function of the volume injected or as a function of the water-cut (Fig. 8.9).

8.4 WATERFLOOD

This is the earliest process (late nineteenth century), and still the most widely used. Its goal is to increase recovery, and also to accelerate production, or, more specifically, to slow down production decline. The means employed is often the maintenance of pressure. Flooding may be of the dispersed type in the oil zone, or of the peripheral type in an existing aquifer.

8.4.1 Technical and Economic Aspects

8.4.1.1 Technical Aspect

With waterflood, the mobility ratio M is often favorable for a light oil (low-viscosity oil) and not too unfavorable for a heavier oil. The efficiency, i.e. the recovery, is therefore high or medium.

As to the sources of water, they are usually aquifer levels at shallow depth, seawater in offshore drilling, or surface water onshore (lakes, rivers).

Waterflood is favorable for heterogeneous reservoirs in which the rock is water wet, which is often the case except for some carbonate reservoirs.

Imbibition plays a significant role if the flow rates are not too high. The water must also be injectable: sufficient permeability and **compatibility** with the formation water. This is because the mixture of injected water with the water in place could produce insoluble precipitates ($BaSO_4$) and plug the wells.

8.4.1.2 Economic Aspect

Investments are generally higher for waterflood than for gas flood. This is because the number of waterflood wells is larger than for gas, since the mobility of water and hence its injectivity are lower. However, the flow rate also depends on the injection pressure.

8.4.2 Time and Start of Flooding

It is not advisable to flood too early, because a minimum production record is generally necessary to identify the natural drive mechanisms of the reservoir and hence the activity of the aquifer (if any).

Some authors have pointed out the potentially beneficial effect of initial saturation with free gas on oil recovery ($P < P_b$). This implies **an optimal gas saturation**, due to the presence of a residual gas saturation S_{gr} decreasing the residual oil saturation S_{or}. Yet an unfavorable effect also occurs in this case ($P < P_b$): the oil flow rates decrease, because S_g and k_{rg} increase and kro decreases. Hence the question is complex and there is no all-purpose answer.

8.4.3 Implementation

The following conditions are necessary for waterflood:

(a) Sufficient water supply in terms of quantity, quality and regularity. **Water treatment** installations (oxygen, prevention of incompatibility with formation water and with the rock, filtration, elimination of bacteria).
(b) Suitable completion of the injection wells and possibly improvement in the injectivity index (tests). It is sometimes also necessary to close the more permeable zones to prevent the premature water breakthrough at the production wells (vertical efficiency).
(c) **Pumping** installations (if necessary).
(d) Monitor the flooding and sweeping system: radioactive tracers, for example, to identify the well causing water breakthrough.

8.5 Gas Injection (Nonmiscible)

The injection of production gas dates nearly as far back as that of water. It was popular for some time in the USA for shallow reservoirs (1000 to 2000 m) implying low gas recompression costs.

Gas injection now has a smaller field of application because reservoir gases are now upgraded and find other uses than injection, **except** in desert or remote areas (and sometimes **offshore**).

This section discusses only gas that is immiscible with the reservoir oil. Gas injection is mainly attractive for a light oil (mobility ratio M not too high).

8.5.1 Technical Aspect

The sweep efficiency is generally **much lower** than that of water (because $M > 1$). Injection is performed either in the gas cap (local) or directly into the oil (dispersed). The injected gas nearly always consists of hydrocarbons:

reservoir production gas. In the frequent case where there is no outside source of gas, the aim is to slow down the drop in pressure, but this comes nowhere close to maintaining reservoir pressure.

Gas injection is nevertheless advantageous compared to water:

(a) If there is a **gas cap**.
(b) If the **oil is light** (the solution GOR is high and the oil viscosity is low).
(c) If the **permeability is high**.

These conditions imply good vertical sweep of the oil by the gas cap, and recovery is good (particularly for **reef-type** reservoirs).

The gas injected into an oil reservoir can also be recovered subsequently (production GOR), after breakthrough.

8.5.2 Economic Aspect

Since the small increase in recovery generally precludes the drilling of many new wells, production wells are often converted (subject to changes in perforations for injection in a gas cap). For injection in the oil zone, the tendency is to adopt a nine-spot pattern. This requires **fewer injection wells** than water-flood, because the injectivity is greater ($M > 1$). However, the gas recompression costs are often high.

8.5.3 Implementation

(a) Well cleaning, casing inspection: do not select a well producing water (it is useless to inject gas into water!).
(b) Well injectivity test.
(c) Closure of zones with preferential paths.
(d) Gas treatment to remove H_2S, CO_2, O_2 and H_2O: corrosion and risk of precipitates (hydrates) liable to plug the lines.
(e) Compression: for reasons of flexibility, it is preferable to install several small compressors rather than one large one. Reciprocating compressors are generally used, driven by gas engines.
(f) Monitoring: use of radioactive markers.

8.5.4 Comparison of Waterflood and Gas Injection

Injection	**Water**		**Gas**
Displaced oil →	**Light**	**Fairly heavy**	
Mobility ratio	M < 1	M > 1	M >> 1
Efficiency	Good	Medium	Poor
Encroachment.............	Stable	± Stable	± Unstable
Imbibition	Favorable	Favorable	–
Number of injectors (1).............	High	Medium	Low
Investment (2)...........	± High	± Medium	± Low

(1) The number of injectors depends on the mobility ratio M, and also on the injection pressure.

(2) Additional investment due to injection compared with investment for primary recovery.

Note that these comparative details are provided for information only, and may vary considerably from one reservoir to another.

8.6 GAS CYCLING IN RETROGRADE CONDENSATE GAS RESERVOIRS

The reinjection of the light components of the produced gas after treatment may have beneficial technico-economic effects, because it allows **better recovery of condensate**, which is a highly valuable product. Cycling of dry gas tends to have a dual effect:

(a) Limit the average pressure drop and hence the drop in production of condensates due to the fact that the percentage of heavy products in

the effluent decreases with the pressure (P < retrograde dew-point pressure).
(b) Revaporize the condensate deposited in the reservoir zones subjected to injection.

The injection wells are sufficiently far from the production wells to avoid any premature breakthrough of dry gas.

Thanks to cycling, condensate recovery often exceeds 60%.

Example:

At North Brae (UK, North Sea), cycling is total. Production for 400 Mcfd of treated gas is 75,000 bpd of condensate (3.2 Mt/year), and the reinjection of 360 Mcfd (90% of the gas produced) will last for nine years: the reservoir will then produce by decompression of the gas and also of the natural gasoline at a declining rate.

Total condensate recovery is estimated to be twice as high as in primary recovery:

(a) Cycling: 65% (9 years).
(b) Cycling + decompression: 82%.
(c) Primary recovery: 35 to 40%.

This reservoir is especially rich in condensate: GPM ~ 900 g/m^3.

8.7 ENHANCED OIL RECOVERY

8.7.1 General Introduction

Waterflood and immiscible gas injection in an oil reservoir often yield only average recovery ratios (25 to 50%). It was therefore logical to try to **improve these recovery ratios**, and research has been intensified in this field in the past twenty years.

Recovery with conventional injection systems is imperfect for two technical reasons:

(a) Incomplete sweep of the reservoir space (macroscopic trapping).
(b) Trapping of residual oil by capillarity in the swept zones (microscopic trapping).

What can be done to enhance recovery?

Any development of more effective processes must therefore meet the following requirements:

(a) **Improve spatial sweep** by reducing the mobility ratio between the two fluids, especially by increasing the viscosity of the injected fluid (or by reducing that of the oil).

(b) **Reduce, or better, eliminate capillary forces** by achieving the miscibility of the two fluids.

The method is even more efficient if both factors can be dealt with simultaneously.

Three methods are distinguished: **miscible, chemical and thermal.**

The following table shows the main fluids and/or products injected, and the mechanisms acting on trapping:

Trapping	Mechanism	Product	Method
Action on microscopic trapping	Miscibility	Carbon dioxide Hydrocarbon gases	Miscible
		Microemulsions	Chemical
	Lowering of interfacial tension	Surfactants	
Action on macroscopic trapping	Increase of water viscosity	Polymers	
	Reduction of crude oil viscosity	Carbon dioxide	Miscible
		Steam injection *in situ* combustion	Thermal

These three enhanced recovery methods cause a significant change in the physicochemical state of at least part of the fluids present.

This is why basic and applied research is being conducted, jointly with the universities, to achieve a better understanding of the mechanisms involved in porous media. Figure 8.11 illustrates (in its own way!) the efforts made in this area.

Drawing by Mick With the kind authorization of *SNEA(P)*

Fig. 8.11

8.7.2 Miscible Methods

8.7.2.1 Miscible Displacements

To improve the oil recovery, it is important to use an injected fluid that is miscible with the oil. Since the interfacial forces are eliminated, there is no residual oil saturation S_{or} and the displacement efficiency is:

$$E_D = \frac{1 - S_{wi} - S_{or}}{1 - S_{wi}} \rightarrow 100\%$$

hence a considerable improvement in recovery, which may reach 30 to 40% for injected gas miscible with the oil in place.

Note that two fluids are miscible if they can blend to form a **single** fluid. Water and whisky, for example... whereas water and oil are not miscible.

8.7.2.2 Two Types of Miscibility

In practice, some hydrocarbon components display the property of being miscible with others: C_2 to C_6 and especially **C_3 and C_4** (intermediates). This is termed **natural or absolute miscibility**. It is convenient to represent the components and any composition of a mixture by a **ternary diagram** (Fig. 8.12) which shows a two-phase zone (Fig. 8.13) whose size varies for different reservoir pressures and temperatures.

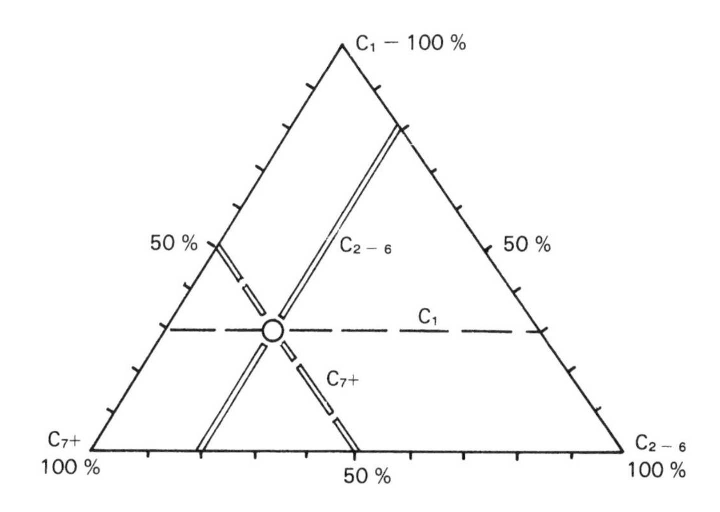

Example:
C_1 = 30 %
C_{2-6} = 20 %
C_{7+} = 50 %

Fig. 8.12

Ternary diagram.

The boundary of the two-phase zone consists of the combination of the bubble-point and dew-point curves which meet at the critical point. .

This boundary and the tangent to the critical point serve to distinguish four zones, including three one-phase zones, on this type of diagram: a gas zone G toward point C_1, an oil zone H toward point C_7+, and supercritical zone S toward point C_{2-6}.

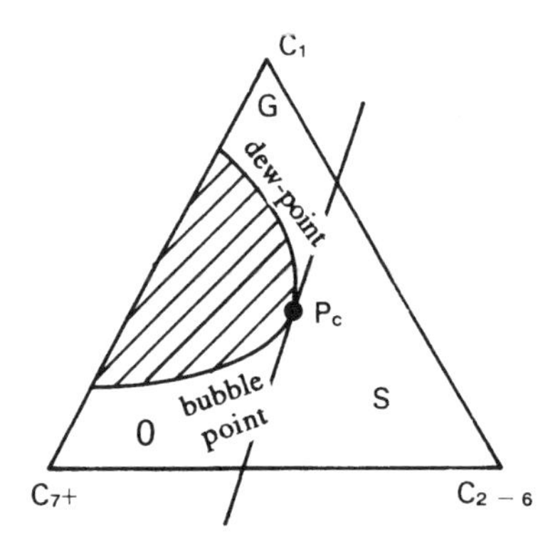

Fig. 8.13a Two-phase zone.

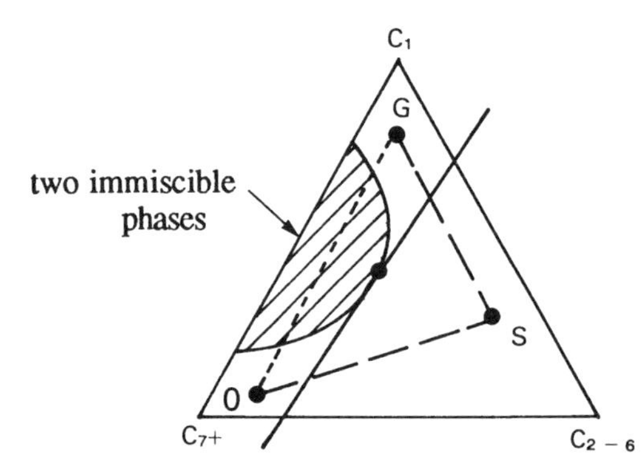

Fig. 8.13b Total miscibility.

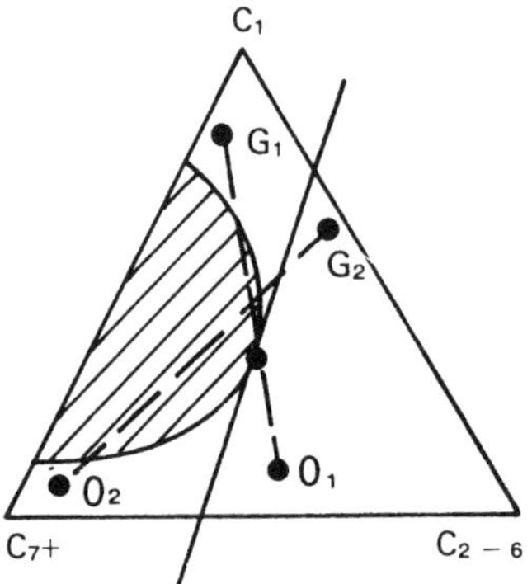

Fig. 8.13c Dynamic miscibility.

A. Natural Miscibility

Any hydrocarbon of which the point indicating its composition lies in the supercritical zone is naturally miscible with a gas and an oil represented on the corresponding diagram at the reservoir pressure P and temperature T (diagram (a) in Fig. 8.13).

This is because any composition resulting from the mixtures is located on segment SG or on segment SH, which **remain outside the two-phase zone** (diagram (b) in Fig. 8.13).

If a **slug** (limited volume) of C_3-C_4 is injected, followed by dry gas (essentially methane), miscibility is obtained between the oil and the C_3-C_4 and between the C_3-C_4 and the dry gas. This slug must be sufficiently large to form a barrier between the two fluids, and its volume is generally about 2 to 10% of the pore volume concerned. Diagram (d) in Fig. 8.13 shows a miscible displacement of this type.

B. Dynamic Miscibility

In certain high-pressure and temperature conditons, some gases become miscible with the oil during displacement. This is termed dynamic (or thermodynamic) miscibility. Two cases can be distinguished:

(a) **Injection of lean gas in a high-pressure light oil**: this process is called high-pressure gas drive. The oil enriches the gas in the mixing zone (diagram (c) in Fig. 8.13, pair G_1, H_1). A rather high reservoir pressure is necessary, in order to have a reduced two-phase zone, and hence a greater possibility of achieving miscibility.

(b) **Injection of wet gas** (or enriched with C_3-C_4) in a fairly heavy oil, a process called condensing gas drive: the gas enriches the oil in the mixing zone (diagram (c) in Fig. 8.13, pair G_2, H_2).

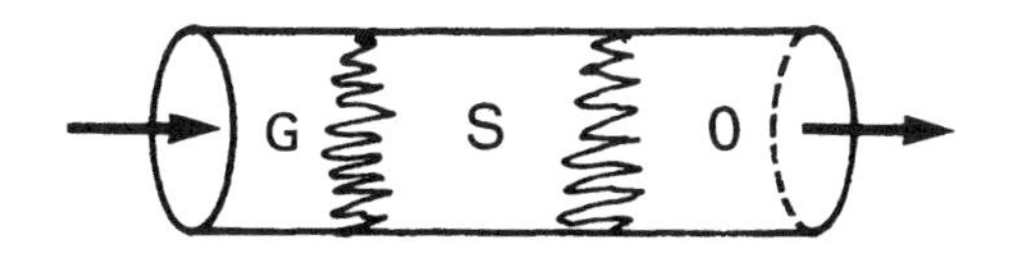

Fig. 8.13d Slug injection.

In both cases, if the pressure, temperature and compositions of the oil and gas are suitable, an exchange of components between the two phases produces a "mixing zone" that is miscible with both fluids.

Only the first case offers favorable natural data, namely light oil at high pressure, and it suffices to inject lean gas, i.e. methane. This applies, for example, to Hassi Messaoud (production gas injected at a wellhead pressure of $P_t \sim 400$ bar).

In the other cases, by contrast, it is necessary to inject either a slug that **is expensive**, or enriched gas (also expensive), or to use rich gas, but to the detriment of the marketing of the LPG[1] (especially C_3-C_4). Hence this situation requires a technico-economic study.

A **minimum** pressure exists to obtain a miscibility condition in the reservoir for fluids of a given composition. This is defined as the **miscibility pressure** ($P_m > 250$ bar for C_2 to C_6).

Note also the use of a **(methyl) alcohol slug followed by water**, which is miscible with both oil and water. This process is expensive and very little used today.

8.7.2.3 CO_2 Injection

In terms of dynamic miscibility, carbon dioxide displays properties similar to those of the intermediates (C_2 to C_6) of the hydrocarbon chain. Its action considerably decreases capillary forces. It is more or less **miscible with the oil in place**, and with the gas and water.

Moreover, CO_2, which dissolves in the **oil**, also increases the volume of the oil considerably (20 to 100%) and **significantly lowers its viscosity**. This is also valid for **heavy oils**, even if CO_2 is not miscible with them (pressure too low).

Thus CO_2 offers an advantage compared to C_2 to C_6 for medium-pressure reservoirs. This is because the miscibility pressure is lower, generally between 130 and 200 bar (instead of $P_m > 250$ bar for C_2 to C_6).

This process, which appears promising, is very often limited by the high cost of the product. In the USA, however, the existence of enormous natural

1. LPG, liquefied petroleum gas.

reservoirs of carbon dioxide allows its transport by pipeline and injection in oil reservoirs on a large scale, in suitable economic conditions.

An **anecdote** concerning the **price** of CO_2: CO_2 injection in the Coulommes field (see the table in Section 8.7.5) had to be terminated by 15 May of the same year, because the sale of CO_2 was then reserved for... carbonated beverages!

8.7.3 Chemical Methods

This term derives from the fact that these processes involve the addition of chemicals to the injected water (or more rarely to gas, producing foams).

Two processes are essentially employed today:

(a) Microemulsions.
(b) Polymers.

8.7.3.1 Microemulsions

The goal is to improve the **displacement efficiency** (microscopic trapping) by injecting a fluid miscible with the oil in place.

A microemulsion consists of a mixture of oil (from production), water and surfactants, which offer the advantage of achieving miscibility, plus alcohols employed as stabilizers (propanol, butanol, etc.). The surfactants used are petroleum **sulfonates**.

Figure 8.14a shows the ternary structure of microemulsions for three typical compositions.

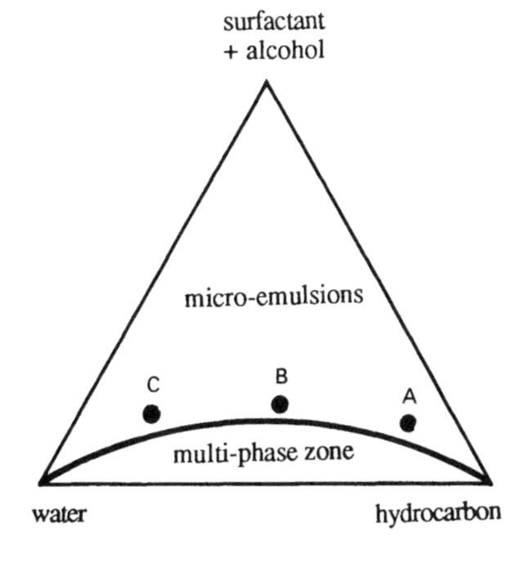

typical micro-emulsion compositions (%)			
	A	B	C
surfactant	10	13	10
alcohol	5	7	5
hydrocarbon	70	40	20
water	15	40	65

Fig. 8.14a

Microemulsion ternary structure.

Recovery is significantly enhanced, but this method is costly because the volume of the microemulsion slug must be more than 2 to 3% of the pore volume to be effective (heterogeneities cause its destruction), and it contains a high proportion of oil.

8.7.3.2 Solutions of Polymers in Water

The goal is to improve **the sweep efficiency by raising the viscosity of the water**, which has the effect of reducing the mobility ratio and making it favorable ($M < 1$). The front is then stabilized, achieving better sweep. The viscosity of the water can thus be increased fifty times or more.

Water-soluble polymers are used (polyacrylamides, polysaccharides, etc.), with molecular weights in the millions. The concentrations range from 100 to 400 ppm. However, these polymers are fragile and very sensitive to the salinity of the pore water (limit = 50 g/l).

The cost of the product (300 to 600 F/t) and that of the installations are relatively low, but the additional recovery is limited.

8.7.3.3 Microemulsions + Polymers

The use of a microemulsion slug followed by a water slug containing polymers yields excellent results. This process has been used in a "pilot" at Châteaurenard (south Paris basin) operated by *SNEA(P)*, with the assistance of *IFP* (Fig. 8.14b).

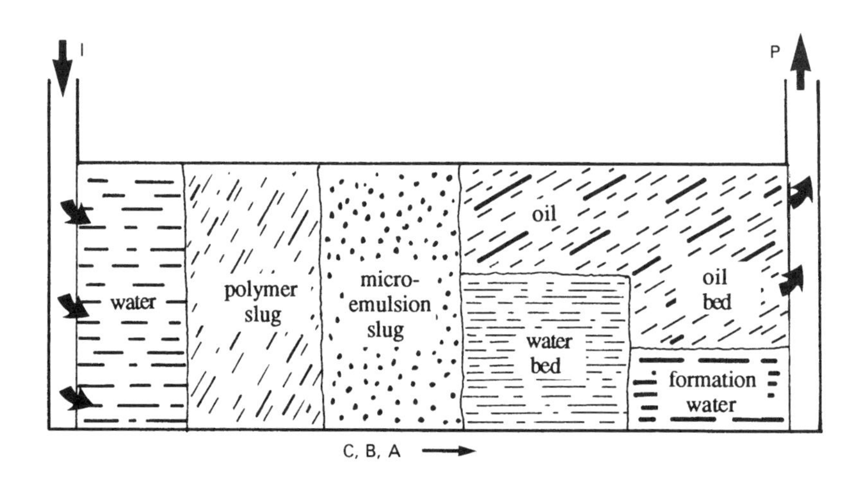

Fig. 8.14b Principle of the method.

In this first pilot, involving a five-spot pattern with 100 m sides, for a reservoir consisting of thin sands, **recovery was 70%**. Undoubtedly a record. Yet the price was very high, according to the specialists concerned. However, the primary aim of a pilot project is essentially technical.

A second semi-industrial pilot, over an area of 30 hectares, initiated in 1983, was designed to cut costs while anticipating relatively high recovery.

8.7.4 Thermal Methods

8.7.4.1 Heavy Oils, and Principle of Thermal Methods

Heavy oil is characterized by high viscosity (table below), which considerably reduces the efficiency of conventional production methods.

Main property ***in situ* viscosity (cP)**	**Type of crude oil**	**Secondary property** **Gravity (°API)**
	Conventional crude oil	
50 to 100		20
	Heavy oil: · Not producible or producible with difficulty by conventional processes. · Movable in reservoir conditions. · Low primary flow rates.	
10,000		~ 10 to 12
	Tar sands: · Not movable in reservoir conditions. · Zero primary flow rates.	

Heavy oil represents a considerable potential, about 200 billion tons in place. Yet its production is extremely small, about 1% of world production.

Raising the temperature in a reservoir boosts the production of oil which it contains, because the viscosity decreases as the temperature rises. Hence the already old idea of supplying heat energy to the reservoirs in which the oil is heavy and viscous (Fig. 8.15).

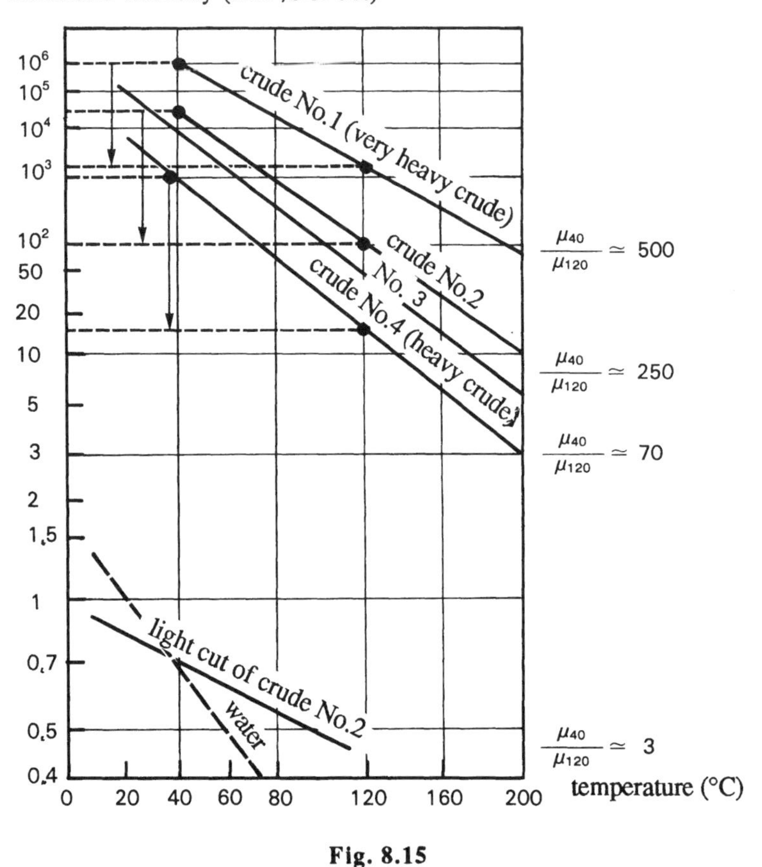

Fig. 8.15

Viscosity of petroleum products and water (*IFP* results).

The viscosity can be reduced by a factor of 10, 50 or even 100!

In addition to the decrease in μ_o, the heat causes swelling of the fluid and cracking of the heavy components, improving the fluidity.

Thermal methods are also the only ones feasible for oil with viscosity > 10 Po (or 1 Pa/s) and a reservoir depth < 1500 m.

Two methods are employed:

(a) Steam injection.
(b) *In situ* combustion.

8.7.4.2 Steam Injection

Hot water flood has also been employed, but its use is very limited. As to steam injection, two processes are employed:

(a) Stimulation.
(b) Injection with drainage between the injection and production wells.

A. Well Stimulation (Huff and Puff)

This is a rather early technique, used particularly in California:

(a) **Heating** of the well (or stimulation) by steam injection,
(b) **Alternate injection** (or heating) and production cycles (Fig. 8.16a).

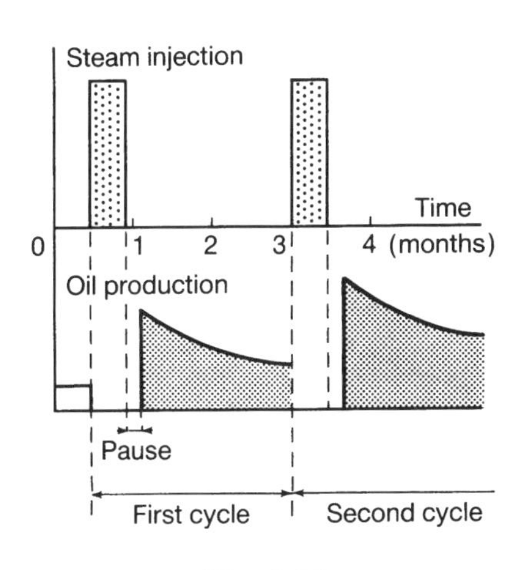

Fig. 8.16a

"Huff and Puff".

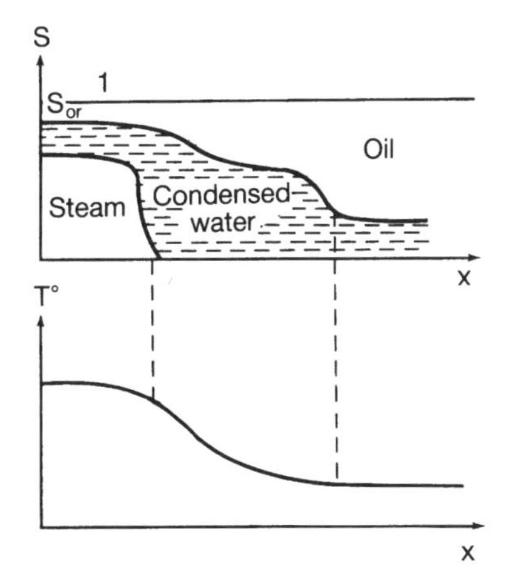

Fig. 8.16b

Steam injection.

After production of part of the condensed water, oil production begins at a much higher level than that obtained before steam injection. The flow rate then decreases since the temperature falls in the reservoir by thermal diffusion, which conditions the duration of production.

After a number of cycles, production falls, because only mechanical energy, associated with natural depletion, allows continued flow to the well.

The PI (productivity index) is often improved by 200 to 500%.

B. Steam Injection (Fig. 8.16b)

Alternating production/injection wells.

A cold water front is obtained followed by hot water and steam.

Many applications exist in both cases, particularly in the United States and Venezuela. In France, Lacq Supérieur was a positive experience.

- **Implementation:** Cyclic, continuous injection.
- **Recovery:** 30 to 60% oil in place.
- **Area of use limited:**
  - Depth < 1000 m.
  - Thickness > 10 m.
  - Oil/steam ≥ 0.15 m^3/t.
- **High investments.**
- **Energy consumption:** 30 to 50%.
- **Application problems:**
  - Completion.
  - Surface installations.
  - Processing of crude and treatment of smoke.

Among all the enhanced recovery methods employed, steam injection is by far the most widely used.

8.7.4.3 *In Situ* Combustion (Fig. 8.17)

This may be forward or reverse.

A. Forward Combustion

The fire is started downhole at the point where air is injected. A combustion front is therefore located in the bed, ahead of which a hot zone causes the vaporization and cracking of the molecules in place. In front of this hot zone is a cold zone where the vapors are recondensed. As the front advances, it finds only coke to burn. In the combustion zone, the temperature may exceed 600°C.

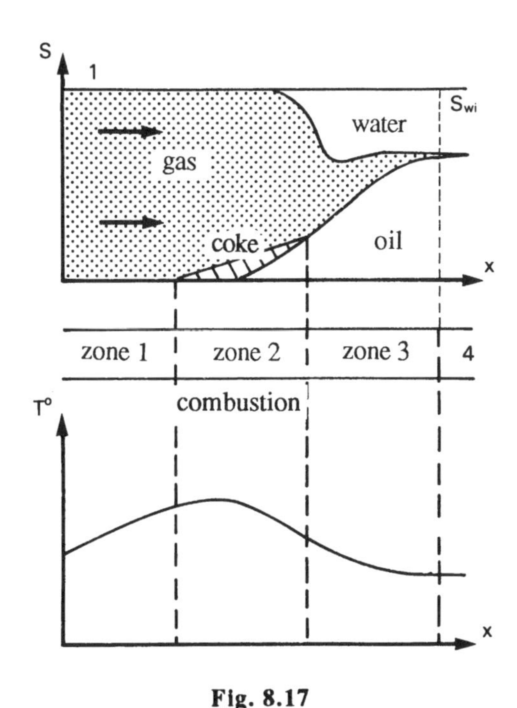

Fig. 8.17

In situ combustion.

Forward combustion is used, for example, in Romania (Suplacu de Barcau field) where *IFP* cooperates in the production of a heavy oil reservoir at a depth of about 50 to 100 m. Recovery anticipated by natural drive mechanisms is 9%. With combustion, the present recovery is > 40% and is ultimately expected to be 50%.

— **Implementation:** Dry, wet combustion.
— **Recovery:** 30 to 60% oil in place.
— **Range of use:**

- Depth < 1500 m.
- Thickness > 3 m.
- Coke > 15 kg/m^3.
- Air/oil < 4000 m^3/m^3.

— **High investments.**
— **Energy consumption:**
  - Surface: 5 to 15%.
  - Reservoir: 10 to 25%.
— **Application problems:**
  - Ignition.
  - Completion.
  - Processing of crude and treatment of smoke.
  - Process control.

B. Reverse Combustion

Air is injected until it reaches the production well. The fire is ignited by an electrical system downhole. Hence countercurrent displacement occurs of the combustion front and the hydrocarbons. This method is very little used because it is difficult to apply.

8.7.5 Enhanced Oil Recovery Pilot Flood

These are generally limited areas of reservoirs subjected to experimental flooding to assess a given process: for example, a five-spot pattern with an injection well at the center. An additional infield well is sometimes drilled to observe the results faster.

At the present time, about 400 "pilots" using enhanced flood are active throughout the world, compared with 50 fifteen years ago. But their number has tended to decline since the drop in crude oil prices in 1986. The following table lists a number of projects conducted by French companies.

8.8 CONCLUSIONS

Secondary and enhanced oil recovery is applied to a large number of reservoir types, except for very active water-drive oil reservoirs, and dry and wet gas reservoirs. In the USA, it is estimated today that an additional 35% of oil is produced by secondary and enhanced recovery, **essentially waterflood.**

Reservoir	Operator	Method	Reservoir type	Average depth	Average thick-ness	Oil		Operations
						API	Reser-voir viscosity	
Lacq Sup (France)	*SNEA(P)*	Steam injection	Porous and fractured limestone	650 m	100 m	22°	20 cP	Pilot started in 1977. One injection well. Oil gain: 12,000 t/year. Oil/steam ratio: 0.18 t/year. Start of industrial development: September 1981.
Château-renard (France)	*SNEA(P)*	Micro-emulsions + polymers	Sands	600 m	5 m	26°	40 cP	Pilot started in 1978. One injection well. Micro-emulsion slug: 10% V_p. Water + polymers: 100% V_p. Recovery: 70%. Very costly production.
Coulom-mes (France)	*Petrorep + IFP + CFP + SNEA(P)*	CO_2	Limestone (fractured)	1850 m	50 m	32°	5 cP	Pilot carried out in 1984. One injection well. 2600 t of CO_2 then 900 t of nitrogen. Good results, but CO_2 very expensive: 700 F/t.
Suplacu de Barcau (Romania)	State of Romania *IFP* /Coop contract	*In situ* com-bustion	Unconso-lidated sands	50 to 200 m	10 m	16°	2000 cP	Industrial development since 1976. 205 production wells and 38 injection wells. Current production: 300,000 t oil/year. Air/oil ratio: 2000 std.m^3/m^3.

The present trend is towards earlier application (better net present worth), although this does not always represent the optimum recovery anticipated. Yet a sufficient production record (one to three years) is often necessary to gain minimum knowledge of the reservoir.

As a rule, a number of injection patterns are investigated by simulation as a function of:

(a) The type of energy injected: water or gas, enhanced processes, or steam or combustion for heavy oils.
(b) The volume of fluid injected.
(c) The well pattern,

and a comparison is made among these various arrangements, with a study on primary recovery as a reference.

Furthermore, the **impact of enhanced recovery** (thermal, miscible, chemical) **is in fact very limited**. Its share of world production is less than 2%. Yet it must be kept in mind that each per cent of additional recovery on a world scale corresponds to about two years of present hydrocarbon consumption. This has led to intensified basic research, whereas applications on fields are barely advancing, given the present price of crude oil (in 1991) (see table of additional oil production costs).

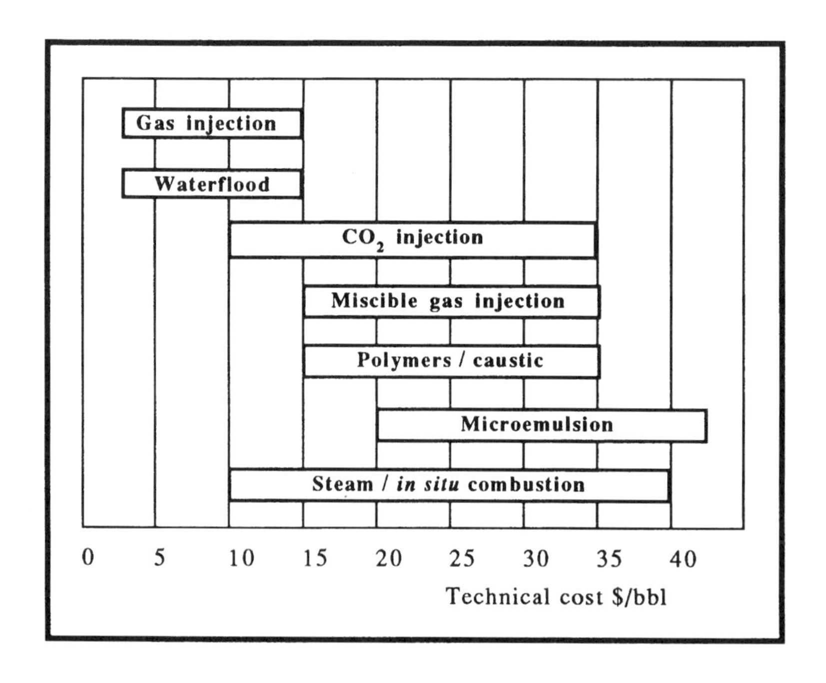

Chapter 9

RESERVOIR SIMULATION MODELS

9.1 ROLE OF MODELS

How can the performance of reservoirs be simulated with different production and development schemes?

How can results be obtained sufficiently fast, at the start of life of the reservoir, to take the best decision?

Simulation models are designed to answer these questions.

It is necessary to analyze the influence of the following main parameters on the recovery ratio and on the economics:

(a) Number of wells.
(b) Well layout.
(c) Flow rates.
(d) Schedule for bringing on stream.
(e) Influence of a potential aquifer.
(f) Secondary and enhanced recovery (one or more methods).

Furthermore, a model can help to obtain a slightly different image of the reservoir from the one obtained initially, by comparison between the operating results and those simulated by the model, after a production period.

9.2 DIFFERENT TYPES OF MODELS

There are three types of models:

(a) **Material balances**: these are the simplest. They represent the reservoir as a **cell**, passing through a series of equilibrium states. Water for example, penetrates into the cell if there is an active aquifer or water flooding (Chapter 7).

It is stated that the fluids in place at a given time (initial fluids minus produced fluids) occupy the available volume in the pores. These material balances are mainly used for preliminary rough calculations before going on to more advanced models, particularly when a new reservoir is discovered. They are run on microcomputers.

(b) **Numerical models**: these serve to subdivide the reservoir into blocks (or cells) in each of which are quantities of fluids subject to the laws of fluid mechanics. Their essential advantage is to be able to represent the variations in characteristics of the reservoir, the fluids, the flow rates to the wells and the pressure in space.

At present, with the spread of computers, numerical models have assumed a considerable lead. This is the **standard tool**.

(c) **Petrophysical models** (scale models of impregnated rocks in the laboratory): these help to investigate certain specific mechanisms, for example, distorsion of fronts, fingering.

Note incidentally that a few decades ago, electrical models (with resistors-capacitors) were used to process a number of specific problems.

9.3 NUMERICAL MODELS

The development of numerical simulation models is connected with that of information processing. In fact, the resolution of equations governing the fluids in a reservoir demands powerful computation resources. These models are in widespread use today. They are essentially data-processing software products.

9.3.1 Principles

The complexity of a model varies mainly with the type of fluid concerned. The earliest models, developed some twenty years ago, called **"Black Oil"**, simulate the production of conventional oil. They have been improved gradually and are still often employed.

The system of equations consists of the law of conservation of mass, the law of flow (Darcy), of capillary relationships and relative permeabilities, the saturation balance, and thermodynamic laws.

The approximation methods used to resolve the system are often of the **IMPES** type (**IM**plicit in **P**ressure, **E**xplicit in **S**aturation). IMPES leads to the resolution of a linear pressure system with dimensions ≅ number of blocks, the saturations then being calculated directly block by block.

The pressure approximation is fairly correct, whereas that of the saturations is affected by sometimes troublesome numerical dispersion. It can be reduced by modifying the curves of k_r which are then called pseudopermeabilities.

The programs used are large and contain approximately 30,000 Fortran instructions.

9.3.2 Modeling and Use

Depending on the type of reservoir (simple or complex structure, degree of heterogeneity, number of phases present — gas, oil, water), various types of simulation models, with varying degrees of complexity and cost, can be employed. The user's art consists in finding a valid compromise between the number of imaginary cases to be processed and the cost of each experiment, which is related to the size and accuracy of the model.

The number of blocks, and the time steps, vary widely, from a few dozen blocks to some 10,000, and from a few thousandths of a day to several months respectively.

Depending on each case, simple models "X.2C" can be used (movement of two fluids or components in one direction, for example) or complex models "XYZ.3C" (movement of three fluids in three directions).

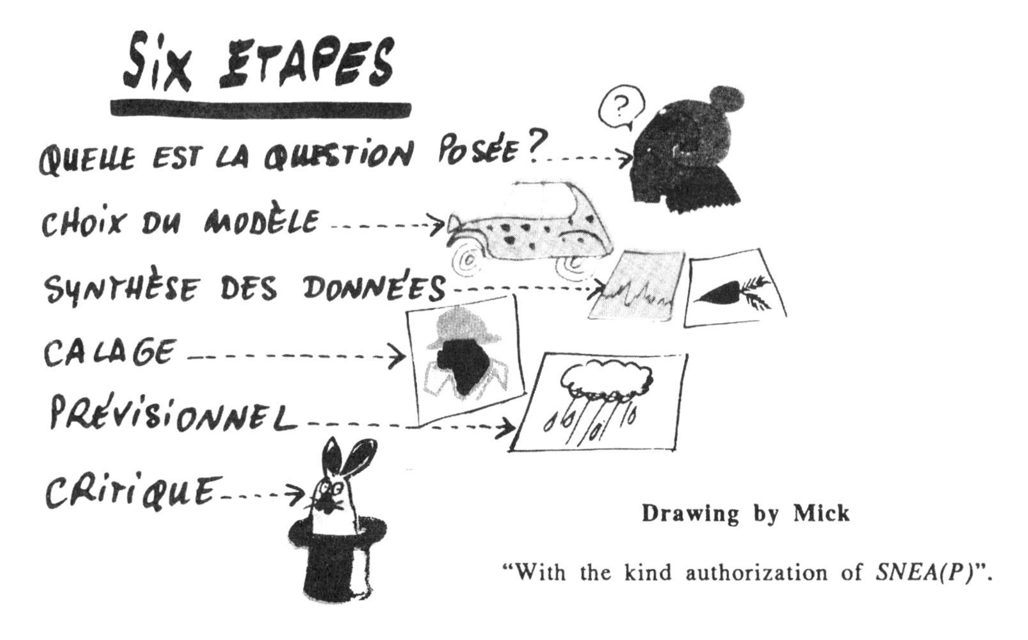

Drawing by Mick

"With the kind authorization of *SNEA(P)*".

An example of simulation is shown in Figs 9.1 to 9.4.

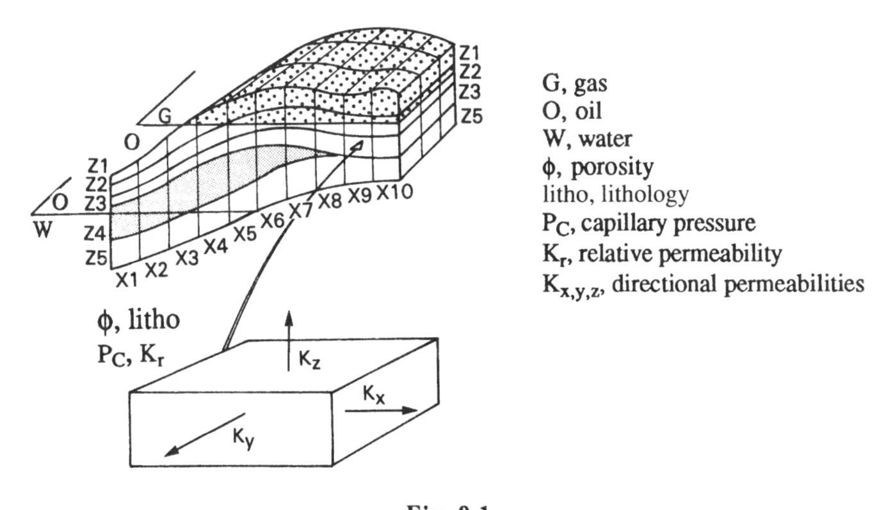

Fig. 9.1

Mathematical reservoir simulation model.

Commensurate with their ability to make production forecasts, these models also sometimes help to improve the understanding of the reservoir, because they have to **simulate its production history: history matching**. The user is therefore required to "find" unknown values of the parameters in zones without wells, or to modify the relative permeabilities for example, in order to obtain the best "fit" of the imaginary cases with the actual behavior of the reservoir. Hence these models have a twofold objective: **forecast studies and phenomenological studies**. In the latter case, a specific model is also used: an XZ "slice" for the analysis of segregation, or a circular radial model RZ for coning, for example.

Upper layer

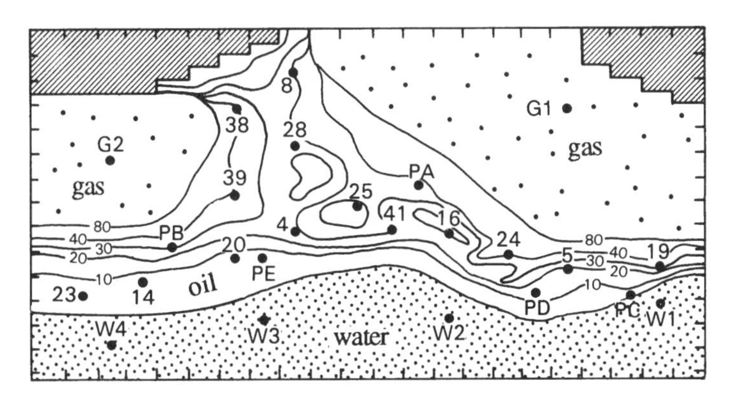

Lower layer

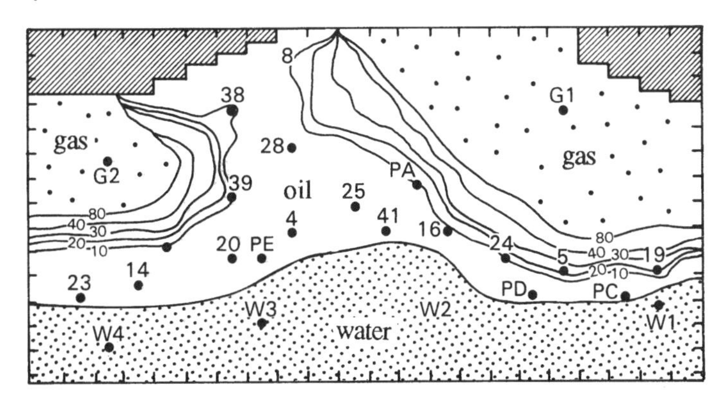

Fig. 9.2

Example of gas isosaturation curves (two layers).

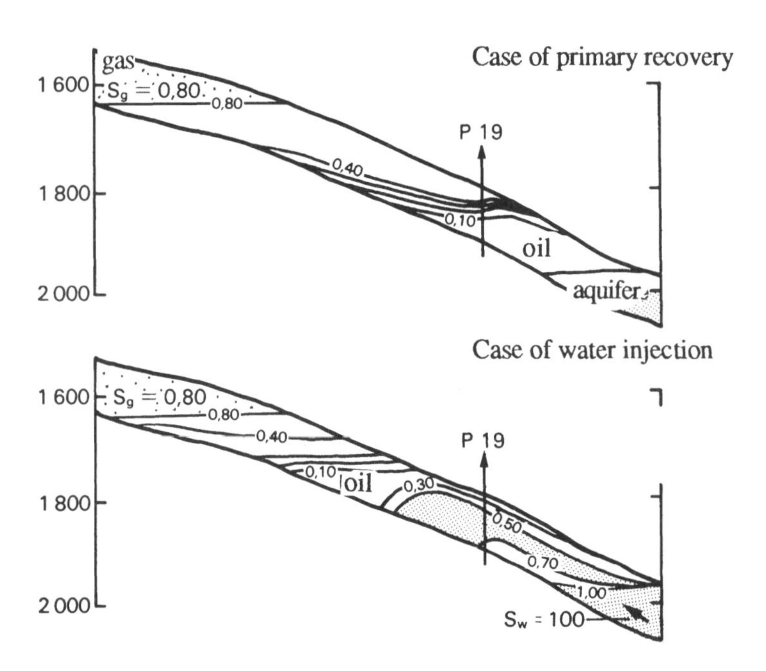

Fig. 9.3

Isosaturation cross-sections.

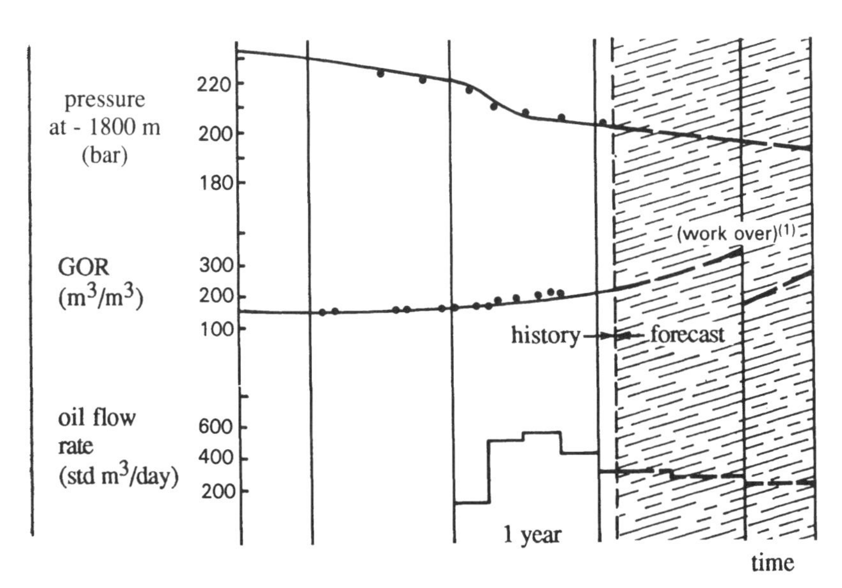

(1)Work over: resumption of wells (modification of perforated zones).

Fig. 9.4

Example of forecast analysis after history matching
(results for one well).

The simulation is carried out on computers ranging from medium capacity, such as the VAX 830, to very high capacity, like the CRAY 1 vector computer.

A "rule of thumb" gives an **operating cost** of about 1 French centime (0.2 cent) per block and per time step. In actual fact, the prices vary considerably according to the type of simulation.

Here are the approximate figures concerning the studies of an average reservoir on the simulator:

(a) Time required for a study: 6 months to one year.
(b) Computer time on supercomputer: 10 to 50 h (CRAY).
(c) Cost of the study: FF 10,000,000 ($ 2,000,000).
(d) Revision of the study every two to three years (or continuous study for large reservoirs).

9.3.3 Specialized Models

Other more specialized models have been developed to deal with specific cases. These include the following:

9.3.3.1 Compositional and Miscible Models (Fig. 9.5)

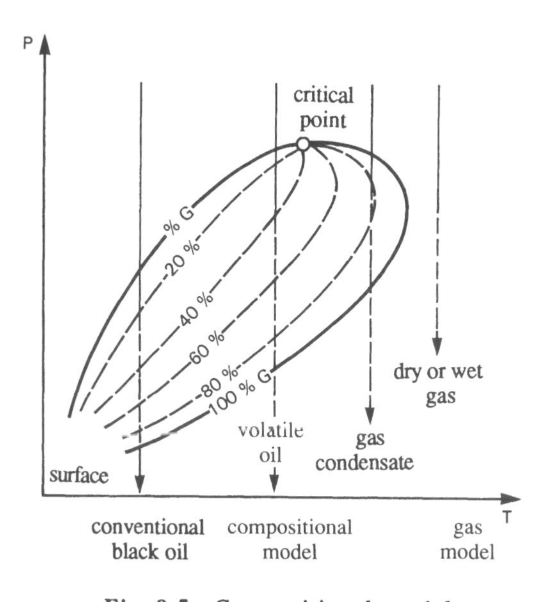

Fig. 9.5 Compositional model.

The oil and gas are each represented by several components in the compositional model. This serves to represent the changes in the fluids more accurately for volatile oils, condensate gases and gas injection.

The **miscible** model offers a better representation of the flows of injected fluids that are miscible with the fluid in place.

9.3.3.2 Chemical Models

These allow the simulation of the injection of polymers and surfactants in particular.

9.3.3.3 Thermal Models

Steam injection and *in situ* combustion models exist. Added to the equations already discussed is a temperature equation. The vaporization of water and (possibly) of oil must also be accounted for. The running costs are rather high.

9.3.3.4 Fractured Models

The representation of the medium is more complex. One of the formulations adopted is that of a **dual porosity system**: one block is associated with a pressure and a porosity in the fracture, and another for those in the matrix rock. The equations are coupled by a rock/fracture transfer term governing the flow from the rock to the fracture as a function of the rock/fracture pressure differences.

This is undoubtedly the most difficult type of reservoir to model. For the time being, the model is still a subject of research as well as a fully operational tool. This results from the many uncertainties existing in the physics of the flows and in the detailed description of this double-facetted medium.

One of the main difficulties stems from the complexity — and ignorance — of the fracture network (Fig. 9.6). Another, perhaps the most important, stems from the mediocre definition of the driving mechanisms concerning exchanges between rock and fractures.

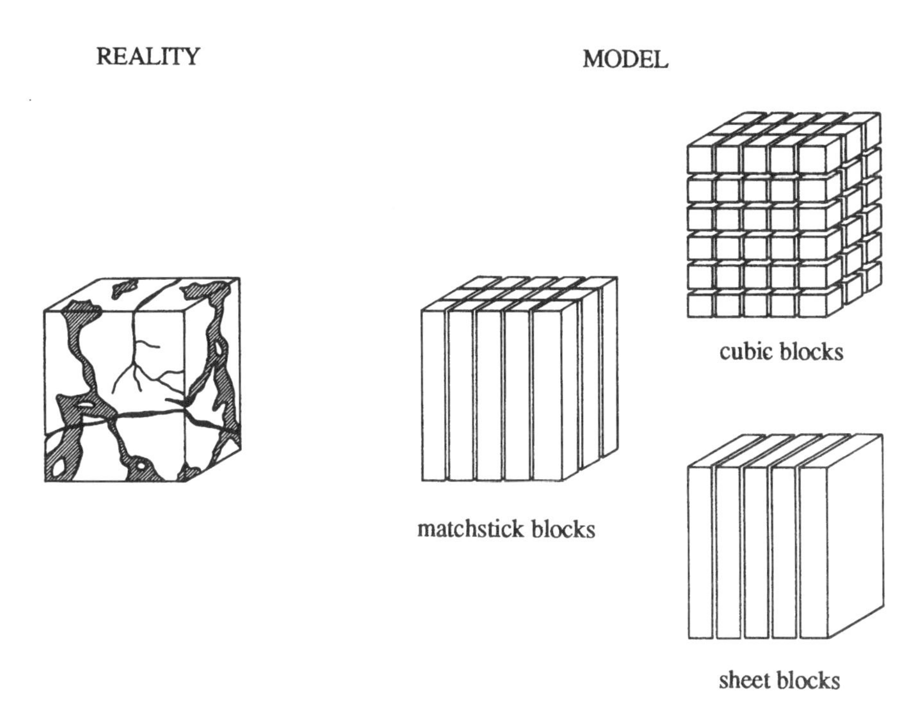

Fig. 9.6

Modeling of fractured environment.

9.3.4 Recent Model

A recent French model, **SCORE** (Simulateur **CO**mpositionnel de **RE**servoir, compositional reservoir simulator), developed by *IFP*, *SNEA(P)*, *CFP* and *GdF*, is chiefly designed to cover routine cases with options (compositional, chemical, etc.). Thus a reservoir can be dealt with by a single model, but this will not always replace the ultraspecialized models.

In this model, the data provided upstream are obtained from many disciplines: geology, geophysics, drilling, production, reservoir laboratories, etc.

Formerly, reports or drawings were used, and the data had to be acquired manually, but automated raw data processing programs are being increasingly developed, as well as interactive modes of access to the simulator.

Downstream, the reservoir simulator furnishes results not only to the reservoir engineer, but also to the production and engineering staff, as for the sizing of surface installations (gathering system, separators, compressors). And it obviously furnishes crude oil or gas production forecasts, which enable the economists, using their own models, to examine the profitability of the investments and to draw up financial forecasts, leading to the optimal development of the field.

We can conclude that a modern reservoir simulator such as SCORE is a significant contribution of information processing to the petroleum industry.

Chapter 10

PRODUCTION AND DEVELOPMENT OF A FIELD

10.1 DEVELOPMENT PROJECT

Article 81 of the French Mining Code stipulates that:

"All mine operator are required to apply most appropriate proven methods in producing and developing the deposits so as to maximize the deposits' final yield in accordance with economic conditions."

The choice of a development scheme is thus an approach that is both technical and economic.

The profitability of the development of a field is generally calculated over a period of ten to fifteen years. The front-end investments (wells, piping, production facilities, pipelines), which are **short-term investments, become profitable through the long-term production of hydrocarbons**.

It is therefore necessary to predict the future production of a field, over a period of many years. It immediately appears that this forecasting is not an easy thing. The reservoirs are often complex, and their development and production depend on the number of wells and on possible enhanced recovery.

A decision must first be made concerning the number and location of the wells to be drilled for natural depletion, and then one (or more) type of secondary or enhanced recovery can be introduced, leading to the **comparison of**

several hypothetical cases. Forecast results are obtained by the use of reservoir models which reproduce with only a relative degree of accuracy the architecture of the reservoir and its drive mechanisms (and **after** simulation of the production history once production has begun).

This enables the **development** of the field by **optimizing** the recovery of the hydrocarbons in accordance with the economic conditions of the time.

Once a reservoir has been discovered, several **appraisal** wells are generally drilled to obtain more information about the reservoir (this is less common offshore, because of the need to limit the number of clusterized well platforms). A development project is then drawn up, often preceded by a preliminary project using very simple models, and taking account of the small amount of information available initially.

The general flow chart of a project is shown in Fig. 10.1. This study takes place according to the following general procedure.

10.1.1 Analytical Phase

This phase involves the compilation of information with a critical analysis of the data. We already showed in Section 4 of Chapter 4 that the very limited understanding of a reservoir, in the appraisal phase, gives rise to significant uncertainties in the values (petrophysical, fluids, etc.) proposed to define the image of the reservoir. Moreover, the drive mechanisms, and especially the potential activity of an aquifer, are poorly or very poorly known.

The critical choice of the data is therefore fundamentally important in reservoir engineering. We shall briefly indicate the most important points:

(a) **Volumes in place**: be prudent in estimations, apply probabilistic methods. Then check the figures found as soon as the production history permits, by simulation model.

(b) **Reservoir image, compartmentalization, heterogeneities**: note that the faults with a throw smaller than the thickness of the reservoir rarely create impervious barriers. Similarly, small shale beds in a shale/sandstone series do not necessarily extend over the area of a reservoir, and do not always constitute a vertical barrier. The analysis of the well pressures usually helps to dispel this uncertainty (but rarely at the start of production).

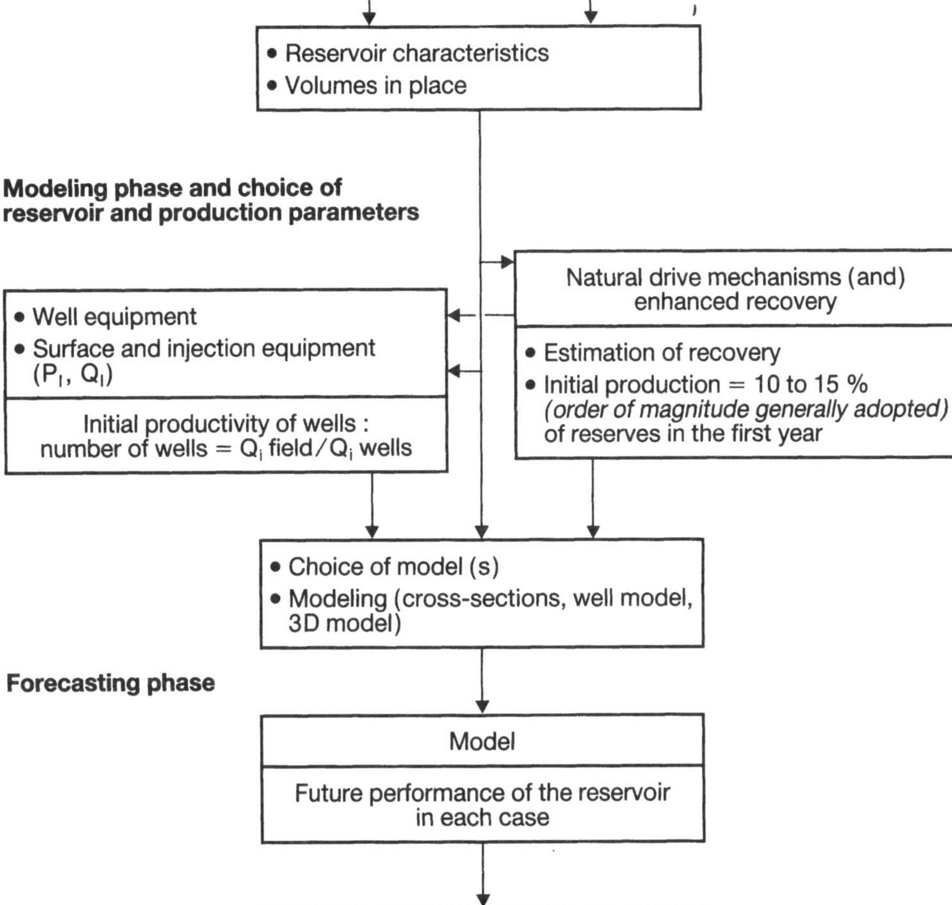

Fig. 10.1

General flow chart of a project.

(c) **Permeability**: production tests give the best (overall) values. The core permeabilities are very useful and help to provide a vertical log and a fairly good idea of the anisotropy ratio k_v/k_h.

(d) **Position of fluids**: the interfaces are determined by logs, for the wells crossing through them. Otherwise, capillary pressure measurements in the laboratory can yield some indication to be used with caution (possible inversion of wettability, especially for carbonates).

(e) **Relative permeabilities**: obtained on samples, they cannot be expected to represent the flows at the scale of a reservoir. An attempt can be made to "fit" them as soon as a sufficient two-phase production history is available (one to two years). However, the key points of these curves are only $k_{ro}(S_{wi})$ or $k_{rg}(S_{wi})$ and $k_{rw}(S_{or})$ or $k_{rg}(S_{or})$.

(f) **PVT analysis**: in each case, the measurements must be adapted to the drive mechanisms of the reservoir and to the production method (process plant). The analysis must be able to reconstruct the thermodynamic changes in the fluids (composite gas-release curves).

(g) **Well productivity**: this is a major uncertainty in the assessment of the reservoir. Hence the great importance of appraisal wells, which can reveal very wide variations in well potential. During this period, simple Buckley-Leverett type calculations can provide invaluable data about the water and/or gas influxes. A coning study is also generally necessary.

(h) **Drive mechanisms**: these must be determined as early as possible. Is there a gas cap for an oil reservoir? The comparison between the initial pressure and the bubble-point pressure (PVT) can give the answer, in the absence of a well that has crossed through the gas/oil interface. Yet caution is necessary, since some reservoirs have a widely variable bubble point.

When there is an **aquifer**, its characteristics must be optimized as soon as permitted by the production history. At the very outset of production, the uncertainty may be large (and recovery may vary by a factor of ten depending on the capacity of the aquifer). Avoid analogies with neighboring fields, which could spring unpleasant surprises.

10.1.2 Modeling Phase

This consists of the interpretation of the data and the attempt to construct a system (model) whose behavior reconstructs that of the actual reservoir:

(a) Geometry and internal architecture of the reservoir (reservoir geology).
(b) Choice of possible drive mechanisms.
(c) Choice of operating conditions (and constraints): productivity and completion of the wells, artificial lift techniques, surface conditions.
(d) Choice of the number and location of producing wells, and possibly injection wells.

These two phases are the most difficult... and the most important.

10.1.3 Forecasting Phase

This involves the calculation of the production forecasts in each case selected. Several alternatives are available for each step of the project: volumes in place, drive mechanisms (more or less active aquifer, for example), operating conditions, number of wells, secondary recovery (water or gas injection), etc.

It can be seen that a project is composed of several alternatives which are compared, and for which **economic optimization** is sought.

Thus a study is a multidisciplinary effort and, focused around the reservoir engineer, enlists the services of the reservoir geologist, the data processing specialist, the producer and the economist.

During the field development period, and even during most of the life of the reservoir, these study phases are revised and supplemented thanks to the know-how acquired subsequently (Fig. 10.2).

10.2 ECONOMIC CONCEPTS USED IN THE PRODUCTION AND DEVELOPMENT OF A FIELD

10.2.1 General Introduction

The economic optimum for the development of a field is identified by the quantified comparison of the different development alternatives examined. For

GRANTING OF PERMIT	EXPLORATION	DEVELOPMENT	PRODUCTION
1			
2	Geology and geophysics		
3	Geology and geophysics; Exploration and appraisal well		
4	Geology and geophysics; Exploration and appraisal well		
5	Exploration and appraisal well		
6	Exploration and appraisal well	Reservoir Engineering and simulations	
DECISION TO DEVELOP 7		Reservoir Engineering and simulations; Development and transport installations	
8		Reservoir Engineering and simulations; Development and transport installations; Geophysics; Drilling of development wells	
9		Reservoir Engineering and simulations; Development and transport installations; Geophysics; Drilling of development wells	
10		Reservoir Engineering and simulations; Geophysics; Drilling of development wells	Production buildup
11			Production buildup
12			Reservoir Engineering (continued); Maximum production
13			Reservoir Engineering (continued); Maximum production
14			Reservoir Engineering (continued); Maximum production
15			Reservoir Engineering (continued); Maximum production
16			Reservoir Engineering (continued); Maximum production
17			Reservoir Engineering (continued); Maximum production
18			(continued); Decline
19			(continued); Decline
20			(continued); Decline
21			(continued); Decline
22			(continued); Decline
23			(continued); Decline
24			(continued); Decline
25			(continued); Decline
26 years			
...			
...			
ABANDONMENT			
SHARE OF TECHNICAL COST	10 to 20 %	40 to 60 %	20 to 50 %

Fig. 10.2

each case, the expenditures (investments and costs) and receipts can be illustrated on the graph in Fig. 10.3 below.

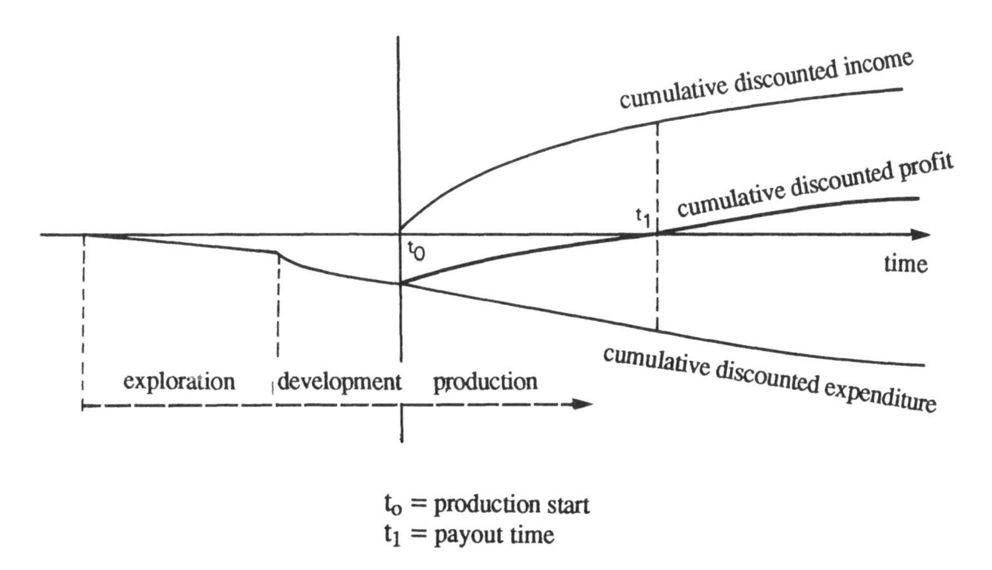

t_o = production start
t_1 = payout time

Fig. 10.3

Time t_o is generally taken as the reference time for discounting.

Expenditures:

Investment (exploration, drilling, production).

\+ Operating costs (salaries, operations, overheads).

\+ Taxes and royalties paid to the producing country.

Receipts:

Selling price of oil and/or gas.

The expenditures, receipts and profits are said to be **discounted** when adjusted in time to their present value. For example, a receipt recoverable in n years has the following value at time t_o:

$$R_o = R_n / (1 + i)^n$$

The discount rate i is often higher than the prevailing bank lending rates (concept of risks incurred). The calculations are carried out for a maxi-mum period of ten to fifteen years.

The **rate of return** is defined which corresponds to a payout time equal to the duration of the operation. If the rate of return is not significantly higher than the interest rate normally set by the banks, the project is not viable (excessively long payout time).

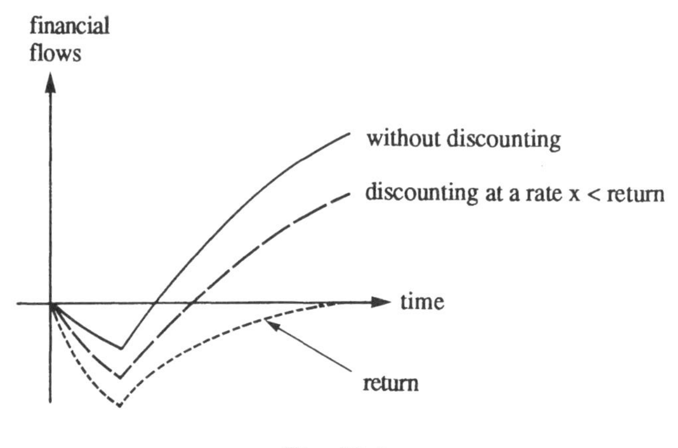

Fig. 10.4

The ratio P/I (profit/investment) expresses the **capital gain** on investment.

The economic optimum corresponds to the highest possible indexes above, with a short payout time. It also depends on the financial health of the operating company and on the concept of risks incurred due to the location of the field.

10.2.2 Different Cases Examined and Decision

The alternatives available to the engineers must be evaluated with the help of economists. The objective of economic studies is in fact to clarify the decisions by re-placing them within the strategic guidelines of the company, adapted to the situation considered. Meanwhile, the needs and requirements of the states in which the activity takes place must be borne in mind, along with complex analyses using several criteria and highlighting the sensitivity of the results to potential changes in key factors.

The main parameters analyzed are the following:

(a) Parameter 1: Uncertainty in the size of the accumulation.
(b) Parameter 2: Search for the best recovery method.
(c) Parameter 3: Comparison of different development schedules.
(d) Parameter 4: Influence of inflation and borrowing.
(e) Parameter 5: Influence of taxation.

10.2.2.1 Uncertainty in the Size of the Accumulation

Every study must take account of several assumptions concerning the amount of the volumes in place. Fairly often, three figures are taken into account, low, medium and high, depending on the data available (Section 4 of Chapter 4). This also applies to reserves. If a single assumption has been made on the productivity of the wells, this accordingly makes three base cases to be examined (with a varying number of wells and production profiles).

10.2.2.2 Search for the Best Recovery Method

A comparison can be made of two development methods with a different number of wells, as well as between water and/or gas injection and simple natural depletion.

10.2.2.3 Comparison of Different Development Schedules

For example, two development schedules with different production rates can be analyzed. In offshore operations, with two drilling/production platforms with simultaneous or staggered development may be compared.

10.2.2.4 Influence of Inflation and Borrowing

The variation in the inflation rate has a negative effect on profitability. The effect of borrowing on profitability is also analyzed.

10.2.2.5 Influence of Taxation

Development may be more or less attractive depending on the taxation conditions applied.

Also worth mentioning are the political aspects, which may or may not favor the production of a specific field, or a given type of development.

10.3 RESEARCH DIRECTIONS IN RESERVOIR ENGINEERING

Reservoir engineering is still a young science, but one that has reached adulthood for one or two decades. It is a whole set of techniques evolving towards a better understanding of reservoirs and utilizes a host of other technologies and calculation methods.

Here are somc current research directions:

(a) Hydrocarbon potential of basins.
(b) Reservoir seismic prospecting.
(c) More systematic study of heterogeneities.
(d) Logs.
(e) Type curves and well models.
(f) Two-phase flow in reservoir conditions.
(g) Drainage by horizontal wells.
(h) Greater flexibility in models.
(i) Better understanding of uncertainties.

In the 1970s and early 1980s, emphasis was placed on modeling, hand in hand with rapid developments in data processing. At present, stress is placed rather on data acquisition, particularly with reservoir seismic prospecting and the analysis of heterogeneities. But this is very relative. And, to end with, worth noting is the increasingly thorough computerization of data acquisition and calculation methods.

Chapter 11

TYPICAL FIELDS

11.1 OIL AND GAS FIELDS IN FRANCE

The French subsoil contains some hundred oil and gas fields (about 75 and 15 respectively). These fields are nearly all medium sized or very small, apart from Lacq (gas), which boasts international size (Fig. 11.2).

Proven reserves are 25 Mt for **oil** and 37 Gm^3 for **gas**.

Oil production topped the three-million-ton mark for the first time in 1987 (3.2 Mt), after a low in 1975 (1 Mt) following a first peak of 3 Mt in 1965. This is due to the recent discoveries in the Paris Basin (Chaunoy, Villeperdue, Vert le Grand, etc.) and, to a lesser degree, in Aquitaine (Vic Bilh, Lagrave, etc.). Cumulative production was 77 Mt in January 1991.

By contrast, **gas** production is decreasing, with the decline of **Lacq** and Meillon Saint Faust. Commercialized cumulative production was 187 Gm^3 in January 1991 (Fig. 11.1).

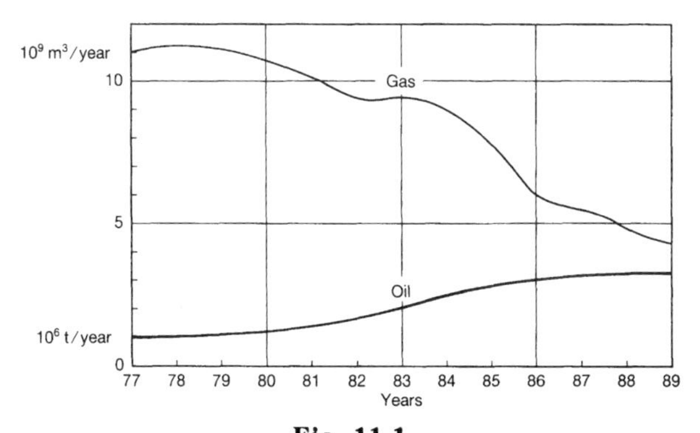

Fig. 11.1

French oil and gas production.

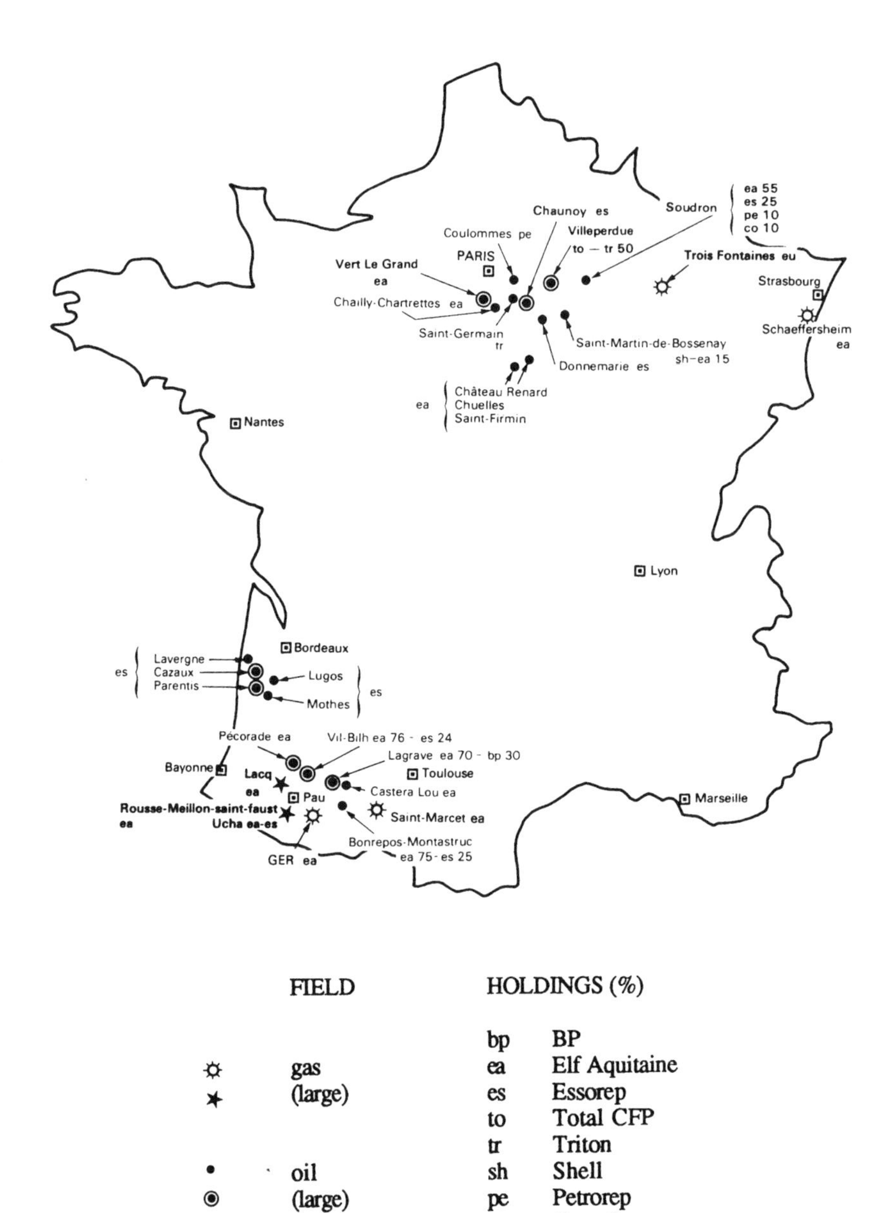

Fig. 11.2

Main oil and gas fields in France.

11.2 EXAMPLES OF FIELDS

Four fields are described in the following pages, with their essential features, **Lacq** and **Frigg** (gas), and **Vic Bilh** and **Villeperdue** (oil).

11.2.1 The Lacq Gas Field

The discovery of Lacq was the first major success of French petroleum exploration. *SNPA*, which focused its prospecting in the southern Aquitaine region, discovered a small oil field at Lacq in 1949, and the vast gas field in 1951 (around 700 m and 3500 m depth respectively) (Figs 11.3, 11.4 and 11.5).

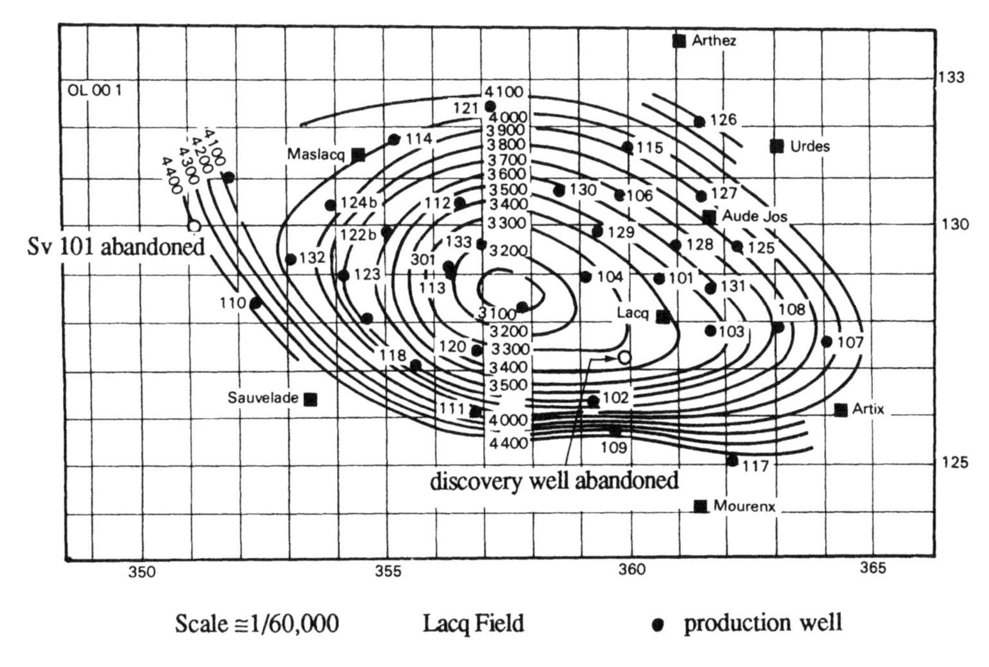

Fig. 11.3

Wells and isobaths at the top of the Neocomian.

This discovery immediately raised a basic technical problem, because the gas was extremely corrosive due to its high H_2S and CO_2 content

(Fig. 11.6), and tended to crack the tubing very rapidly. Production of this gas accordingly appeared unfeasible.

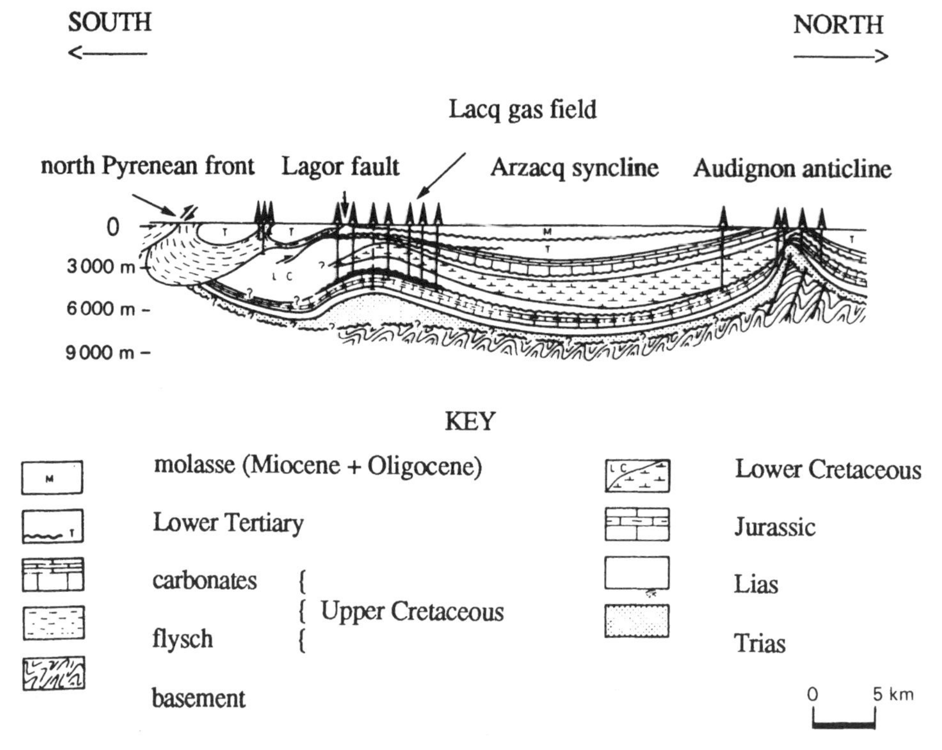

Fig. 11.4 Schematic cross-section through Lacq and Audignon.

This challenge was soon taken up by *SNPA*, which asked the Pompey and Vallourec steelworks to develop a special steel, which was achieved in 1954. Thus, thanks to the tenacity of the oil men and steel manufacturers, a major field, considered unexploitable, was ultimately produced.

This field earned a fortune for *SNPA*, which became *SN Elf Aquitaine* in 1976, and also helped to develop the region of Pau.

Gas production from Lacq, which, in 1963, accounted for nearly all the natural gas consumed in France, now only covers 20% of French needs (Fig. 11.7). But, because of the H_2S content of the gas, *SNEA(P)* incidentally became the world's leading sulfur producer.

The main characteristics of the field are given in Table 11.1.

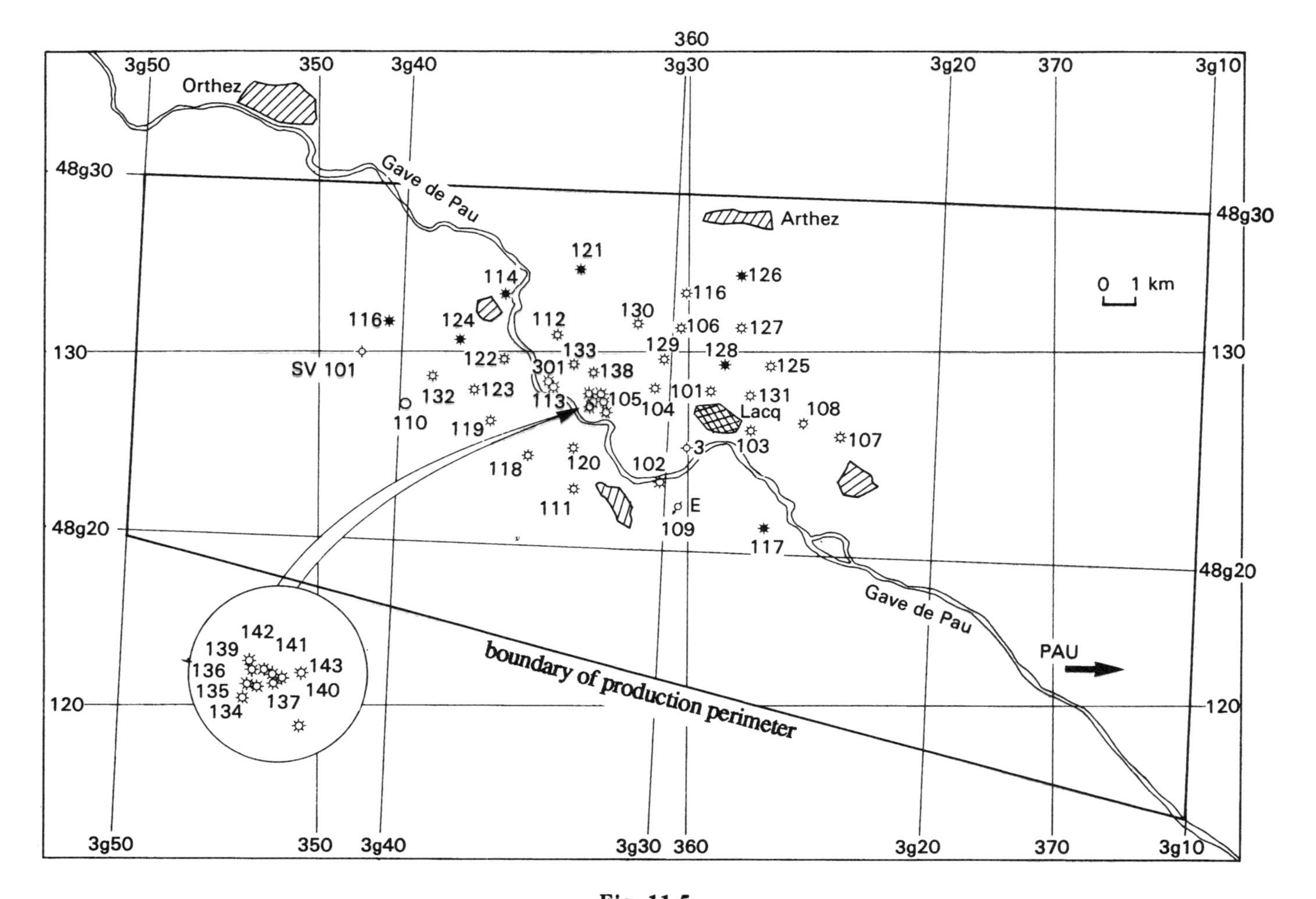

Fig. 11.5

Lacq lease perimeter. Deep Lacq field.

Hydrocarbons = 74.2%	
Methane	69.0%
Ethane	3.0%
Propane	0.9%
Butane	0.5%
Pentane	0.2%
Hexanes and +	0.6%
Sour Gases = 24.8%	
Hydrogen sulfide	15.3%
Mercaptans	0.2%
Carbon dioxide	9.3%
Water = 1.0%	
Water	1.0%

Fig. 11.6

Molecular composition.

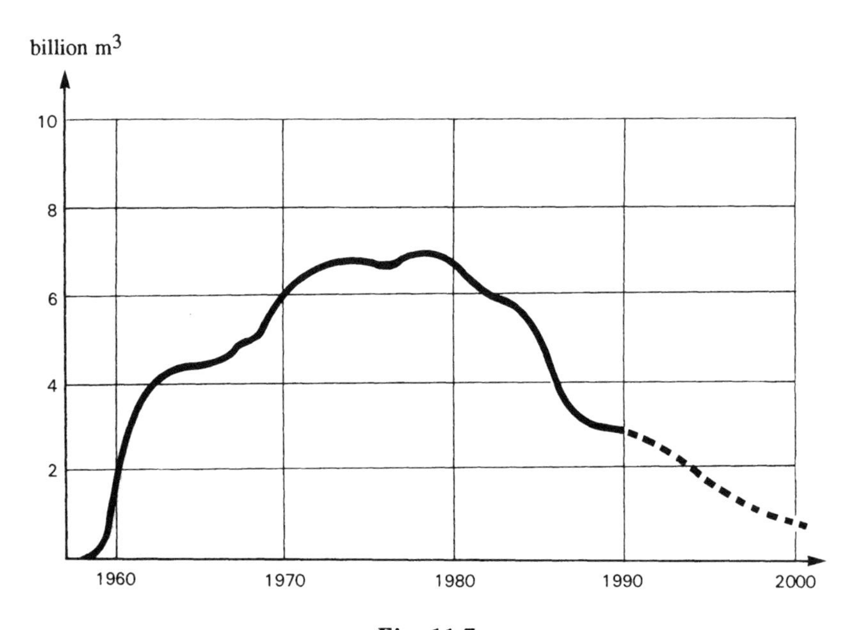

Fig. 11.7

Annual gas production of the Lacq plant.

Table 11.1

LACQ (Deep)

Country:	France
Company:	*SNEA(P)*
Date/place:	December 1951/Aquitaine
Production:	3 Gm^3/year
Trap:	Anticline
Rock/age:	Dolomitic limestone / Portlandian (Jurassic)
Depth:	3200 to 4200 m
Number of reservoirs:	1
Area (i):	120 km^2
Thickness:	≈ 1000 m
Useful thickness:	200 m
Gas column:	> 1000 m
Caprock:	Marls
Gas in place:	260 Gm^3

RESERVOIR CHARACTERISTICS

Levels:	1
Thickness:	1000 m
Porosity:	3% (1 to 20%)
Permeability:	0.1 mD + fractures
S_{wi}:	15 to 50%
Type of sedimentation:	Dolomitic limestone

GAS CHARACTERISTICS

Initial pressure:	670 bar at – 3700 m
Type of gas:	Wet
Content in GPM:	24 g/m^3
Gas gravity:	0.8
Temperature:	125°C at – 3700 m
Gas/water interface:	Compact at base

PRODUCTION CHARACTERISTICS

Well production:	350,000 m^3/day
Number of wells:	46
Production rate:	3 Gm^3 of gas/year in decline 200,000 m^3 of condensate/year
Cumulative production:	211 Gm^3 (January 1991)

ENVIRONMENT

Location map:	Lacq perimeter
Other levels:	Upper Lacq (oil, Senonian at – 600 m)
Other fields:	Meillon Rousse (gas)

RECOVERY

Reserves:	235 Gm^3
Mechanisms:	One-phase gas drive (anticipated R%: 90%)

11.2.2 The Frigg Field

The **Frigg** field is three-fourths owned by the two companies *SNEA(P)* and *Total CFP*. It was discovered in 1971, and came on stream in 1977. This time interval corresponded partly to a period of study of the delimitation of the dividing line between the Norwegian and British waters, so that the decision to develop was taken only in 1974; and partly to the actual development period, with an imposing set of platforms: the Seven Isles of Frigg. Capital outlay for the development of this field was considerable, about 20 billion francs.

Expanding knowledge about Frigg:

The initial surveys indicated a rather homogeneous reservoir (Frigg sands) separated by a tuff and shale barrier (Balder) from the underlying entirely aquifer Cod formation (Fig. 11.9).

For recovery, the major problem concerned the activity of the aquifer. The question arose whether or not the regional Cod aquifer activated the gas field. The answer was found fairly soon, and simulations demonstrated aquifer activity corresponding to "windows" in the tuff. An observation well drilled in the NE zone indicated a slight rise of the water level (O/W and G/O).

The producing wells were drilled in clusters from two drilling/production platforms (2 x 25 wells) (Fig. 11.8). The wells penetrated only 50 to 60 m down from the top of the reservoir, in order to avoid water coning. The well spacing was 250 m. Production is shown in Fig. 11.10.

Since the simulation model indicated an average water rise of 40 m, one well was deepened in 1984 and indicated a 55 m rise in the water level at this point. This strong aquifer activity could be highly detrimental, due to premature influxes in the wells, and due to the large volumes of gas trapped by the water. Drastic measures were taken to deal with the problem.

Several wells were deepened, and others were drilled between the platforms and the boundaries of the gas zone, in order to identify the possible vertical barriers and "windows". This reassessment survey conducted in 1984/1986 (midlife of the field) included 1000 m of core drilling and 5500 m of detailed 2-D/3-D seismic shooting.

The results considerably altered the initial image of the field:

(a) Much more heterogeneous reservoir.

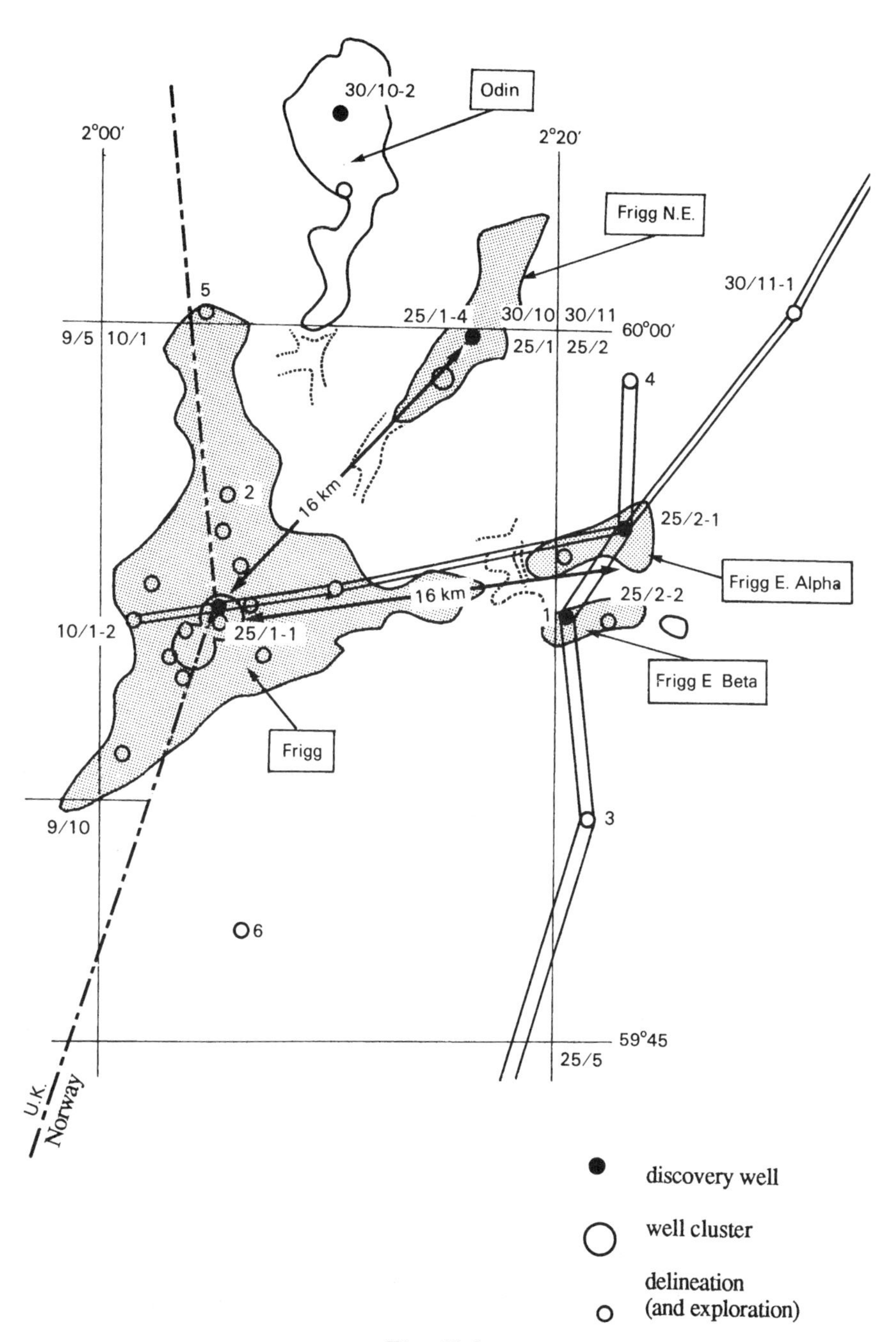

Fig. 11.8

Frigg and satellites.

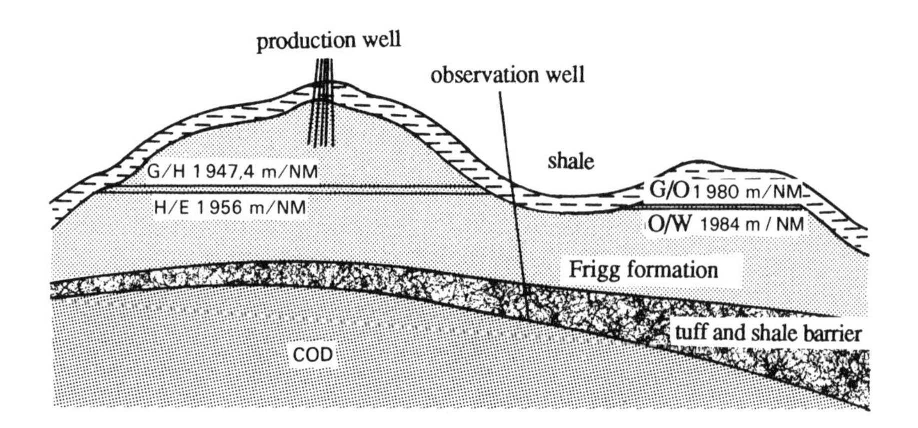

Fig. 11.9

Observation well.

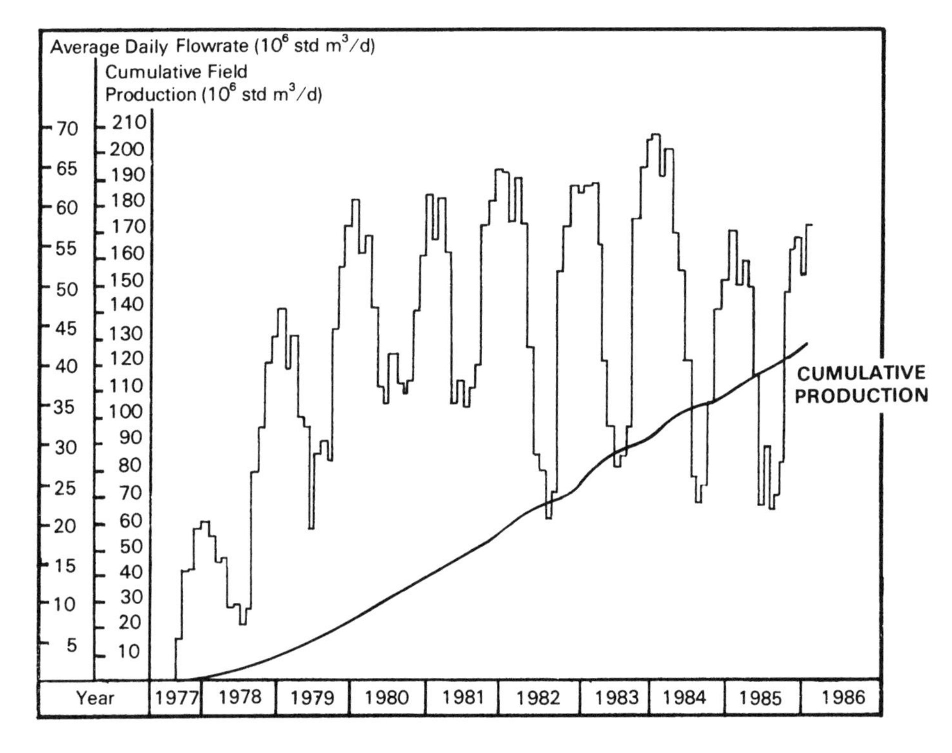

Fig. 11.10

Frigg production case history.

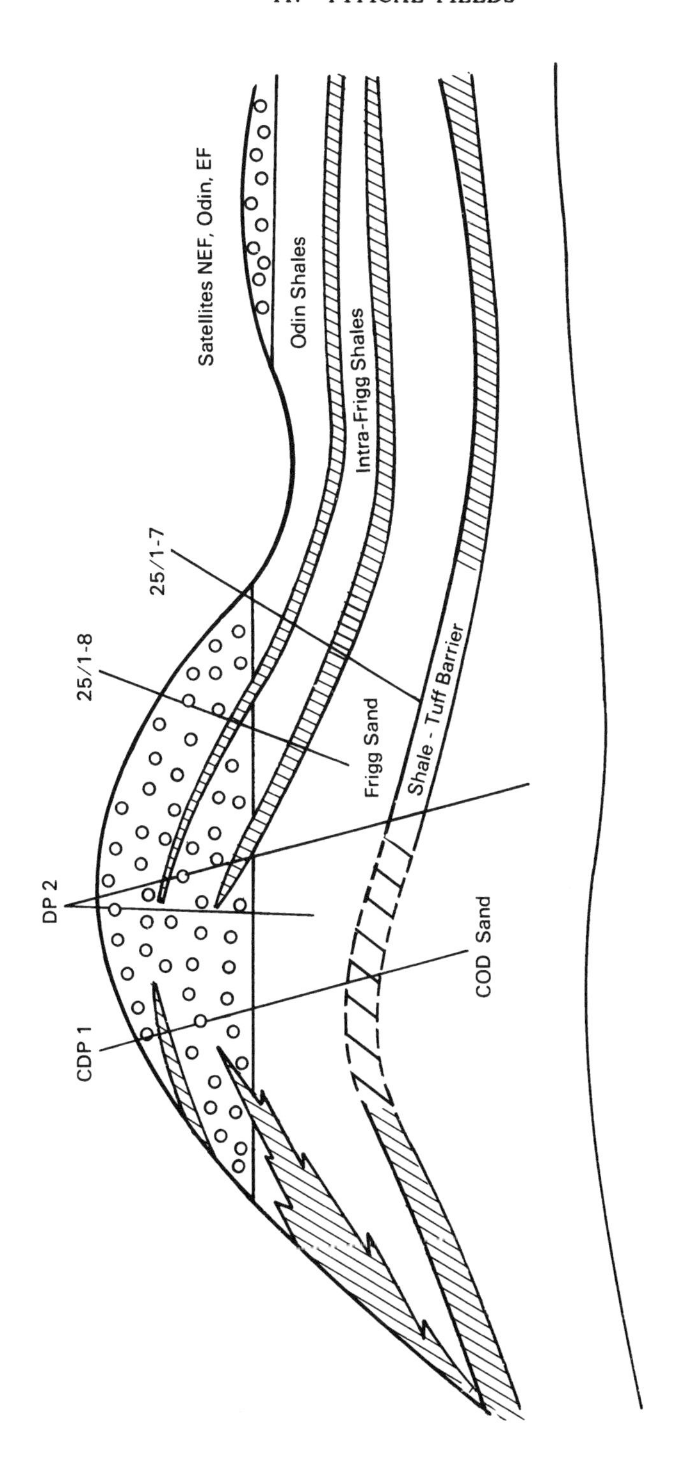

Fig. 11.11
Geological model.

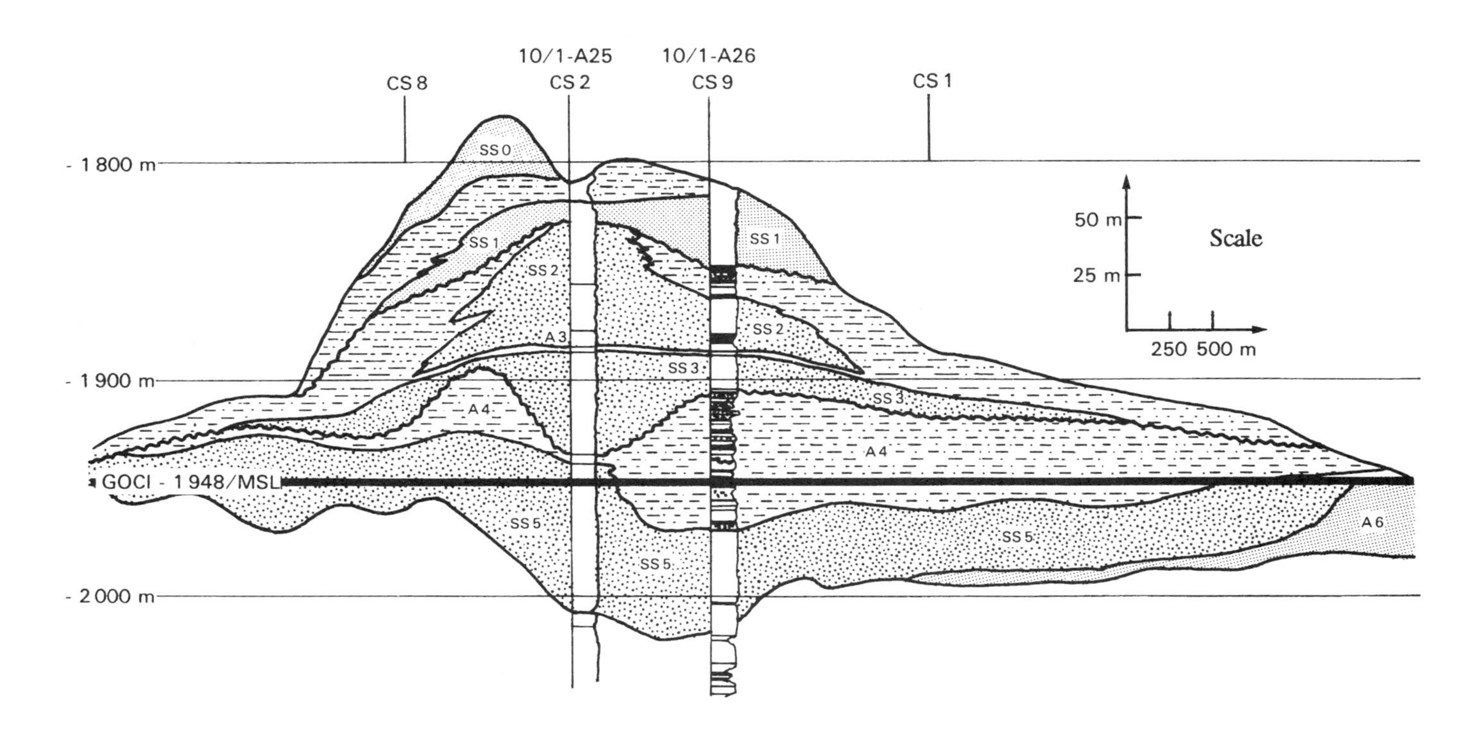

Fig. 11.12
Depth cross-section.

Table 11.2

FRIGG

Country:	Norway/United Kingdom	
Companies:	*SNEA(P)* (41%) *Total* (21%) *Norsk Hydro* and *Statoil* (38%)	Norway
	Total (55.5%) *SNEA(P)* (44.5%)	United Kingdom
Date/place:	July 1971 / North Sea	
Production:	Since Sept. 1977: 15 Gm^3/year	
Trap:	Anticline	
Rock/age:	Sand/Eocene	
Depth:	1900 m	
Number of reservoirs:	1	
Area (i):	100 km^2	
Thickness:	200 to 250 m	
Gas column:	160 m	
Caprock:	Shaly	
Gas in place:	265 Gm^3 + 75 Gm^3 (satellites)	

RESERVOIR CHARACTERISTICS

Levels:	3 (separated by two discontinuous shaly levels)
Thickness:	200 to 250 m
Porosity:	29%
Permeability:	1300 mD
S_{wi}:	10%
Type of sedimentation:	More or less consolidated sands

GAS CHARACTERISTICS

Initial pressure:	196.9 bar at – 1900 m
Type of gas:	Condensate
Content in GPM:	4.3 g/m^3
Viscosity (P_i):	0.02 cP
Temperature:	60°C
Gas/oil interface:	– 1947 m/SL (with an oil disc 9 m thick)

PRODUCTION CHARACTERISTICS

Levels:	3
Well production:	2 to 4 Mm^3/day
Well head pressure:	70 bar
Skin/treatment:	Variable
Number of wells:	≈ 60
Production rate:	15 Gm^3/year
Cumulative production:	193 Gm^3 (January 1991)

ENVIRONMENT

Location map:	"Central" North Sea
Other levels:	Cod (underlying aquifer)
Other fields:	Heimdal, Odin (near northeast Frigg)

RECOVERY

Reserves:	≈ 200 Gm^3 (61% Norway and 39% United Kingdom)
Mechanisms:	Gas expansion and water drive (anticipated R%: 70%)

(b) Containing more or less continuous shale levels (Figs 11.11 and 11.12),

hence the extremely variable water inflows in the different zones, since the presence or absence of shale barriers governed the rise in the water level.

At present, a team of reservoir engineers and reservoir geologists has been assigned to refine the data and simulate the reservoir, in order to predict the gas production capacity in the forthcoming years more accurately.

The characteristics are given in Table 11.2.

11.2.3 The Vic Bilh Field

This field is located 30 km northeast of Pau (Figs 11.13, 11.14 and 11.15). Its technical characteristics are given in Table 11.3. Note that Vic Bilh means "old country" in the Bearn dialect.

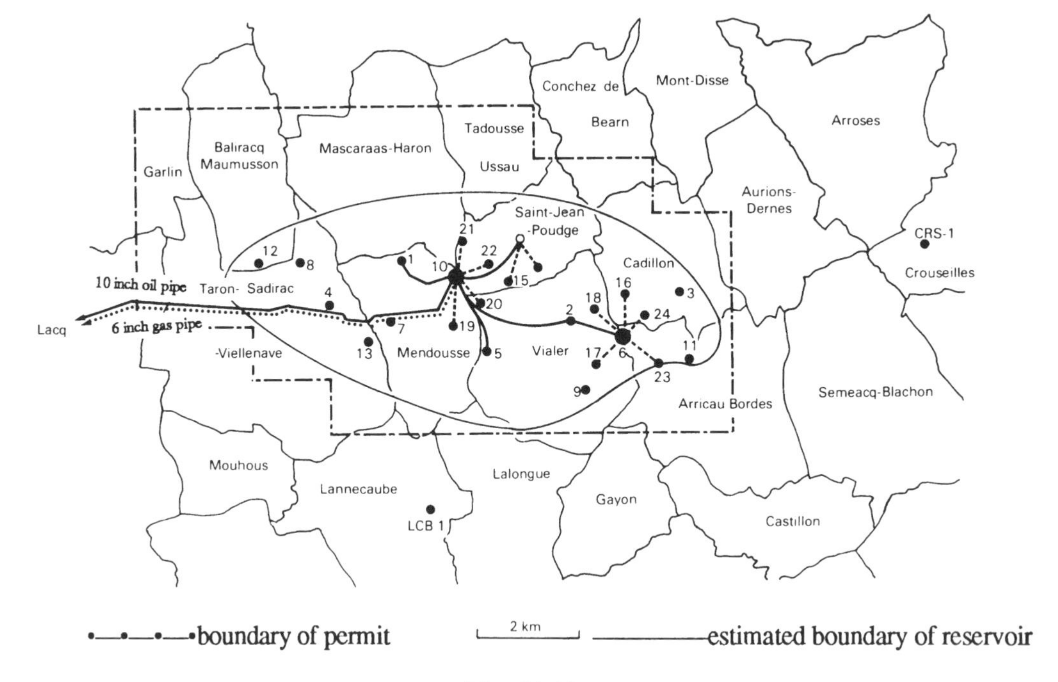

Fig. 11.13

Vic Bilh field.

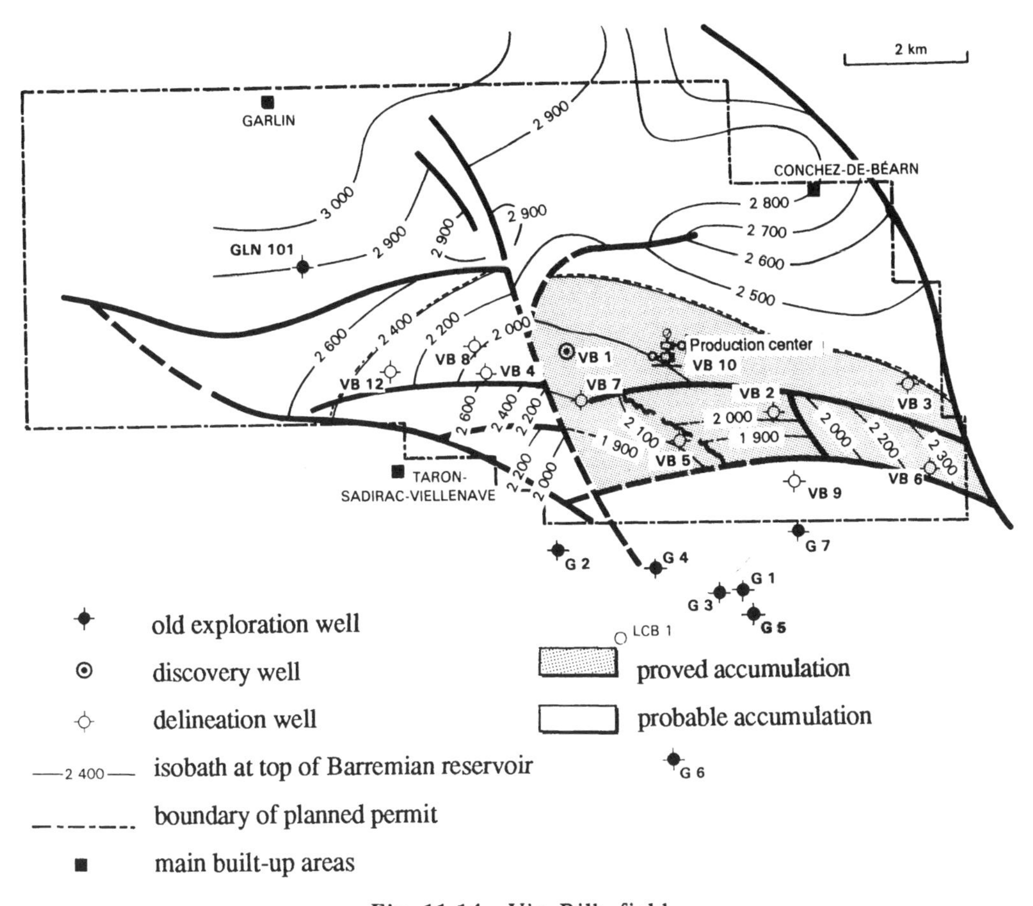

Fig. 11.14 Vic Bilh field.

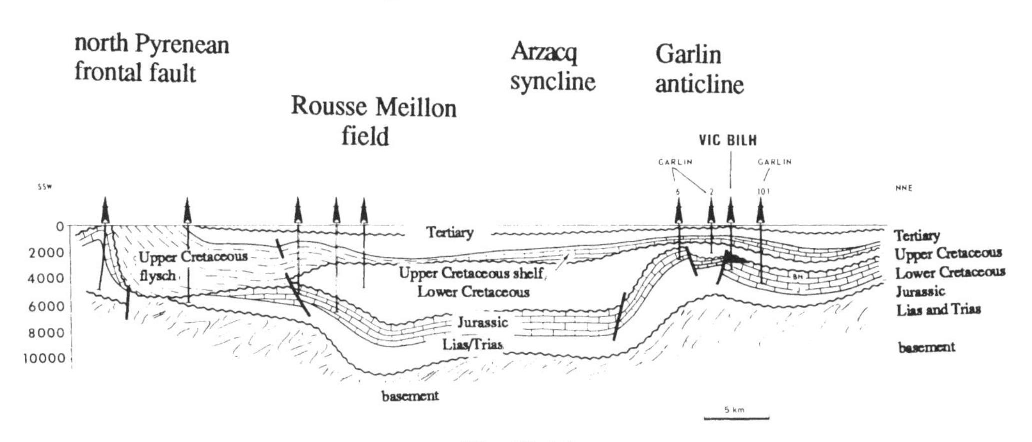

Fig. 11.15

Schematic geological cross-section across the Arzacq syncline passing through the Rousse Meillon and Vic Bilh fields.

Table 11.3

VIC BILH

Country:	France
Company:	*SNEA(P) / Essorep* (76%/24%)
Date/place:	June 1979 / North of Pau
Production:	200,000 t/year
Trap:	Faulted anticline
Rock/age:	Limestone / Barremian / Berriasian (Cretaceous)
Depth:	2250 to 2700 m
Number of reservoirs:	Complex
Area (i):	16 km^2
Thickness:	400 m
Useful thickness:	Variable
Oil column:	400 m
Oil in place:	26 Mt

RESERVOIR CHARACTERISTICS

Levels:	Several
Thickness:	400 m
Porosity:	< 15%
Permeability:	0.1 to 5 mD + fractures
S_{wi}:	25 to 35%

OIL CHARACTERISTICS, highly variable

Initial pressure:	Variable
Bubble point:	Identical (or <)
B_{oi} / R_{si}:	Variable / 0 to 80 m^3/m^3
Viscosity (P_i):	Very variable
API gravity:	5°/28° API
Oil/water interface:	– 2430 m
Gas/water interface:	– 2030 m

PRODUCTION CHARACTERISTICS

Well production:	10 to 200 m^3/day
Production index:	1 to 100 m^3/day.bar
Number of wells:	≈ 50
Activated production:	Viscous oil is "lifted" by very light oil (fluxing effect)
Production rate:	200,000 t/year
Cumulative production:	2.85 Mt (January 1991)

ENVIRONMENT

Location map:	Lease
Other levels:	None
Other fields:	Lagrave, Pécorade

RECOVERY

Reserves:	≈ 5 Mt
Mechanisms:	Dissolved gas + more or less active aquifer (anticipated R%: 20%)
Water flood:	Planned

This field was very difficult to discover because it is located in a structurally very complicated area. The first drilling operations in this zone began in 1945. It was discovered in 1979 by a high-resolution seismic survey.

This is a complex reservoir formed of several levels, and the oil may be light (upper part) or very viscous (lower part). This results in variable production per well, ranging from 10 to 200 m^3/day.

11.2.4 The Villeperdue Field

This field is located 100 km east of Paris (Fig. 11.16). Its technical characteristics are given in Table 11.4.

This is an early discovery made in 1959 by *RAP*, non-commercial at the time, and resumed by *Triton* and *Total* in 1982.

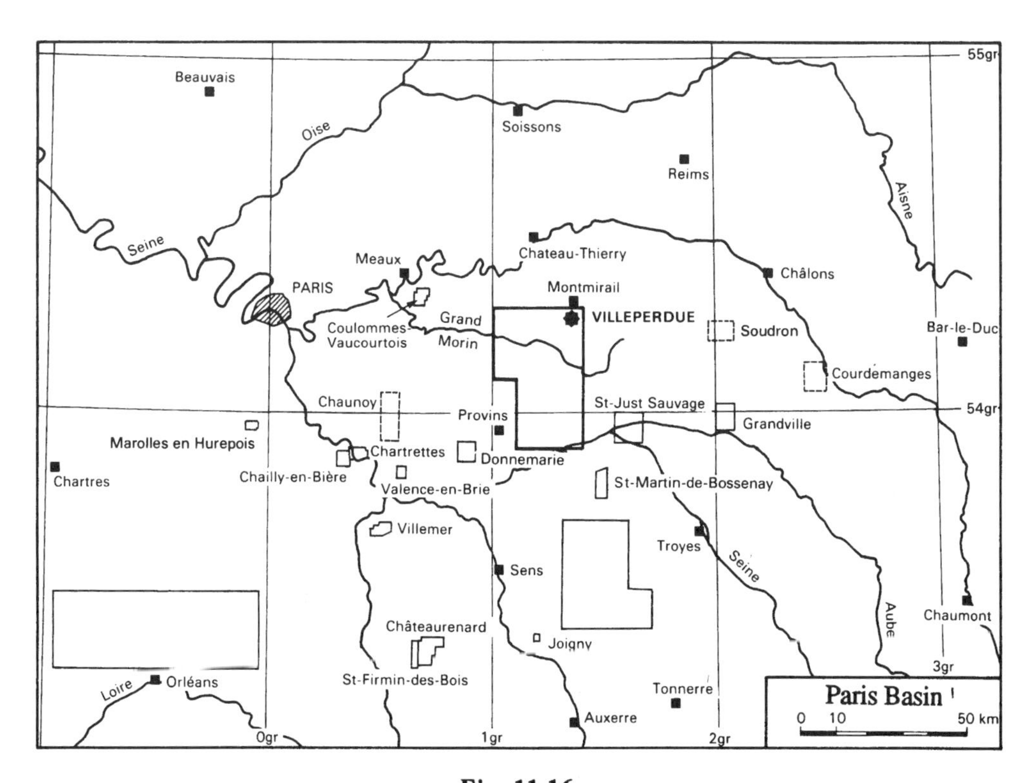

Fig. 11.16

Geographic location.

Table 11.4

VILLEPERDUE

Country:	France
Company:	*Total* (50%), *Triton* (50%)
Date/place:	1981 (1959) / Villeperdue Montmirail (Brie)
Production:	500,000 t/year
Trap:	Anticline / Stratigraphic
Rock/age:	Carbonates / Lower Callovian (Dogger)
Depth:	1850 m
Number of reservoirs:	1
Area (i):	83 km^2
Thickness:	30 m
Useful thickness:	20 m
Oil column:	42 m
Caprock:	Marls
Oil in place:	70 Mm^3

RESERVOIR CHARACTERISTICS

Levels:	2 (R_1 and R_2)
Thickness:	8 to 12 m
Porosity:	12% and 12% (8 to 18%)
Permeability:	Low to good, > 10 mD
S_{wi}:	40%
Type of sedimentation:	Oolitic limestone

OIL CHARACTERISTICS

Initial pressure:	175 bar
Bubble point:	6 bar
GOR:	3 m^3/m^3
Viscosity (P_i):	3.5 cP
API gravity :	34.4°
Oil/water interface:	– 1680 m

PRODUCTION CHARACTERISTICS

Levels:	2 (R_1 and R_2)
Well production:	20 m^3/day (3 to 80 m^3/day)
Production index:	0.1 to 2 m^3/day bar
Number of wells:	≈ 125
Production activated by pumping	
Production rate:	500,000 t/year
Cumulative production:	3.40 Mt (January 1991)

ENVIRONMENT

Location map:	Paris Basin
Other levels:	Rhaetic to the east
Other fields:	Dogger and Triassic

RECOVERY

Reserves:	≈ 10 Mt
Mechanisms:	Expansion of oil + pore water + rock (anticipated R%: 15%)
Water flood:	700,000 m^3/year (1988)

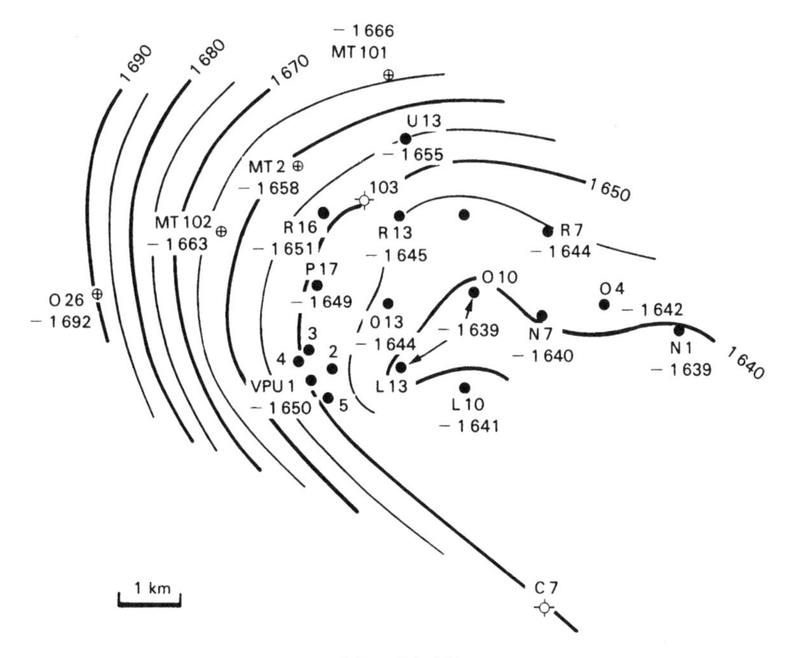

Fig. 11.17

Isobaths at top of Dogger.

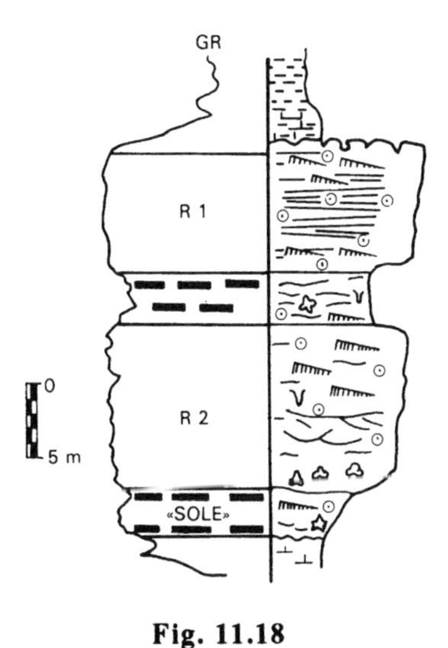

Fig. 11.18

The reservoir.

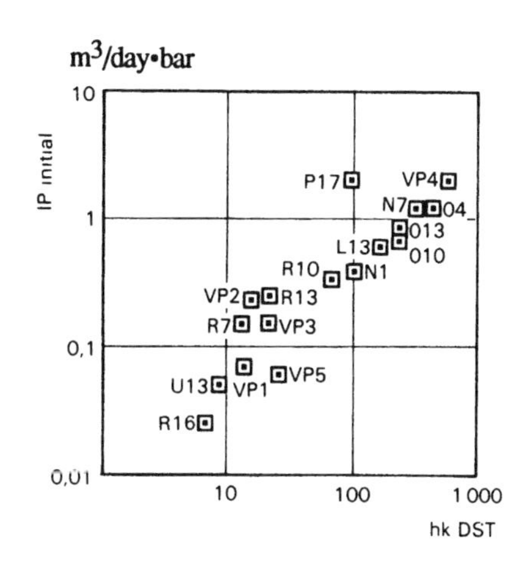

Fig. 11.19

Variation of hK in the field.

This field is distinguished by considerable heterogeneity and by a structural map that is difficult to clarify by seismic prospecting (Figs 11.17, 11.18 and 11.19).

The development strategy therefore consisted of a step-by-step process by successive sectors, in a grid with a well spacing of 1200 m, whereas the final grid probably adopted will be 400 m if the technical conditions (reservoir characteristics) and economic conditions are favorable.

At the end of 1988, the field was not yet fully explored, especially toward the east, where the structural closure had not yet been surveyed, but the facies deteriorated in that direction. This accordingly indicated a composite structural and stratigraphic trap.

REFERENCES

GENERAL

Frick, T.C., "Petroleum Production Handbook", 2., *Society of Petroleum Engineers of AIME* (Dallas), 1962.

Smith, C.R., *"Mechanics of Secondary Oil Recovery"*, Robert E. Krieger Publishing Company (New York), 1975.

Chapter 1
RESERVOIR GEOLOGY AND GEOPHYSICS

Bitot, P. *(SNEA(P))*, "Eléments de géologie de production"[1], *ENSPM* lectures, ref. 32 837, 1985.

Guillemot, J. *(ENSPM)*, *"Eléments de Géologie"*, Editions Technip, 1986.

Mari, J.L. *(IFP)*, "Les différents aspects de la sismique de gisement", *ENSPM* lectures, ref. 34250, 1986.

Chapter 2
CHARACTERIZATION OF RESERVOIR ROCKS

Mondrain, P. *(ENSPM)*, "Caractéristiques des roches réservoirs", *ENSPM* lectures, ref. 32519, 1985.

Monicard, R. *(IFP)*, *"Caractéristiques des roches réservoirs. Analyse des carottes*[1]*"*, Editions Technip, 1975.

1. Recommended reading matter for detailed study.

Chapter 3
FLUIDS AND PVT STUDIES

Gravier, J.F. *(Total CFP)*, *"Propriétés des fluides de gisement*[1]*"*, Editions Technip, 1986.

Mondrain, P. *(ENSPM)*, "Introduction du PVT dans la thermodynamique classique", *ENSPM* lectures, ref. 26506.

Chapter 4
VOLUMETRIC EVALUATION OF OIL AND GAS IN PLACE

Groult, J. *(Total CFP)*, Notes for *ENSPM* lectures on geology, geophysics and production geology, ref. 23973, 1976.

Leroy, G. *(SNEA(P))*, "Cours de géologie de production"[1], *ENSPM* lectures, ref. 24 429, 1976.

Chapter 5
ONE-PHASE FLUID MECHANICS AND WELL TEST INTERPRETATION

Daviau, F. *(SNEA(P))*, "Eléments de base pour l'interprétation des essais de puits", *ENSPM* lectures, ref. 31747, 1983.

Daviau, F. *(SNEA(P))*, *"Interprétation des essais de puits: les méthodes nouvelles*[1]*"*, Editions Technip, 1986.

Mondrain, P. and Cossé, R. *(ENSPM)*, "Eléments de mécanique de fluides monophasiques et interprétation des essais de puits", *ENSPM* lectures, ref. 30947, 1983.

Chapter 6
MULTIPHASE FLOW

ARTEP, *"Influence de l'étude des condtions aux limites sur le phénomène du cône d'eau en régime supercritique"*, Editions Technip, 1978.

Craig, F.F., "The Reservoir Engineering Aspects of Water Flooding", *Doherty series*, **3**, 1971.

Cuiec, L. *(IFP)*, "Mouillabilité et réservoirs pétroliers", *Revue de l'IFP*, 1986.

Reiss, L.H. *(SNEA(P))*, "Mécanique des fluides polyphasiques en milieu poreux"[1], *ENSPM* lectures, ref. 26713, 1978.

1. Recommended reading matter for detailed study.

Chapter 7
PRIMARY RECOVERY
ESTIMATION OF RESERVES

COSSE, R. *(ENSPM)*, «Drainage naturel. Estimation des réserves». *ENSPM* lectures, réf. 35614, 1987.

CESEG report *(ENSPM)*, «Le pétrole dans le monde. Eléments statistiques». 1990.

REISS, L.H. *(SNEA(P))*, *Reservoir engineering en milieu fissuré* [1]. Editions Technip (Collection *ENSPM*), 1980.

GIGER, F. and COMBE, J. *(IFP)*, and REISS, L.H. *(SNEA(P))*, «L'intérêt du forage horizontal pour l'exploitation des gisements d'hydrocarbures». *Revue de l'IFP*, 1983.

REISS, L.H. *(SNEA(P))*, «Production from horizontal wells after 5 years». *J.P.T.*, November 1987.

Chapter 8
SECONDARY AND ENHANCED RECOVERY

COSSE, R. *(ENSPM)*, «Méthodes classiques de récupération assistée: injection d'eau et de gaz»[1], *ENSPM* lectures, réf. 33293, 1985.

LATIL, M. *(IFP)*, *Récupération assistée*. Editions Technip, 1975.

BARDON, C. *(IFP)*, «La récupération améliorée» (report), 1983.

Chapter 9
RESERVOIR SIMULATION MODELS

CILIGOT-TRAVAIN, G. *(SNEA(P))* and LEMONNIER, P. *(IFP)*, «Modèles de simulation numérique de gisement» (report). 1986.

COMBE, J. and LATIL, M. *(IFP)*, «L'Optimisation de l'exploitation pétrolier grâce au simulateur de gisement Score» (report). 1985.

CILIGOT-TRAVAIN, G. *(SNEA(P))*, «Les modèles de gisements»[1], *ENSPM* lectures (updated version to be published shortly).

1. Recommended reading matter for detailed study.

Chapter 10
OILFIELD DEVELOPMENT

COTTIN, R.H. *(SNEA(P))*, «Elaboration d'un projet de gisement»[1], *ENSPM* lectures, réf. 28944, 1980.

IFP L'exploitation des gisements d'hydrocarbures. Editions Technip (Collection Recherches et Témoignages), 1974.

BERTRAND, G. (*DHYCA*, Service Conservation des Gisements), «Développement optimal d'un gisement» (report). 1985.

Chapter 11
TYPICAL FIELDS

DHYCA (Service Conservation des Gisements) Annual Report, 1990 .

IFP, «Le fichier des champs pétroliers en France». Central Library, 1990.

SNEA(P), reports on the Frigg and Vic-Bilh fields.

Total CFP, report on the Villeperdue field.

1. Recommended reading matter for detailed study.

INDEX

Accumulation, 21, 115
Acidizing, 146
Acoustic impedance, 27
Afterflood, 145
Anticline, 23
Apraisal well, 310
Aquifer, 214, 228
Area
  drainage, 136
  injection, 271
Average characteristics, 125

Black oil, 301
Bounded reservoir, 138
Breakthrough, 190, 208
Bright spot, 27, 29
Bubble point, 79, 214
Buckley-leverett theory, 190
Build-up, 142
Buoyancy, 36

Capillarity, 50, 179
Capillarity imbibition, 187
Capillarity migration, 56
Capillarity pressure, 52, 54
Cap-rock, 21, 121
Carbonate, 22
Chalk, 23
Chemical methods, 289
Clapeyron diagram, 78
Combustion in situ, 294
Completion, 134
Composite well log, 17
Compressibility, 38, 91, 102, 163, 215
Compressibility factor Z, 86, 89
Condensate gas, 241
Condensation retrograde, 81
Coning
  bottom, 205
  edge, 206
Constant pressure outer boundary, 137
Core analysis, 15, 35, 60
Correlation, 17
Critical flow rate, 216
Critical point, 79
Critical speed, 203

Damage, 143
Darcy's law, 38, 135
Decline laws, 244
Deliverability curve, 153
Density log, 68
Depletion
  natural, 178, 211
Derivative curve, 164
Development, 309
Dew point, 79

Diffusivity, 134
Dissolved gas drive, 226
Dolomite, 23
Drive mechanisms, 211

Efficiency
areal sweep, 272
displacement, 275
vertical, 274
Electric log, 65
Encroachment, 201
Enhanced oil recovery, 261, 282
Equation of state, 86, 135
Expansion, 212
Extension well, 116
Exudation, 252

Faciès, 17, 119
Fault, 16, 119
Field, 309
Fingering, 201, 204
Five-spot pattern, 268
Flow rate, 139, 141, 196
Fluid interface, 122
Fluid summation, 37
Forecasting, 313
Formation volume factor
gas, 88
oil, 91
Fractured formation, 245
Fracturing, 249
Front, 189
Frontal displacement, 189

Gamma ray log, 66
Gas
condensate, 83
dry or wet, 83
natural, 86
Gas-cap, 213
Gas-cap drive, 234
Gas cycling, 281
Gas injection, 279
Gas in place, 115
Gas-oil ratio, 91
Geological model, 17
Geology, 15
Geophysics, 15
Gravity, 192, 214
Gross pay, 119

Heterogeneities, 216, 310
History matching, 303
Horizontal drain hole, 254
Horner's method, 143
Hot water, 293
Huff and puff, 293
Hydrate, 102, 288
Hydrocarbon, 18, 81
Hydrodynamism, 55

Image of the reservoir, 15
Imbibition, 187
Impregnated rock, 119
Improvement, 146, 282
Income, 315
Injection, 267, 272
Instability, 201
Interfacial tension, 52
Interpolation, 217
Isobath, 25, 122
Isochrone, 25
Isopach, 123
Invaded zone, 66, 144
Investments, 315

Jamin effect, 180
Jurin's law, 52

Karst, 23, 258
Kerogen, 18

Layer-bed, 21
Limestone, 23
Logging, 62

Material balance, 218, 223
Microresistivity log, 66
Migration, 19, 20
Miscibility, 284
Miscible slug, 287
Microemulsion, 289
Mixture, 77, 80
Mobility ratio, 201, 213
Model, 218, 300
Multiphase flow, 177, 216

Net pay, 119
Neutron log, 68
Nonflowing well, 150

Oil, 91
- heavy, 291
- residual, 181
- volatile, 236

Oil in place, 115
Oil window, 18
One phase, 84
Optimization, 310, 316
Organic matter, 18

Partial penetration effect, 146
Pattern, 267, 271
Payout time, 315
Perforation effect, 146
Permeability, 33, 38
- absolute, 43
- average, 42
- relative, 43, 182

Permeameter, 45
Petrophysics, 33
Pilot flood, 296
Physical constants, 87
Plug
- formation, 278

Pool, 21
Pore, 33
Porosity, 33
Pressure
- build-up, 142
- drawdown, 138
- flowing, 139
- geostatic, 38
- initial, 139
- pseudo, 152
- reservoir, 139, 160
- static, 133, 139
- well head, 155

Primary recovery, 211
Probabilistic methods, 128
Probabilities
- objective, 129
- subjective, 129

Produced water, 186, 209
Production capacity, 133
Production data, 16
Production forecast, 224, 302
Production log, 74
Productivity index, 148
Pumping, 150
PVT studies, 77, 134

Quadratic equation, 152
Quick look, 71

Radius, 136
Rate of return, 316
Recovery, 211, 219
Reserves, 115, 211, 220
Reservoir, 3, 21, 119
Resistivity log, 65
"Rock eval" method, 19

Sampling, 134
Sandstone, 21

Saturated oil, 82
Saturation, 49
critical gas, 184
discontinuity, 193
irreducible, 55, 183
residual, 181
Scanner, 61
Secondary recovery, 211, 261
Segregation, 236
Seismic survey, 25, 26
Semi steady-state flow, 137
Shale, 19, 22, 65
Skin effect, 133, 144
Simulation
model, 299
Sonic log, 68
Source rock, 18
Spontaneous potential, 65
Standard conditions, 88, 119
Steady-state flow, 137
Steam injection, 293
Stimulation, 146
Storage coefficient, 164
Stratigraphic scale, 31
Surface tension, 52
Surfactant, 289
Sweeping, 263, 276

Tar sands, 291
Test
back pressure, 156
drill stem, 159
initial, 154
interference, 158
isochronal, 156
periodical, 157
Thermal methods, 291
Tilted interface, 55
Tongue, 201, 203
Transient flow, 137, 153
Transfert function, 254
Transition zone, 54
Trap, 23
Two phases, 91, 149
Type curve, 163

Undersaturated oil, 82, 212

Viscosity
gas, 90
oil, 99
water, 102
Volume in place, 311

Water
formation, 101
Water cut, 191
Water drive, 229
Waterflood, 277
Water influx, 232
Water-oil ratio, 191
Weldge tangent, 194
Well log, 63
Well model, 167
Well site, 267
Well test, 133
Wettability, 50

Imprimé en France par EUROPE MEDIA DUPLICATION S.A.S.
53110 Lassay-les-Châteaux
N° 13528 – Dépôt légal : avril 2005
N° d'éditeur : 1114